ELECTROANALYTICAL CHEMISTRY

VOLUME 23

edited by Allen J. Bard
and Cynthia G. Zoski

CRC Press
Taylor & Francis Group
Boca Raton London New York

CRC Press is an imprint of the
Taylor & Francis Group, an **informa** business

CRC Press
Taylor & Francis Group
6000 Broken Sound Parkway NW, Suite 300
Boca Raton, FL 33487-2742

First issued in paperback 2019

© 2010 by Taylor and Francis Group, LLC
CRC Press is an imprint of Taylor & Francis Group, an Informa business

No claim to original U.S. Government works

ISBN-13: 978-1-4200-8485-6 (hbk)
ISBN-13: 978-0-367-38433-3 (pbk)

Visit the Taylor & Francis Web site at
http://www.taylorandfrancis.com

and the CRC Press Web site at
http://www.crcpress.com

Introduction to the Series

This series is designed to provide authoritative reviews in the field of modern electroanalytical chemistry defined in its broadest sense. Coverage is comprehensive and critical. Enough space is devoted to each chapter of each volume so that derivations of fundamental equations, detailed descriptions of apparatus and techniques, and complete discussions of important articles can be provided, so that the chapters may be useful without repeated reference to the periodical literature. Chapters vary in length and subject area. Some are reviews of recent developments and applications of well-established techniques, whereas others contain discussion of the background and problems in areas still being investigated extensively and in which many statements may still be tentative. Finally, chapters on techniques generally outside the scope of electroanalytical chemistry, but which can be applied fruitfully to electrochemical problems, are included.

Electroanalytical chemists and others are concerned not only with the application of new and classical techniques to analytical problems but also with the fundamental theoretical principles upon which these techniques are based. Electroanalytical techniques are proving useful in such diverse fields as electroorganic synthesis, fuel cell studies, and radical ion formation, as well as with such problems as the kinetics and mechanisms of electrode reactions, and the effects of electrode surface phenomena, adsorption, and the electrical double layer on electrode reactions.

It is hoped that the series is proving useful to the specialist and nonspecialist alike—that it provides a background and a starting point for graduate students undertaking research in the areas mentioned, and that it also proves valuable to practicing analytical chemists interested in learning about and applying electroanalytical techniques. Furthermore, electrochemists and industrial chemists with problems of electrosynthesis, electroplating, corrosion, and fuel cells, as well as other chemists wishing to apply electrochemical techniques to chemical problems, may find useful material in these volumes.

Allen J. Bard

Cynthia G. Zoski

Contents of Other Series Volumes

VOLUME 21

VOLUME 22

Contents

Contributors to Volume 23

Shaowei Chen
Department of Chemistry
  and Biochemistry
University of California
Santa Cruz, California

Hubert H. Girault
Casali Institute of Applied Chemistry
Ecole Polytechnique Federale de
  Lausanne
Lausanne, Switzerland
and
Laboratoire D'Electrochimie
  Physique et Analytique
Lausanne, Switzerland

Philippe Hapiot
Sciences Chimiques de Rennes
CNRS, Campus de Beaulieu
Université de Rennes
Rennes, France

Ovadia Lev
Casali Institute of Applied
  Chemistry
The Hebrew University
  of Jerusalem
Jerusalem, Israel

Srinivasan Sampath
Department of Inorganic
  and Physical Chemistry
Indian Institute of Science
Bangalore, India

Jacques Simonet
Sciences Chimiques de Rennes
Université de Rennes
Rennes, France

1 Electrochemistry at Liquid–Liquid Interfaces

Hubert H. Girault

CONTENTS

1.1   INTRODUCTION

In 1989, a review with the same title was published in this series [1]. Indeed, at that time, electrochemistry at polarized liquid–liquid interfaces had undergone a second youth with the pioneering work of C. Gavach et al. in France and J. Koryta, Z. Samec et al. in what was then Czechoslovakia, and M. Senda et al. in Japan. A legacy of J. Koryta is the acronym that is now widely used even outside the chemistry community, namely, ITIES, which stands for Interface between Two Immiscible Electrolyte Solutions [2]. This first review, written in two parts, respectively, in 1985 and 1989, was dedicated first to a historical perspective dating back to the end of the nineteenth century, to a presentation of the thermodynamics of interfacial polarization, including electrocapillary phenomena, and to an introduction to the different charge transfer processes, namely, ion-transfer, assisted-ion-transfer, and electron-transfer reactions. The key advantage in preparing a second review nearly two decades later is to realize the extent of many developments that have in fact taken place during this period. Indeed, in 1989, we had very little information on the interface structure apart from that derived from thermodynamic analyses—no molecular dynamics yet, no x-ray reflectivity yet, and no surface-sensitive spectroscopic techniques yet. In fact, it sounds like 1989 was a very long time ago. For ion-transfer reactions, it is clear that the rate constants reported over the years have increased regularly as the methods and instrumentation have improved, yielding better-quality data, but more important, new theories have been developed that shed a new light on the reaction mechanism. In the field of assisted-ion-transfer reactions, a major development has been the concept of ionic partition diagrams that is widely used to report the lipophilicity, that is, the $\log P$, of ionizable molecules, particularly those of therapeutic importance. From a technological viewpoint, one can cite the introduction of micro-ITIES that can now be used in conjunction with Scanning Electrochemical Microscopy (SECM), and, of course, the development of a full range of spectroelectrochemical techniques such as voltabsorptometry, voltfluorimetry, potential-modulated absorbance and fluorescence, and nonlinear optical methods.

The classical electrochemical methodologies have been applied outside the classical water–nitrobenzene (NB) or water–1,2-dichloroethane (DCE) interface; new solvent systems have been investigated; and, in particular, Kakiuchi et al. have demonstrated that organic electrolyte solutions can be replaced by ionic liquids, also called Room Temperature Molten Salts (RTMS).

Already back in 1989, functionalizing the interface had started, mainly with phospholipid monolayers. Since then, many other types of functionalizations have been studied, for example, with metallic or semiconducting nanoparticles, or with molecular catalysts, or even with dyes for photosensitization.

The present review is not an exhaustive account of the nearly one thousand references on electrochemistry at liquid–liquid interfaces that have appeared over the years; it presents a self-standing overview of the aspect of electrochemistry that many still consider as exotic, ranging from basic principles to recent trends. Indeed, to classically trained electrochemists, the concepts of the 4-electrode potentiostat, of concomitant ion- and electron-transfer reactions, and of electro-capillarity without mercury, are sometimes difficult to explain. I hope that this chapter will help those not familiar with the field to appreciate the diversity that soft molecular interfaces can provide. From an experimental viewpoint, these molecular interfaces present a key advantage. They are easy to prepare and provide highly reproducible results. Just mix two immiscible liquids and wait for an interface to form—no electrode polishing, no tedious single crystal preparation.

1.2 INTERFACIAL STRUCTURE AND DYNAMICS

The structure of a liquid–liquid interface is difficult to define because, by definition, we deal with a dynamic molecular interface with thermal fluctuations. Our knowledge to date stems mainly from molecular dynamic calculations, from capacitance and surface tension measurements, and from some experimental spectroscopic investigations.

1.2.1 MOLECULAR DYNAMICS

1.2.1.1 Bare Water–Solvent Interfaces

Over the past two decades, molecular dynamics has provided not only a pictorial view of the interfaces that unfortunately cannot experimentally be imaged as solid electrodes by microscopic techniques but also some new concepts regarding, in particular, surface dynamics. Following the pioneering Monte-Carlo simulation study of the water–benzene interface by Linse [3], molecular dynamic studies of ITIES were actively pursued by Benjamin who studied first the structure of the H_2O–1,2 DCE interface [4], and who wrote two excellent reviews in 1996–97 [5,6]. In the beginning, most simulations were aimed at establishing density profiles and surface roughness, but with new methodologies appearing, such as the use of bivariate representations [7], or the dropball method to determine surface roughness [8] together with the use of larger sets of simulated molecules and longer run times, the description of the interface has become more detailed [9]. The main conclusion of the earlier work was an interface that affects the molecular organization of the adjacent phases, that is, relatively sharp at the molecular level but with corrugations caused by thermal fluctuations and capillary waves. The density profiles obtained by slicing the system were showing oscillations extending to the bulk, but it was difficult to distinguish oscillations in the interfacial plane from those perpendicular to it. Regarding the hydrogen-bonding organization, the consensus was that interfacial water molecules tend to arrange themselves so as to maximize the number of hydrogen bonds and to minimize their potential energy. More recently, Benjamin [10] has shown that hydrogen bond networks depend strongly on the nature of the organic solvent, but that, generally, hydrogen bond lifetimes are longer at the interface compared to the bulk, especially for solvent pairs where water fingers are likely to form. The different lifetimes that were obtained at the Gibbs dividing plane $\tau_{w\text{-DCE}} = 15$ ps, $\tau_{w\text{-NB}} = 10$ ps, and $\tau_{w\text{-CCl4}} = 7$ ps are indeed longer than the bulk value of about 5 ps.

The early work on water molecule orientation was carried out with a monovariate analysis of the dipole vector versus the plane of the interface, but Jedlovszky et al. [7] developed, in 2002, a bivariate representation to show that, for the H_2O–DCE system, two preferential orientations of the water molecules dominate: One with a parallel alignment of the molecular plane with the interface, and another with a perpendicular alignment of the molecular plane with a hydrogen atom pointing directly to the organic phase and with the molecular dipole vector pointing about 30° toward the organic phase. The first orientation was prevalent throughout most of the interfacial region and in the subsurface water layer adjacent to the interface, while the second occurred only for those molecules penetrating deep into the organic phase. This distribution characteristic seems rather general as it has been observed for different systems [11,12].

Liquid–liquid interfaces for ITIES research are limited by the choice of the organic solvent that must, of course, be immiscible with water and able to dissolve electrolytes. As a consequence, electrochemistry at ITIES is often limited to the H_2O–NB, H_2O–1,2-DCE, water–heptanone, and water–2-nitrophenyloctylether (NPOE) systems, the last two having been developed for their low toxicity.

In the case of the H_2O–NB interface, first studied by Michael and Benjamin [13] and recently revisited by Jorge et al. [12], the interface can be viewed as relatively sharp on the molecular scale but with some thermal fluctuations. This recent work suggests the existence of two tightly packed interfacial layers with both molecular planes parallel to the interface and restricted mobility on the normal axis—one water layer on the aqueous side and one nitrobenzene layer on the organic side.

Since the early work of Benjamin [4], the water–1,2-DCE interface has received a lot of attention. In particular, Benjamin et al. have shown that the presence of a static electric field tends to broaden the interface and decrease the surface tension by increasing the amplitude of finger-like distortions without strongly affecting the local microscopic structure or dynamics [14]. Later, this conclusion has been supported by the mean-field (Poisson–Boltzmann) calculations of Daikhin et al. [61]. More recently, the group of Richmond has combined molecular dynamics and sum frequency generation to probe and characterize this interface in a self-consistent manner, where molecular dynamic simulations are performed to generate computational spectral intensities of the H_2O–CCl_4 and H_2O–DCE interfaces that can be compared to experimental data. These calculations yield spectral profiles that depend both on frequency and interfacial depth. In 2004 [15], Walker et al. could conclude on the broad nature of the H_2O–DCE interface. Indeed, the interface was found to show spectral characteristics of a mixed-phase interfacial region consisting of randomly oriented water molecules with a broad distribution of interactions with DCE and other water molecules, thereby corroborating the concept of mixed-solvent layer introduced by Girault and Schiffrin in 1983 [16]. In 2007 [17], Walker and Richmond confirmed that the width of the H_2O–DCE interface was much broader than that of the H_2O–CCl_4 system. However, despite this diffuse structure, water molecules present throughout the interfacial region show a high degree of net orientation. These simulations can identify some water molecules present in the organic phase with an orientation sensitive to their degree of immersion. Molecules closest to the interface direct their OH bonds toward water, while those further away direct their OH bonds toward DCE [18].

The H_2O–hetpa-2-one interface has been studied by Fernandes et al. [19] who have shown that the interface is molecularly sharp and corrugated by capillary waves. The organic molecules in direct contact with the aqueous phase behave as amphiphilic molecules, with their polar heads toward the aqueous phase and the nonpolar chain into the bulk of the organic phase. As with surfactant molecules, the second layer reverses its orientation, forming a bilayer structure, but this ordering was found to vanish quickly already at the third layer. These bilayer structures have also been observed by Wang et al. [20] for the water–hexanol interface, for which they also remark that relatively static waves corrugate the inner part of the interface considerably more than that for the water–hexane interface, and that the relatively important water solubility in hexanol occurs in hydrogen-bonded cages formed by the OH groups of the alcohol.

The H_2O–NPOE interface was very recently simulated by Jorge et al. [21] who have shown that the presence of an alkyl chain in NPOE introduces an added degree of hydrophobicity compared to the H_2O–NB interface, resulting in an increase of interfacial tension. Also, interfacial NPOE molecules appear less organized than nitrobenzene molecules.

1.2.1.2 Aqueous Ion Solvation at the Interface

Apart from the simulation of purely molecular interfaces between two pure solvents, molecular dynamics has been very useful in apprehending aqueous ion solvation in the interfacial region. The landmark paper in this field was a publication by I. Benjamin who showed how the presence of a cation in the interfacial region perturbs the interfacial structure, the ion–dipole interactions creating water fingers when the ion enters the organic phase [22]. This concept was confirmed in subsequent calculations for anions such as chloride [23]. In 1999, Schweighofer and Benjamin studied the transfer of tetramethylammonium (TMA^+) at the H_2O–NB interface [24]. This paper presented some interesting conclusions. First, unlike alkali-metal ions such as Na^+, TMA^+ undergoes a complete change of solvation shells. The potential of mean force calculated corroborates well the electrostatic continuum model of Kharkats and Ulstrup [25] for the solvation energy profile. Finally, the dynamics of TMA^+ transfer under the influence of an electric field follows Stokes's law relating the drift velocity and electric field strength.

In 2002, dos Santos and Gomes published a study on calcium ion transfer across the H_2O–NB interface [26]. They observed a monotonous increase in the potential of mean force, that is, the solvation energy as the ion crosses the interface, and the process was found to be nonactivated. During the transfer from water to nitrobenzene, the first hydration shell remains intact, whereas the second hydration shell loses its water molecules.

In a recent publication by Wick and Dang [27], the excess concentration of cations (i.e., Na^+ and Cs^+) and of anions (i.e., chloride) at the H_2O–1,2-DCE interface has been studied, showing that these cations have a positive excess concentration and a lower potential energy at the interface, which is not observed for other solvents such as CCl_4, while Cl^- has a negative excess concentration although it shows a positive value at the liquid water–vapor and at the H_2O–CCl_4 interfaces. These authors argue that the uniqueness of the H_2O–DCE interface stems from the average interfacial 1,2-DCE molecule orientation, resulting in favorable cation interactions but unfavorable Cl^- interactions.

Back in 1983, the concept of mixed solvent layer [16] resulted from the determination of water surface excess concentrations at different interfaces by interfacial tension measurements that showed that, in the case of the H_2O–DCE interface, and unlike the liquid water–vapor or the water–heptane interfaces, the water excess concentration was less than a monolayer as expected for aqueous 1:1 electrolyte. The molecular dynamics results of Wick and Dang seem therefore to corroborate this early concept of interfacial structure in the presence of electrolytes in the aqueous phase.

One should also mention the work of Jorge et al. who have studied ion solvation at the H_2O–NPOE interface [28].

1.2.1.3 Lipophilic Ion Solvation at the Interface

Regarding the possible adsorption of lipophilic ions, different studies have been performed to see whether these lipophilic ions are preferentially located at the interface. For example, Chevrot et al. have studied the widely used cobalt bis(dicarbollide) anions $[(B_9C_2H_8Cl_3)_{(2)}Co]^-$, CCD^-, at the water-nitrobenzene [29] and at the water–chloroform interface [30]. This anion adsorbs at the former and very much at the latter, acting as a surfactant despite its nonamphiphilic nature. These authors attribute the excellent extracting properties of CCD^- to this specific adsorption as illustrated in Figure 1.1.

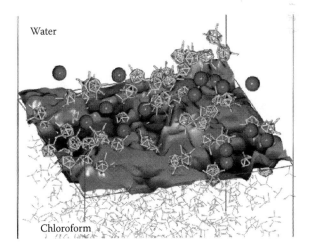

FIGURE 1.1 Distribution of all CCD^- anions and of Cs^+ ions within 10 Å from the interface. The surface of the interface is color coded as a function of its z-position. (Chevrot, G., R. Schurhammer, and G. Wipff, 2006, *J Phys Chem B*, Vol. 110, p. 9488. Used with permission.)

1.2.1.4 Water–Ionic Liquid Interfaces

The interface between water and the ionic liquid made of 1-butyl-3-methylimida-zolium cations (BMI^+) and bis(trifluoromethylsulfonyl)-imide anions (Tf_2N^-) has been recently simulated by Sieffert and Wipff [31]. Performing demixing experiments, these authors found that "the randomly mixed liquids separate much more slowly (in 20 to 40 ns) than classical water–oil mixtures do (typically, in less than 1 ns), finally leading to distinct nanoscopic phases separated by an interface." The width of the interface was found to be sharper than that calculated when using another anion, namely, PF_6^- [32].

In summary, molecular dynamics has confirmed what was expected from surface excess concentration measurements—that the more miscible the solvents, the rougher the interface, the lower the interfacial tension. It has also confirmed that lipophilic ions are specifically adsorbed on the organic side of the interface. In addition, it has introduced the concept of water protrusions or water fingers in the organic phase. It has clearly shown that the presence of ionic species enhances the formation of protrusions. In the case of solvents with a hydrocarbon chain such as hexanol, heptanone, or nitrophenyloctyle-ther, molecular dynamics has demonstrated the layering of the first organic solvent molecules.

1.2.2 SPECTROSCOPIC STUDIES

1.2.2.1 Roughness Measurement

In 1995, Michael and Benjamin had suggested that picosecond time-resolved fluorescence following the excitation of amphiphilic solutes adsorbed at the interface could be used to probe the width of the interface [33]. In 1999, Ishizaka et al. performed the first experiment to probe the interfacial roughness of the $H_2O–CCl_4$ and the H_2O–1,2-DCE interface by measuring the dynamic fluorescence anisotropy of sulforhodamine 101 (SR101) using time-resolved total internal reflection (TIR) fluorimetry [34].

If the roughness of the interface is comparable to the molecular size of SR101, its rotational motions are strongly restricted in the interfacial layer, and its emission dipole moment is within the X-Y plane of the interface. In such a case, the time profile of the total fluorescence intensity of the interfacial dye should be proportional to $I_{//}(t)+I_{\perp}(t)$, where $I_{//}(t)$ and $I_{\perp}(t)$ represent the fluorescence decays with emission polarization parallel and perpendicular to the direction of excitation polarization, respectively. When the angle of the emission polarizer is set at $45°$ with respect to the direction of excitation polarization (magic angle), fluorescence anisotropy is canceled, and the TIR fluorescence decay is given by a single-exponential function. If the interfacial layer is thick or rough, the interfacial molecules are weakly oriented; the rotational motions of SR101 take place rather freely, similar to those in a bulk phase. In this case, the total fluorescence intensity must be proportional to $I_{//}(t)+2I_{\perp}(t)$, and the magic angle must be equal to $54.7°$. The magic-angle dependence revealed that rotational reorientation of SR101 at the $H_2O–CCl_4$ interface was restricted in the two-dimensional plane of

TABLE 1.1

Data for the Time-Resolved Fluorescence Anisotropy of SR101

Organic Phase	Interfacial Tension/ mN·m⁻¹	Magic Angle	Fractal Dimension	$E_T(30)$/ kcal·mol⁻¹
Cyclohexane	51	45°	1.90	30.9
CCl₄	45	45°	1.93	32.4
Toluene	33	~45°	2.13	33.9
Chlorobenzene	37	~45°	2.20	36.8
O-Dichlorobenzene	39	45 to ~ 54.7°	2.30	38
1,2-Dichloroethane	28	~54.7°	2.48	41.3

Source: Ishizaka, S., H. B. Kim, and N. Kitamura, *Anal Chem*, Vol. 73, 2001, p. 2421.

the interface, while at the H_2O–DCE interface it took place rather freely as in an isotropic medium.

Furthermore, energy transfer dynamics measurements between SR101 and another dye, Acid Blue 1 (AB1), at the H_2O–CCl_4 or H_2O–DCE interface were measured. The fluorescence dynamics are given by

$$I_D(t) = A \exp\left[-\left(\frac{t}{\tau_D}\right) - P\left(\frac{t}{\tau_D}\right)^{d/6} \right] \qquad (1.1)$$

where A is a preexponential factor, and τ_D is the excited-state lifetime of the dye SR101 in the absence of the dye AB1. P is a parameter proportional to the probability that AB1 resides within the critical energy transfer distance R_0 of the excited donor, and d is called the fractal dimension. It should be around 2 for a planar geometry and 3 for a bulk geometry. The value d was found to be equal to 2 and 2.5 for the H_2O–CCl_4 and H_2O–DCE interfaces, respectively. This also indicated that the H_2O–CCl_4 interface was sharp with respect to the molecular size of SR101 (about 1 nm), while the H_2O–DCE interface was relatively rough compared to the H_2O–CCl_4 interface. In 2001, Ishizaka et al. extended this study of roughness measurements to different solvent pairs, and the main results are given in Table 1.1 [35].

Actually, in 2004, Kornyshev and Urbakh proposed a theoretical model to show that the dependence of the direct energy transfer signal on the potential drop across the interface can give valuable information about the interfacial dynamic corrugations and pattern formation on the length scales between 1 and 10 nm [36].

1.2.2.2 Polarity Study

The polarity of a liquid–liquid interface is an important factor to consider for heterogeneous reaction kinetics, as the solvent environments at the interface are different from those in bulk media. In 1998, Wang et al. reported a second harmonic generation (SHG) spectroscopic study on the polarities of H_2O–DCE and H_2O–CB interfaces by using N,N-diethyl-p-nitroaniline (DEPNA) as a probe [37]. According to their study, interfacial polarity can be considered equal to the arithmetic average of the polarity of the adjoining bulk phases, indicating that long-range solute–solvent interactions determine the difference in the excited and ground-state solvation energies of the interfacial molecules rather than local interactions.

Ishizaka et al. have also studied the polarity of a liquid–liquid interface but by time-resolved TIR fluorimetry [35]. In bulk solutions, the nonradiative decay rate constant of the polarity sensitive probe sulforhodamine B (SRB) increased with an increase in a solvent polarity parameter $[E_T(30)]$, and this relationship was used to estimate the polarities of water–oil interfaces. The nonradiative decay is given by this pseudoempirical equation:

$$k_{nr} \propto \exp\left[-\left(\frac{\beta}{RT}+\kappa\right)\left(E_T(30)-30\right)\right]\exp\left[-\frac{\Delta G^0_{A^*B^*}}{RT}\right] \tag{1.2}$$

where β and κ are constants, and $\Delta G^0_{A^*B^*}$ is the Gibbs energy difference between the fluorescent molecules A* and B* in a nonpolar solvent. A* can only decay radiatively to the ground state, S_0, and is in rapid equilibrium with a nonemissive state, B*, which can only decay to S_0 via internal conversion. $E_T(30)$ is an empirical parameter often used to indicate the polarity of a solvent. It is based on the absorption spectra of the solvatochromic dye known as Dimroth–Reichardt's betaine and calculated from the spectral data as follows:

$$E_T(30)(kcal\cdot mol^{-1}) = hc\tilde{v}_{max}N_A = \frac{28591}{\lambda_{max}}(nm) \tag{1.3}$$

where $\tilde{v}_{max}$ is the wavenumber and λ_{max} the maximum wavelength of the intramolecular charge-transfer π–π* absorption band of Dimroth–Reichardt's negatively solvatochromic pyridinium N-phenolate betaine dye [38].

For an oil phase of a relatively low polarity $[E_T(30) < 35 \text{ kcal}\cdot\text{mol}^{-1}]$, the polarity of the water–oil interface agreed with that of the arithmetic average of the polarities of the two phases, as predicted by Wang et al. [37]. For o-dichlorobenzene and 1,2-DCE of relatively high polarity $[E_T(30) > 35 \text{ kcal}\cdot\text{mol}^{-1}]$, the interfacial polarity determined by TIR spectroscopy was lower than the average value. The results were discussed in terms of orientation of the probe molecules at the interface as shown in Figure 1.2.

Another approach proposed by Steel and Walker is based on the concept of molecular rulers [39]. These rulers are solvatochromic surfactants composed of an anionic sulfate group attached to a hydrophobic, solvatochromic probe by alkyl spacers of different lengths. The probe is p-nitroanisole, an aromatic solute whose bulk solution excitation wavelength monotonically shifts by more than 20 nm from 293 to 316 nm as the solvent polarity or static dielectric constant varies from 2 for cyclohexane to 78 for water. To measure only the absorbance of

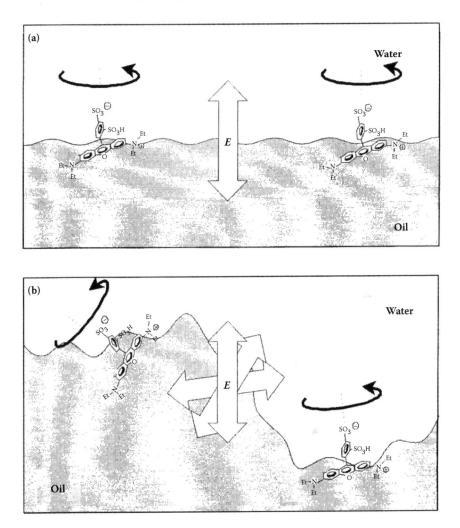

FIGURE 1.2 Schematic illustrations of (a) sharp and (b) rough water–oil interfaces. E denotes the direction of the electric field generated across the water–oil interface. (Ishizaka, S., H. B. Kim, and N. Kitamura, 2001, *Anal Chem*, Vol. 73, p. 2421. Used with permission.)

the probe located at the interface, the authors have used surface second-harmonic generation (SSHG). The resonance maximum for the probe alone adsorbed at the water–cyclohexane interface was found at 308 nm, consistent with the proposition of Eisenthal that the local dielectric environment can be represented by averaged contributions from the adjacent phases [37]. In the case of the molecular ruler, the resonance maximum shifted to that of the cyclohexane limit when the spacer was varied from C_2 to C_6. In the case of the water–octanol interface, the molecular ruler was located in the alkane layer of the first octanol layer, the same as that reported for by molecular dynamics for the water–hexanol interface [20].

1.2.2.3   Interfacial Acid–Base Equilibria

Surface second harmonic generation (SSHG) is very useful to measure the spectrum of interfacial species and therefore to do a pH titration in the interfacial layer that is not centrosymmetric. The concept of measuring a surface pK_a was introduced by Zhao et al. [40] who showed that the surface pK_a of p-hexadecylaniline was 3.6 compared to a bulk value of 5.3, indicating that the interface prefers to accommodate neutral rather than charged species.

In 1997, Tamburello et al. also measured the surface concentrations of different forms of eosin B at the air–water interface [41]. Two surface pK_a values were measured to be 4.0 and 4.2, that is, larger values than the bulk values of 2.2 and 3.7, respectively. These shifts indicate that the neutral and the monoanionic forms of eosin B are favored at the interface compared to the monoanionic and the dianionic forms, respectively.

Similar experiments were also carried out at ITIES, first by the group of Higgins and Corn [42] who studied the pH dependence of the adsorption of amphoteric surfactants such as 2-(n-octadecylamino) naphthalene-6-sulfonate (ONS) at the polarized H_2O–DCE interface and observed the polarization dependence of the protonation. A more thorough study was carried out on 4-(4′-dodecyloxyazobenzene) benzoic acid [43,44]. In 2004, Pant et al. used the same technique to monitor the acid–base properties of Coumarin 343 (C343) at the H_2O–DCE interface [45]. A pH-dependent aggregation was observed: at pH values smaller than 8, C343 adsorbs in J-aggregated protonated form; at pH = 9–10, C343 adsorbs in both protonated and deprotonated forms; and at pH = 11, C343 adsorbs in H-aggregated deprotonated form at the interface. The observed large shift in pK_a value of C343 at the interface is attributed to intramolecular hydrogen bonding along with the aggregation of dye molecules. Surface tension data show a weak adsorption of C343 at the interface for pH = 11 and pH = 3 and a strong adsorption at the intermediate pH values, reaching a maximum at pH = 10, which is consistent with the SHG data.

In summary, spectroscopic studies of the properties of liquid–liquid interfaces have corroborated the conclusions drawn from molecular dynamics, namely, that the H_2O–CCl_4 interface is much sharper than the H_2O–DCE interface. Additionally, it is clear that the interfacial polarity can be considered as the average of the polarities of the two solvents. Finally, it is worth pointing out that spectroscopic data can be directly compared to molecular dynamics calculations to extract structural information as recently reviewed by Benjamin [46].

1.2.3   Polarized ITIES

1.2.3.1   Potential Window

An ITIES is, by definition, the interface between two immiscible electrolyte solutions. As for an electrode–electrolyte interface, we can distinguish polarizable and polarized interfaces. A polarizable interface usually separates a very

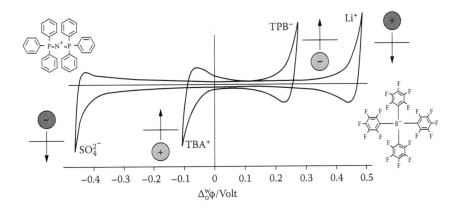

FIGURE 1.3 Potential window for a system comprising $LiSO_4$ in water and TBATPB, that is, tetrabutylammonium tetraphenylborate in 1,2-DCE limited, respectively, by the transfer of TBA$^+$ and TPB$^-$ from the organic to the aqueous phase as illustrated and BTTPATFPFB, that is, bis(triphenylphosphoranylidene)ammonium tetrakis(pentafluorophenyl)borate as drawn. In that case, the potential window is limited, respectively, by the transfer of the aqueous ions sulfate and lithium to the organic phase as illustrated.

hydrophilic salt in water, such as lithium chloride (LiCl), and a very lipophilic salt in the organic phase, such as tetraheptylammonium tetra-kis-4-chlorophenylborate (THA$^+$TPBCl$^-$), or a very hydrophobic ionic liquid such as 1-octyl-3-methylimidazolium bis(nonafluorobutylsulfonyl)imide (C$_8$mim$^+$C$_4$C$_4$N$^-$) [47]. Figure 1.3 illustrates potential windows for polarizable interfaces.

A system is said to be polarizable if one can change the Galvani potential difference or, in other words, the difference of inner potentials between the two adjacent phases without changing *noticeably* the chemical composition of the respective phases, that is, without *noticeable* electrochemical reactions taking place at the interface. A system is said to be polarized if the distribution of the different charges and redox species between the two phases determine the Galvani potential difference.

In the two cases, it is interesting to know how the electric potential varies from one phase to the next and therefore what the charge distribution is on either side of the interface. At present, molecular dynamics is not powerful enough to treat a system containing enough ions to evaluate the electric potential variation. Theoretical approaches have to more rely on classical models such as the modified Poisson–Boltzmann equation.

1.2.3.2 Capacitance Measurements

The electrical potential distribution has been extensively studied in the 1980s and 1990s by capacitance measurements, as excellently reviewed in 1998 by Samec [48]. The first model of polarized ITIES is that of Verwey–Niessen [49], dating back to 1939, of two back-to-back diffuse layers, and then adapted by Gavach et al. [50,51] by considering the presence of an inner layer. The key problem with this inner layer concept is that its thickness should be measurable with a Parsons–Zobel plot [52] with a positive intercept when plotting the reciprocal of the experimental capacitance data versus the reciprocal of the capacitance from the diffuse layers, also called the Gouy–Chapman capacitance. At the mercury–electrolyte interface, the intercept is always found to be positive, but at ITIES, the intercept is negative, which means that either the diffuse layers interpenetrate or that some ion-pairing takes place between ions from each phase, or that a mixed solvent layer must be considered. Pereira et al. have indeed observed that the interfacial capacity is always larger than that predicted by the Gouy–Chapman theory [53,54]. Back in 1986, Torrie and Valleau had shown by Monte Carlo simulations that image forces at the boundary between two dielectric media played a nonnegligible role [55]. In 1999, Huber et al. [56] presented simulations with a lattice-gas model, which is useful for modeling the space charge regions, as this approach is midway between molecular dynamics simulations with realistic interaction potentials able to treat only fairly small systems, and analytical models such as the Gouy–Chapman model, which have to rely on mathematical approximations sometimes difficult to verify experimentally. In particular, it is possible to treat ensembles that are sufficiently large to include the space-charge regions. In this way, they were able to show how the nature of the ion influences the capacitance data. The larger the Gibbs energy of transfer of an ion from one solution to the other, the smaller the overlap between the space-charge regions, and the lower the interfacial capacity. The same lattice-gas model had been used previously to show how ion interactions and pairing can explain the asymmetry of the capacity response, depending on the nature of the ions [57]. At the same period, Pecina and Badiali have shown that both ionic adsorption and the roughness of the interface lead to a higher capacity compared with the prediction of the Gouy–Chapman theory. They introduced a correction from a flat geometry involving the interplay between a roughness function in terms of a height–height correlation function of the surface, the Debye lengths of the system, and a length characterizing the adsorption [58,59]. The increase of capacitance with surface roughness was corroborated by Daikhin, Kornyshev and Urbakh [60]. The same authors later showed how capillary waves are affected by interface polarization [61]. For low potential drops across the interface, the mean square height $<\xi^2>$ increases proportional to V^2:

$$<\xi^2> \approx <\xi^2>_{pzc} + \frac{V^2 k_B T}{4\pi\gamma^2} C_{GC} \ln\left(\frac{\kappa_1\kappa_2}{(\kappa_1+\kappa_2)k_{gr}}\right) \tag{1.4}$$

where $<\xi^2>_{pzc}$ is the mean square height at the potential of zero charge, V the applied potential difference, γ the interfacial tension, C_{GC} the Gouy–Chapman capacitance, κ the respective Debye lengths, and k_{gr}^2 the small-wave-vector gravitational cutoff given by $\Delta\rho g/\gamma$, where g is the gravitational acceleration and $\Delta\rho > 0$ is the difference in the densities of two liquids. For higher potential drops, the height of the roughness grows even faster than described by Equation 1.4, and the interface may become unstable.

In 2001, Urbakh et al. summarized a number of outstanding questions [62]:

> Which characteristics of the ionic density profiles determine the observed dependences of the capacitance on the nature of the ions?
> Which parameters control the sign of the deviation of the capacitance from the Gouy–Chapman result and the asymmetry of the capacitance curves as a function of the potential?
> What information on the free energy profile of the ions across the interface can be obtained from the capacitance data?

To answer these questions, these authors developed an analytical model based on a modified nonlinear Poisson–Boltzmann equation, taking into account the overlap of the two back-to-back diffuse layers and the resulting differential equation:

$$\frac{d}{dz}\varepsilon_0(z)\frac{d}{dz}\psi(z)+\frac{4\pi e^2}{k_B T}N_0(\psi(z))=-\frac{d}{dz}\delta\varepsilon(z)\frac{d}{dz}\psi(z)$$

$$-\frac{4\pi e^2}{k_B T}[N(\psi(z))-N_0(\psi(z))]$$

$$(1.5)$$

where $eN_0(\psi(z))$ is the charge density at a sharp interface. If the left-hand term of Equation 1.5 is equal to zero, then we have the classical Gouy–Chapman differential equation for a sharp interface. The right-hand side of the equation provides a correction to the Gouy–Chapman case, considering the diffuseness of the interface and the presence of the adjacent "unfriendly" phase. In this way, with a perturbation approach, they were able to derive an analytical expression for interfacial capacitance.

$$C=\frac{dQ}{dE}=C_{G-C}\left[\frac{dU_0}{dV}+L_1\frac{dU_1}{dV}+L_2\frac{dU_2}{dV}+L_3\frac{dU_3}{dV}\right] \qquad (1.6)$$

where the first term in Equation 1.6 defines the Gouy–Chapman capacitance of two back-to-back ionic double layers separated by a sharp interface, and the three other terms are caused by the overlap of the double layers in the interfacial region and a smooth variation of dielectric properties across the interface. The integral parameters L_1 and L_2 represent length parameters that depend only on the specific interaction of ions of the "first" and "second" salt with the contacting solvents. The integral parameter L_3 is a length parameter that describes the dielectric property profile. The functions U are functions of the overall potential difference V (see [62] for details). This model was then further refined to take into consideration a "mixed boundary layer" where the overlapping of two space-charge regions occurs, and the effects of ion association and adsorption at the interface [63]. In this way, they could derive a more user-friendly equation for capacitance:

$$C=\frac{C_d^1 C_d^2}{C_d^1+C_d^2}\left(1+\frac{e}{C_d^1}\left[\frac{d\Gamma_1^+}{dE}-\frac{d\Gamma_1^-}{dE}\right]-\frac{e}{C_d^2}\left[\frac{d\Gamma_2^+}{dE}-\frac{d\Gamma_2^-}{dE}\right]\right) \qquad (1.7)$$

where C_d^1 and C_d^2 are the diffuse double-layer capacitances in the two adjacent phases, which can be calculated within the Gouy–Chapman theory considering the surface charge density in addition to the charge in the diffuse double layer, and Γ represents the surface concentration in the interfacial microscopic layer where overlap of double layers and ion association occur.

In 2005, an interesting critical summary of the current knowledge on capacitance measurements and potential distribution, together with a new model, was published by Monroe et al. [64]. In this work, the good old Verwey–Niessen theory was extended to allow ionic penetration at the interface. With this adaptation, several features could be accounted for, such as asymmetry and shifts of the capacitance minimum, that could not be described by the classical Gouy–Chapman or Verwey–Niessen theories. Gibbs energies of ion transfer were used as input parameters to describe ionic penetration into the mixed-solvent interfacial layer, and experimental data were successfully reproduced.

Recently, Markin and Volkov have studied the adsorption of ion pairs at ITIES using a generalized Langmuir isotherm that takes into account the limited number of adsorption sites, the final size of molecules, the complex formation at the interface, and the interaction between adsorbed particles, and were able to fit the interfacial tension variation with salt concentrations [65–67].

1.2.3.3 X-ray Reflectivity

In 2006, the first x-ray reflectivity study of an ITIES was published in a series of papers by Luo et al. [68–70]. They studied an interface between a nitrobenzene solution of tetrabutylammonium tetraphenylborate (TBATPB) and an aqueous solution of tetrabutylammonium bromide (TBABr). The concentration of TBABr was varied to control the Galvani potential difference using an experimental setup as shown in Figure 1.4. The ion distributions were predicted by a Poisson–Boltzmann equation, that explicitly includes a free energy profile for ion transfer across the interface described either by a simple analytic form or by a potential of mean force from molecular dynamics simulations.

When varying the temperature of the system, the authors showed clearly that the measured interfacial width differs from the predictions of capillary wave theory with a progressively smaller deviation as the temperature is raised. It was therefore concluded that both molecular layering and dipole ordering parallel to the interface must take place. Either layering or a bending rigidity that can result from dipole ordering could explain these measurements [69]. Regrettably, capacitance measurements were not carried out directly on the systems studied by x-ray reflectivity, which could have led to a direct correlation between capacitance data and values recalculated, for example using Equation 1.6, from the ionic distribution profiles. Very recently, Onuki presented some calculations for a 3-ion system that corroborate the x-ray reflectivity measurements [71]. The calculations of the ion distributions at the interface take into account the solvation and image interactions between ions and solvent and, in the case of two ions, show that hydrophilic and hydrophobic ions tend to undergo a microphase separation at an interface, giving rise to an enlarged electric double layer.

1.2.3.4 Specific Adsorption at ITIES

Following the discovery in 1957 of electroadsorption by Guastalla [72], that is, the lowering of the interfacial tension at an ITIES when surfactants are driven to the interface upon the passage of a current through this interface [1], many groups have studied the potential dependent specific adsorption of ionic surfactants at ITIES. The concept of electrocapillary phenomena at ITIES in the presence of surfactants stems from the pioneering work of Watanabe et al. [73,74], and of Dupeyrat and Nakache [75]. In 1987, Kakiuchi et al. measured electrocapillary curves in the presence of ionic surfactants and showed how the specific adsorption affects the potential distribution as it does on a solid electrode as shown in Figure 1.5 [76].

In 2001, Kakiuchi et al. showed how the relationship between the Gibbs energy of adsorption and transfer for an ionic surfactant leads to the adsorption maximum being attained when the value of the applied potential difference is close to that of the standard ion-transfer potential. This takes place also in the presence of concentration gradients in the vicinity of the interface and predicts that the adsorption is maximum around the half-wave potential in voltammetric measurements [78]. This relationship is also applied in electrocapillary emulsification.

The specific adsorption of ions at the ITIES has been studied by Su et al. who have simulated the capacitance response for different adsorption isotherm conditions [77]. Figure 1.6 illustrates the effect of adsorption on the capacitances and potential distribution in the case of a potential dependent Langmuir isotherm for which the surface coverage is given by

$$\theta = \frac{a_z}{a_{H_2O}} \frac{\exp\left(-\Delta G_a^{\ominus}/RT\right)\exp\left[-zF\left(\phi^2-\phi^w\right)/RT\right]}{1+\frac{a_z}{a_{H_2O}}\exp\left(-\Delta G_a^{\ominus}/RT\right)\exp\left[-zF\left(\phi^2-\phi^w\right)/RT\right]} \qquad (1.8)$$

1.2.3.5 Adsorption and Instability of ITIES

Mechanical instability of ITIES due to adsorption and chemical desorption of surfactants has been known for many years. For example, Arai et al. studied the instability of the water–octanol interface caused by the adsorption of sodium dodecylsulfate by potentiometry and could relate interfacial polarization and oscillations [79]. The early work has been thoroughly reviewed by Kihara and Maeda [80]. In 2001, Maeda et al. demonstrated that self-sustained oscillations could be studied for a system comprising cetyltrimethylammonium in water and picric acid in nitrobenzene [81].

In 2002, Kakiuchi et al. proposed a thermodynamic theory that showed that, upon coupled adsorption and transfer of surfactants, the electrocapillary curve may have a positive curvature, as opposed to a negative curvature in classical electrocapillary theory. This implies that the double-layer capacitance would be

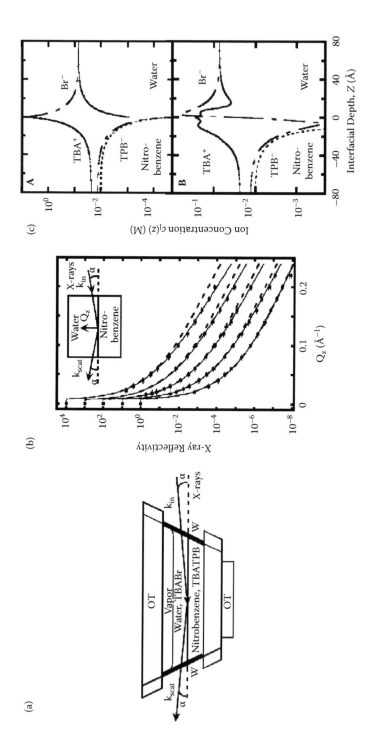

FIGURE 1.4 (a) Cross-sectional view of stainless steel sample cell. W: mylar windows; T: thermistors to measure temperature. The kinematics of surface x-ray reflectivity is also indicated: k_{in} is the incoming x-ray wave vector, k_{scat} is the scattered (reflected) wave vector, α is the angle of incidence and reflection (Luo, G., S. Malkova, J. Yoon, D. G. Schultz, B. Lin, M. Meron, I. Benjamin, P. Vanysek, and M. L. Schlossman, *J Electroanal Chem*, Vol. 593, 2006, p. 142.) (Middle) X-ray reflectivity, $R(Q_z)$, as a function of wave vector transfer Q_z from the interface between 0.01 M TBATPB in nitrobenzene and a TBABr solution in water at five concentrations (0.01, 0.04, 0.05, 0.057, and 0.08 M, bottom to top) at a room temperature of $24°C \pm 0.5°C$. Solid lines are predictions using the potential of mean force from molecular dynamics (MD) simulations. Dashed lines are predicted by the Gouy–Chapman model. No parameters have been adjusted in these two models. Data for different concentrations are offset by factors of 10 (R = 1 at $Q_z = 0$). Error bars are indicated by horizontal lines through the square data points and are usually much smaller than the size of the squares. The points at $Q_z = 0$ are measured from transmission through the bulk aqueous phase. (b) The kinematics of x-ray reflectivity: k_{in}, incoming x-ray wave vector; k_{scat}, scattered wave vector; and α, angles of incidence and reflection. (c) Ion distributions at the interface between a 0.08 M TBABr solution in water and a 0.01 M TBATPB solution in nitrobenzene. Solid lines, TBA+; short–long dashed line, Br−; short dashed line, TPB−. (A) Gouy–Chapman theory. (B) Calculation from MD simulation of the potential of mean force. (Luo, G. M., S. Malkova, J. Yoon, D. G. Schultz, B. H. Lin, M. Meron, I. Benjamin, P. Vanysek, and M. L. Schlossman, 2006, *Science*, Vol. 311, p. 216. Used with permission.)

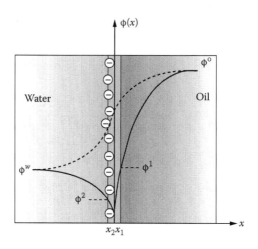

FIGURE 1.5 Schematic representation of the potential profiles across the ITIES in the absence (dashed line) and presence (solid line) of specific adsorption of anionic species from the aqueous phase to the interface. (Su, B., N. Eugster, and H. H. Girault, 2005, *J Electroanal Chem*, Vol. 577, p. 187. Used with permission.)

negative, which is unrealistic, and, as a consequence of this, the interface escapes from the unstable state by dissipating energy through spontaneous emulsification or other Marangoni-type movements of the interface. Unlike classical electro-emulsification in which the interfacial tension is driven close to zero, the criterion of the positive curvature proposed by Kakiuchi can explain the cases of spontaneous emulsification at finite interfacial tension values, and the regular oscillations observed with the transfer of surface-active ions across the electrified liquid–liquid interface [82]. Experimental studies by Kakiuchi's group over the years have confirmed this thermodynamic model and shown in particular the influence of interface size on the instability phenomena [83–87].

More recently, this topic has been revisited by Daikhin and Urbakh [88] who presented a kinetic description of ionic surfactant transfer across an ITIES that includes the charging of the interface, adsorption, and transfer as well as characteristics of the electrical circuit. This model showed that the irregular current oscillations are due to a dynamic instability induced by the interplay between the potential-dependent adsorption and direct transfer across the interface. In particular, this model showed that current anomalies occur in a potential range close to the standard ion-transfer potential.

1.2.3.6 Water–Ionic Liquid Interfaces

The capacitance of the water–ionic liquid interface was recently studied by Ishimatsu et al. [89] who measured electrocapillary curves and showed that the aqueous ions Li^+ and Cl^- are not specifically adsorbed on the aqueous side of the interface. More recently, Yasui et al. [90] have shown ultraslow responses, on the order of minutes, for the variation of the interfacial tension as a function of

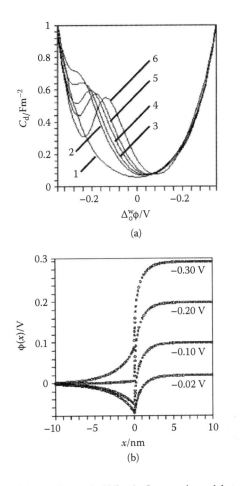

FIGURE 1.6   Differential capacitance (a, b) for the Langmuir model at various values of c_z. Potential profiles at various potential differences for the Langmuir mode. (Su, B., N. Eugster, and H. H. Girault, 2005, *J Electroanal Chem*, Vol. 577, p. 187. Used with permission.)

the applied potential for an interface between water and the ionic liquid, trioctylmethylammonium bis(nonafluorobutanesulfonyl)amide. This ultraslow relaxation was attributed to long-range and collective ordering of ions of the electrical double layer on the ionic liquid side of the interface.

In summary, the understanding of electrical polarization at liquid–liquid interfaces has improved much over the past decades, expanding beyond the classical Gouy–Chapman analysis. After many years of extensive capacitance data measurements for different solvent pairs, electrolytes, and so on, the different theoretical approaches all point toward a three-layer model in which the outer layers are classical diffuse layers that can be treated in a first approximation by the Poisson–Boltzmann equation, and a central layer that ions from both side can penetrate, and where solvent fingering may take place, and where the surface roughness is potential dependent. However, central questions remain: How to define the potential of zero charge? What is the relationship between the potential of zero charge, minimum capacitance, and electrocapillary maximum?

1.3 ION-TRANSFER REACTIONS

1.3.1 THERMODYNAMIC BACKGROUND

1.3.1.1 Nernst Equation for Ion Transfer

The basic equation for an ion-transfer reaction is the Nernst equation that results from the equality of the electrochemical potential of a given ion, i, in the two adjacent phases:

$$\mu_i^{\ominus,w} + RT \ln a_i^w + z_i F \phi^w = \mu_i^{\ominus,o} + RT \ln a_i^o + z_i F \phi^o \tag{1.9}$$

where $\mu_i^\ominus$ is the standard chemical contribution to the electrochemical potential, a the activity, and ϕ the inner potential. The Galvani potential difference, in other words the difference of inner potentials, is then the sum of a standard term and a concentration ratio term:

$$\Delta_o^w \phi = \phi^w - \phi^o = \Delta_o^w \phi_i^\ominus + \frac{RT}{z_i F} \ln \left(\frac{a_i^o}{a_i^w} \right) \tag{1.10}$$

$\Delta_o^w \phi_i^\ominus$ is the standard transfer potential equal to

$$\Delta_o^w \phi_i^\ominus = \frac{\Delta G_{tr,i}^{\ominus,w \to o}}{z_i F} \tag{1.11}$$

and expresses, in a voltage scale, the standard Gibbs energy of transfer $\Delta G_{tr,i}^{\ominus,w \to o}$ from water to the organic solvent. An important consequence of Equation 1.9 is that, contrary to a neutral molecule, the partition constant P_i of an ion is dependent on the Galvani potential difference between the two phases.

$$P_i = \frac{a_i^o}{a_i^w} = \exp \left[\frac{z_i F}{RT} \left(\Delta_o^w \phi - \Delta_o^w \phi_i^\ominus \right) \right] = P_i^\ominus \exp \left[\frac{z_i F}{RT} \Delta_o^w \phi \right] \tag{1.12}$$

$P_i^\ominus$ represents the standard partition constant when the interface is not polarized (potential of zero charge). In this way, when one applies a Galvani potential more positive than the standard value, cations will transfer from the aqueous to the organic phase, and the anions from the organic to the aqueous phase.

1.3.1.2 Distribution Potential for a Single Salt

Let us consider the distribution of a salt C^+A^- between two immiscible liquids. At equilibrium, we have an equality of the electrochemical potentials of both the cation and the anion, and by taking into account the electroneutrality in each phase

($c_{C^+} = c_{A^-}$), we obtain that the distribution of the salt between the two phases polarizes the interface, and that the Galvani potential difference is given by [91]

$$\Delta_o^w \phi = \frac{\Delta_o^w \phi_{C^+}^\ominus + \Delta_o^w \phi_{A^-}^\ominus}{2} + \frac{RT}{2F} \ln\left(\frac{\gamma_{C^+}^o \gamma_{A^-}^w}{\gamma_{C^+}^w \gamma_{A^-}^o} \right) \qquad (1.13)$$

In the case of dilute solutions, the second term in this equation is negligible. The Galvani potential difference imposed by the distribution of a salt is independent of the volume of the phases in contact and is called the distribution potential. In 2004, Markin and Volkov calculated the distribution potential in the case where one phase is of microscopic dimensions comparable to the Debye length, such that the electroneutrality condition is broken [92].

1.3.1.3 Distribution Potential for an Acid

A problem that is often encountered in preparative or pharmacological chemistry is linked to the distribution of acid or base molecules. To treat this problem, let us look at the simple case of the distribution of an acid between two phases (Figure 1.7).

At equilibrium, we have equality of the electrochemical potentials of the two ions A⁻ and H⁺ in the two phases (ignoring the presence of OH⁻ ions), we have the equality of the chemical potentials of the acid AH in the two phases, and we also consider the acid–base equilibria in both phases. As a result, the Galvani potential difference is given by the distribution potential of the A⁻ and H⁺ ions, and the acidity constant in the organic phase is linked to that in the aqueous phase by

$$K_a^o = \frac{a_{A^-}^o a_{H^+}^o}{a_{AH}^o} = K_a^w \frac{P_{A^-} P_{H^+}}{P_{AH}^\ominus} = K_a^w \frac{P_{A^-}^\ominus P_{H^+}^\ominus}{P_{AH}^o} \qquad (1.14)$$

This equation shows that, to calculate the pK_a^o of an acid in the organic phase knowing its pK_a^w value in water, we need to know the standard distribution coefficients of the various species involved.

1.3.1.4 Solvation Energy Profile

In 1991, Kharkats and Ulstrup calculated the Gibbs solvation energy profile for a sharp interface using a simple electrostatic continuum model [25]. The gist

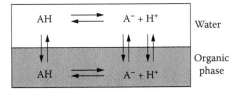

FIGURE 1.7 Distribution of an acid in neutral and ionized form.

of the method is to calculate the electrostatic energy of the ion using volume integrals:

$$W = \frac{1}{2} \iiint_V \vec{D} \cdot \vec{E} \, dv \tag{1.15}$$

By considering the image of an ion of radius a located at a distance h from the other electrolyte, we have the following relations for $h > a$:

$$W = \frac{(ze)^2}{32\pi\varepsilon_o\varepsilon_1 a} \left[4 + 2\left(\frac{\varepsilon_1-\varepsilon_2}{\varepsilon_1+\varepsilon_2}\right)\left(\frac{a}{h}\right) - \frac{1}{12}\left(\frac{\varepsilon_1-\varepsilon_2}{\varepsilon_1+\varepsilon_2}\right)^2\left(\frac{a}{h}\right)^4 \right] \tag{1.16}$$

and for $h \le a$,

$$W_1 = \frac{(ze)^2}{32\pi\varepsilon_o\varepsilon_1 a} \left[\begin{array}{l} 2\left(1+\dfrac{h}{a}\right)+2\left(\dfrac{\varepsilon_1-\varepsilon_2}{\varepsilon_1+\varepsilon_2}\right)\left(2-\dfrac{h}{a}\right) \\[2ex] +\left(\dfrac{\varepsilon_1-\varepsilon_2}{\varepsilon_1+\varepsilon_2}\right)^2\left(\dfrac{[1-(2h/a)][(h/a)+1]}{(2h/a)+1}+\dfrac{\ln[(2h/a)+1]}{2h/a}\right) \end{array} \right]$$

$$+ \frac{\varepsilon_2}{4\pi\varepsilon_o a_1}\left(\frac{ze}{\varepsilon_1+\varepsilon_2}\right)^2\left(1-\frac{h}{a}\right) \tag{1.17}$$

As can be seen in Figure 1.8, the change of the electrostatic contribution to the Gibbs energy is smooth over a distance of a few ionic radii.

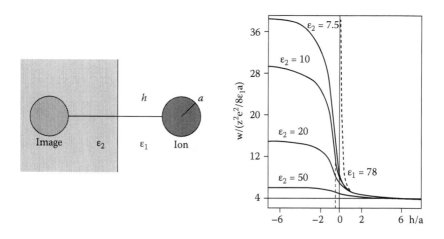

FIGURE 1.8 Left: Ion located near a dielectric boundary. Right: Gibbs solvation energy profile. (Kharkats, Y. I., and J. Ulstrup, 1991, *J Electroanal Chem*, Vol. 308, p. 17. Used with permission.)

1.3.2 Kinetic Measurements

Any kinetic measurements of an electrochemical reaction rate is by no means an easy task if the time scale for diffusion or mass transport in general is of the same order of magnitude as the transfer rate. In the case of liquid–liquid interfaces, the large iR drop associated with the low conductivity of the organic phase and the large double-layer capacitance are additional difficulties. For large macroscopic interfaces, techniques such as cyclic voltammetry or square-wave voltammetry, where it is difficult to decouple the ohmic loss from the limiting kinetic effect, cannot be used reliably. The most popular techniques have been impedance measurements that can separate mass transport from the electrochemical reaction in the frequency domain [93–95], and chronocoulometry where the kinetic data can be obtained by extrapolation of diffusion-controlled behavior [96,97]. The advantage of chronocoulometry, compared to, say, potential-step chronoamperometry, is that the extrapolation is made from a relatively long time domain. Indeed, the integration preserves the kinetic limitations on the ion-transfer current. In fact, there is no ideal technique to measure the kinetics of ion-transfer reactions as impedance measurements rely heavily on fitting the data on a model-equivalent circuit that might not always be appropriate, and as chronocoulometry has difficulty in taking the double layer charging into account. This makes any comparison between different publications difficult, but at least the trends observed with the same technique under similar experimental conditions can provide useful semiquantitative information.

1.3.2.1 Series of Analogous Ions

In the early days, most studies were dedicated to changing some system parameters and using a series of homologous ions. In 1983, Samec et al. showed by convolution potential voltammetry that, for a series of tetraalkylammonium cations, the logarithm of rate constant for the crossing of the inner layer, that is, the rate constant corrected for diffuse-layer effects, was directly proportional to the Gibbs energy of transfer [98]. In 1989, the same group revisited this ion series using impedance measurements at equilibrium potentials and measured much faster apparent rate constants for the tetraalkylammonium and viologen series [94]. Another very important conclusion was that the apparent rate constants were measured to follow a Butler–Volmer law, that is, $\ln k$ was found proportional to the applied voltage, as postulated by Gavach et al. in what can be considered as the first attempt to measure the rate constants of tetraalkylammonium transfer [99]. A year later, Shao et al. confirmed this Butler–Volmer behavior for the transfer of acetylcholine by dissolving the ion in either phase [96]. The Tafel slopes obtained are illustrated in Figure 1.9.

In 1992, Kakiuchi et al. showed, also by impedance measurements, that the apparent rate constants for anions were even faster than for cations of relatively equal radius [100] (Table 1.2). This led the authors to conclude that dielectric friction at the interface must be considered. They also showed than the Goldman-type current-potential relationship was more appropriate than the Butler–Volmer mechanism (see Section 1.3.3.1) [101].

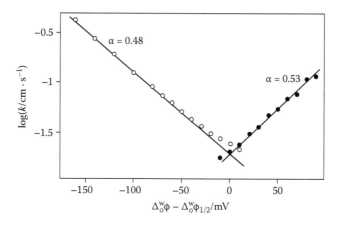

FIGURE 1.9 Tafel plots for acetylcholine transfer (°) from 1,2-DCE to water, and (•) from water to 1,2-DCE. (Adapted from Shao, Y., and H. H. Girault, 1990, *J Electroanal Chem*, Vol. 282, p. 59.)

When analyzing kinetic data for ion-transfer reactions, some groups used different corrections such as the Frumkin correction for the double-layer effects, or the Levich correction for the double-layer corrections under Nernst–Planck transport as proposed by Samec et al. [102]. These corrections have been thoroughly discussed by Murtomaki et al. who have clearly shown that the Frumkin correction can be used for equilibrium measurements, and the Levich correction must be used under faradaic conditions [103].

1.3.2.2   Effect of the Viscosity of the Adjacent Phases

In 1990, Shao and Girault studied the transfer of acetylcholine at the water+sucrose/1,2-DCE interface [96]. By addition of sucrose to the aqueous phase, one can vary its viscosity from 1 to 6 mPa·s without altering much the dielectric constant of the solution. Using chronocoulometry, it was shown that

TABLE 1.2
Standard Transfer Potential and Apparent Ion-Transfer Rate Constant

Ion	Me_4N^+	Et_4N^+	Pr_4N^+	Me_2V^{2+}	Et_2V^{2+}	Pr_2N^+
$\Delta_o^w\phi/mV$	27	−67	−170	−15	−43	−58
$k_{app}/cm\cdot s^{-1}$	0.14	0.09	0.14	0.05	0.05	0.07

Ion	PF_6^-	ClO_4^-	BF_4^-	SCN^-
$E_{1/2}/mV$	343	240	197	161
$k_{app}/cm\cdot s^{-1}$	0.16	0.11	0.17	0.09

Source: Wandlowski, T., V. Maecek, K. Holub, and Z. Samec, *J Phys Chem-Us*, Vol. 93, 1989, p. 8204; Kakiuchi, T., N. J., and M. Senda, *J Electroanal Chem*, Vol. 327, 1992, p. 63.

both the diffusion coefficient of the ion in the aqueous phase and the apparent standard rate constant were inversely proportional to the viscosity of the aqueous phase. Also, the addition of sucrose causes a variation of the Gibbs energy of hydration of acetylcholine, resulting in an observed variation of the Gibbs energy of transfer from the water + sucrose solution to 1,2-DCE. An interesting observation in this work was the linear dependence between $\ln D_{Ac^+}^w$ and the variation of hydration energy $\Delta G_{tr}^{w \to w+s}$. This suggests that the activation energy for diffusion is proportional to the solvation energy. Also, a linear relationship between $\ln k_{Ac^+}^{\ominus}$ and $\Delta G_{tr}^{w+s \to DCE}$ was observed. The direct proportionality of the rate constant for ion-transfer and the diffusion coefficient in the aqueous phase led these authors to conclude that ion-transfer reactions are a special case of ion transport through a heterogeneous solution. In 1995, Kakiuchi's group carried out a similar study at the H_2O–NB interface but found that increasing the viscosity of the aqueous phase had little effect on the rate of the TEA^+ transfer [104]. This suggests that varying the viscosity of the less viscous solution has little effect, as the interfacial viscosity is partially controlled by the adjacent phase of the highest viscosity (NB η = 2.03 mPa·s, DCE η = 0.84 mPa·s, at 20°C). Actually, T. Solomon had proposed that the viscosity data could be interpreted in a framework of Linear Gibbs Energy Relationships [105]. Indeed, viscosity or diffusion can be considered as activated processes and, therefore, variation of viscosity can be interpreted as changes of activation energies for diffusion, themselves proportional to the Gibbs energies of solvation.

1.3.2.3 Effect of the Dielectric Constant of the Adjacent Phases

In 1991, Shao et al. studied the transfer of acetylcholine from an aqueous solution to a mixture NB–CCl_4 [97]. By varying the organic phase composition, one can change the viscosity and the relative permittivity. Again, some linear relationships were observed, in particular, between $\ln k_{Ac^+}^{\ominus}$ and $\Delta G_{tr}^{w \to NB+TCM}$. Rather than considering that $\ln k_{Ac^+}^{\ominus}$ is proportional to the Gibbs energy of transfer (what is called the Brønsted relationship), it was proposed that $\ln k_{Ac^+}^{\ominus}$ is proportional to a linear combination of the Gibbs solvation energies, for example, the half sum in a first approximation.

$$k^{\ominus} = k^0 \exp\left[-\frac{\lambda^w \Delta G_s^w + \lambda^o \Delta G_s^o}{RT} \right] \tag{1.18}$$

In the same way that the Gibbs energy of activation for diffusion is proportional to the solvation energy, it was postulated that the Gibbs energy of activation for transfer is a weighted average of the Gibbs energy of solvation in the adjacent phases. Also in the same way that Eisenthal et al. postulated that the interfacial polarity is the average of the polarity of the two adjacent solvents [37], it was postulated that the activation of ion-transfer reactions depend on long-range interactions from the two electrolytes, and not from short-range local forces. Equation 1.8 is also in agreement with Solomon's approach to the viscosity data [105].

1.3.2.4   The Effect of Temperature

In 1992, Wandlowski et al. made a thorough study of the temperature dependence of the rate of ion-transfer reaction for a series of ammonium and phosphonium salts [95]. The apparent rate constants were found to increase with the temperature, and the authors deduced an activation energy barrier of about 5 kJ·mol⁻¹ that is rather low and that highlights the importance of entropic contribution.

In fact, an experiment that would be worth doing is to measure the pressure dependence for ion-transfer reactions. Indeed, in the 1960s Hills et al. measured the pressure dependence of ionic conductances in both water and NB to measure the volume of activation, and it would be very interesting to know the activation volume for the transfer reactions [106–108].

1.3.2.5   Kinetic Measurements at Micro/Nano ITIES

In the case of micro-ITIES, the rate of mass transport is enhanced when compared to linear diffusion, and higher rate constant values can be reached. To measure the rate of ion-transfer reactions, one can use nano-ITIES as proposed by Cai et al. in 2004 [109]. From a steady-state analysis using 25-nm-radius pipettes, Equation 1.19 yields a rate constant value of about 2.2 cm·s⁻¹ (Figure 1.10):

$$\frac{i}{i_d} = \frac{1}{1 + \left(\frac{1}{\lambda} + \frac{m_f}{m_r}\right)\exp[\alpha n F(E - E^{\ominus\prime})/RT]} \tag{1.19}$$

where m is the mass transfer coefficient for the forward and reverse reactions, and λ the kinetic parameter equal to $k^{\ominus}/m_f$.

In 2007, Rodgers and Amemiya used finite element simulation to fit experimental cyclic voltammograms and study TEA⁺ transfer from the outer aqueous phase into the inner NB phase [110] (Figure 1.11). The voltammogram was found reversible when using a 2.1 µm radius micropipette and no kinetic data could be extracted.

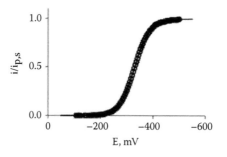

FIGURE 1.10   Background-subtracted steady-state voltammogram (symbols) of TEA⁺ transfer from DCE to water at a 25-nm-radius silanized pipette and the best theoretical fit using Equation 1.19. (Cai, C. X., Y. H. Tong, and M. V. Mirkin, 2004, *J Phys Chem B*, Vol. 108, p. 17872. Used with permission.)

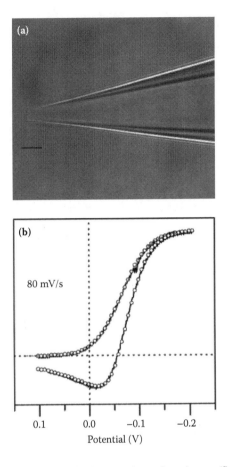

FIGURE 1.11 Optical microscopic image of a micropipette (Scale bar = 10 μm). Background-subtracted CV of simple TEA$^+$ transfer at a 2.1-μm-radius pipette. The vertical lines correspond to $\Delta_w^o \phi_{TEA^+}^{\Theta'}$. The open circles represent simulated CVs. (Rodgers, P. J., and S. Amemiya, 2007, *Anal Chem*, Vol. 79, p. 9276. Used with permission.)

In summary, the measurement of ion-transfer kinetics has been a real experimental challenge. As the techniques have progressed, higher and higher rate constants have been measured. One central question is: Why are the rate constants measured at micro- or nano-interfaces so much higher than those measured on macroscopic interfaces? Perhaps, here, the cotransport of solvent molecules plays a role if the two phases are not equilibrated prior to the experiments. At a large ITIES, we may assume that the two solvents are fully mutually saturated and at equilibrium. The cotransport of water molecules with the ions has little effect on the solvent composition. In the case of nano-ITIES, the current densities are so high that it may be possible that the ion transfer may destroy the interface by emulsification. All in all, in 2009, ion-transfer kinetics remains an open question!

1.3.3 Ion-Transfer Theory

For many years, there had been a lack of appropriate theories for ion-transfer reactions across an ITIES. There were two ways to consider these reactions: either as an activated process as originally proposed by D'Epenoux et al. [111] or as an ion conductivity process as proposed by Shao et al., based on the electrochemical gradient argument [96], by Kontturi et al. using the Nernst–Planck equation [112,113], and as proposed by Kakiuchi following the Goldman model of ion permeation through biological membranes [101,114,115].

1.3.3.1 Butler–Volmer Approach

Using a kinetic formalism, the flux corresponding to the transfer of an ion across a water–organic solvent interface can be written:

$$J = zF[k^{w \to o}c^w - k^{o \to w}c^o] \tag{1.20}$$

where c^w and c^o represent the bulk concentrations. Of course, the key question becomes: How do the rate constants depend on the applied potential?

When at equilibrium the bulk concentrations are equal, the potential difference is equal to the formal potential for the ion-transfer reaction, and the global activation energy barrier is symmetrical [91]. As the potential is varied, the overall driving force $zF(\Delta_o^w \phi - \Delta_o^w \phi^{\ominus'})$ partially lowers the activation energy barrier and, as in the Butler–Volmer mechanism, the current is then given by

$$I = zFAk^{\ominus}c\left[e^{zF\alpha(\Delta_o^w\phi - \Delta_o^w\phi^{\ominus'})} - e^{-zF(1-\alpha)(\Delta_o^w\phi - \Delta_o^w\phi^{\ominus'})}\right] \tag{1.21}$$

where α represents the fraction of the driving force operating on the global activation energy barrier. In the general case, when the concentrations in the adjacent phases are different, this equation becomes

$$I = zFAk^{\ominus}\left[c^w e^{zF\alpha(\Delta_o^w\phi - \Delta_o^w\phi^{\ominus'})} - c^o e^{-zF(1-\alpha)(\Delta_o^w\phi - \Delta_o^w\phi^{\ominus'})}\right] \tag{1.22}$$

Different approaches have been proposed to justify a Butler–Volmer behavior. In 1986, Gurevich and Kharkats proposed a stochastic approach to ion-transfer reactions [116–118]. Here, the main equation is the Langevin equation:

$$\frac{d\mathbf{u}}{dt} = -\beta\mathbf{u} + \frac{1}{m}\left[-\frac{\partial V}{\partial \mathbf{r}} + A(t,\mathbf{r})\right] \tag{1.23}$$

where $\mathbf{u}$ is the velocity vector of the ion, m its mass, $\partial V/\partial \mathbf{r}$ is the force acting on the ion in the regular potential V, and $A(t,\mathbf{r})$ is the stochastic force due to the fluctuations of the condensed phase. The value β is the phenomenological coefficient of friction describing the dissipative processes due to particle–medium interaction. From the Fokker–Planck equation, we obtain two equations that describe the distribution function f:

$$\frac{\partial f}{\partial t} = -\frac{\partial f}{\partial \mathbf{r}} \mathbf{j} \tag{1.24}$$

and

$$\mathbf{j} = -\frac{kT}{\beta m}\left[\frac{\partial f}{\partial \mathbf{r}} + \frac{1}{kT}\frac{\partial V}{\partial \mathbf{r}}\right] \tag{1.25}$$

The key problem to proceed is to know the potential V. In a first approximation, Gurevich and Kharkats used the electrical potential and, using a harmonic model, they end up with an expression very similar to a Butler–Volmer equation. In 1995, Indenbom proposed taking into account the elastic properties of the interface to derive an expression of the activation energy barrier [119].

In 1997, Schmickler published an ion-transfer theory based on a lattice-gas model [120]. The potential of mean force experienced by a single ion in the vicinity of a liquid–liquid interface was then calculated. Typically, it shows a maximum and a minimum, which means that ion transfer is treated as an activated process involving the transfer over a barrier following a Butler–Volmer-type law.

1.3.3.2 Goldman-Type Transfer

The flux of an ion is directly proportional to the gradient of electrochemical potential and can be expressed by the Nernst–Planck equation:

$$\begin{aligned}
\mathbf{J}_i(x) &= -c_i(x)\tilde{u}_i(x)\mathbf{grad}\tilde{\mu}_i(x) \\
&= -RT\tilde{u}_i(x)\mathbf{grad}c_i(x) - z_iFc_i(x)\tilde{u}_i(x)\mathbf{grad}\left(\phi(x) - \phi_i^{\ominus}(x)\right)
\end{aligned} \tag{1.26}$$

with $\phi_i^{\ominus}(x) = -\mu_i^{\ominus}/z_iF$. To facilitate its integration, we can write this equation as

$$\mathbf{J}_i(x) = -RT\tilde{u}_i(x)\mathbf{grad}\left[c_i e^{z_iF\left(\phi(x)-\phi_i^{\ominus}(x)\right)/RT}\right]\cdot e^{-z_iF\left(\phi(x)-\phi_i^{\ominus}(x)\right)/RT} \tag{1.27}$$

such that, integrating at constant flux, gives

$$\mathbf{J}_i\int_w^o \frac{e^{z_iF\left(\phi(x)-\phi_i^{\ominus}(x)\right)/RT}}{\tilde{u}_i(x)}\,\mathrm{d}x = -RT\int_w^o \mathrm{d}\left(c_i e^{z_iF\left(\phi(x)-\phi_i^{\ominus}(x)\right)/RT}\right) \tag{1.28}$$

If we assume as a first approximation that $\phi(x) - \phi_i^\ominus(x)$ varies linearly across the interface, by introducing the constants y_o and y_w,

$$y_o = \frac{z_i F}{RT}\left[\phi(o) - \phi_i^\ominus(o)\right], \quad \text{and} \quad y_w = \frac{z_i F}{RT}\left[\phi(w) - \phi_i^\ominus(w)\right] \qquad (1.29)$$

we have

$$z_i F\left(\phi(x) - \phi_i^\ominus(x)\right)/RT = y_w + (y_o - y_w)\frac{x}{L} \qquad (1.30)$$

Of course, the problem here is that the standard term $\phi_i^\ominus(x) = -\mu_i^\ominus/z_i F$, which expresses the standard chemical potential, varies in a stepwise manner across the mixed-solvent layer, say, 1 nm thick, whereas the potential drop varies in a monotonic way in the absence of specific adsorption across the two back-to-back diffuse layers, say 10 nm thick. As a result, Equation 1.30 is a very rough approximation, as soon as the Gibbs energy of transfer of the ion is larger than ± 5 kJ·mol^{-1}.

The second major assumption of this approach is to assume that the electrochemical mobility is constant throughout the interfacial region, in other words, that the viscous drag is the same in the two solvents, or that the diffusion coefficients are equal in the adjacent phases. Doing so, we have

$$\frac{\mathbf{J}_i}{\tilde{u}_i}e^{y_w}\int_0^L e^{(y_o - y_w)x/L}\,dx = \frac{L\mathbf{J}_i}{\tilde{u}_i}\left[\frac{e^{y_o} - e^{y_w}}{(y_o - y_w)}\right] = -RT\left[c_i^o e^{y_o} - c_i^w e^{y_w}\right] \qquad (1.31)$$

The ionic flux across the interface is then

$$\mathbf{J}_i = \frac{2RT\tilde{u}_i y}{L}\left[\frac{c_i^o - c_i^w e^{2y}}{1 - e^{2y}}\right] = \frac{RT\tilde{u}_i}{L}\frac{y}{\sinh y}\left[c_i^w e^y - c_i^o e^{-y}\right] \qquad (1.32)$$

with

$$y = \frac{y_w - y_o}{2} = \frac{z_i F_i}{2RT}\left(\Delta_o^w\phi - \Delta_o^w\phi_i^\ominus\right) \qquad (1.33)$$

For small values of y, Equation 1.32 reduces to a simple diffusional flux equation:

$$\mathbf{J}_i = \frac{RT\tilde{u}_i}{L}\left[c_i^w - c_i^o\right] \qquad (1.34)$$

The major drawback of this Goldman-type approach is to assume that the diffusion coefficients are equal throughout, when the key to the matter is to understand how it varies from one bulk phase to another, either in a smooth monotonous manner or above an activation barrier.

In 1996, Aoki proposed an ion-transfer kinetic model in which the transfer is controlled both by an activation energy and velocity of the ion in the viscous interface, thereby bridging the kinetic and mass transport models [121]. Here, the velocity of a large spherical ion is considered to be driven by the desolvation energy, the energy of overcoming the interfacial tension between the two liquids, and the electrostatic energy in the double layer, all three giving rise to an activation energy, and the thermal fluctuation. It is retarded by the viscous force, as expressed by the Langevin equation. The kinetic equation is derived from the expressions for the velocity and activation energy through the Boltzmann distribution equation, and can elucidate both the properties of the activation control and viscous control. The logarithmic forward rate constant is approximately linear with the potential difference as shown by Equation 1.35.

$$\ln\left(\frac{k_f}{k_0}\right) = \alpha\zeta(1+2p) + \frac{\sigma_1 - \sigma_2}{2} - \frac{(\sigma_0 - \sigma_1)pw}{a} \tag{1.35}$$

where ζ is the dimensionless potential difference $\zeta = zF(\phi_1 - \phi_2)/RT$, α the proportion of potential drop taking place in phase 1, a is the ionic radius, w is the width of the interface, and the parameters are defined by

$$\sigma_i = \frac{2\pi a^2 u_i}{k_B T} \quad \text{and} \quad p = \frac{\sqrt{k_B T}}{\sqrt{m\xi w}} \tag{1.36}$$

where u_i is the density of solvation by the respective solvent, and ξ is the viscous friction coefficient.

1.3.3.3 Hydrodynamic Approach

In 2001, Ferrigno and Girault proposed a hydrodynamic approach to ion-transfer reactions [122]. In the same way that the friction factor for the diffusion coefficient in the Stokes–Einstein equation can be given by solving the Navier–Stokes equation for a sphere in a laminar flow, the Navier–Stokes equation was solved numerically to account for the passage of a sharp boundary between two continuum media. These data show that the friction coefficient varies from $6\pi\eta^w r_{ion}$ to $6\pi\eta^w r_{ion}$ in a continuous manner over a distance one order of magnitude larger than the size of the ion.

1.3.3.4 Marcus Theory for Ion-Transfer Reactions

In 2000, Marcus proposed an ion-transfer theory based on the recognition that an ion-transfer reaction involves a mechanism for initiating a desolvation from the initial liquid, A, and concerted solvation by the receiving liquid, B [123]. The model involves four steps:

The formation of a protrusion of solvent B penetrating in the solvent A
A bimolecular reaction between the ion and the protrusion
A diffusion of the attached ion through the interface
Detachment of the ion

All in all, the apparent rate k_{rate} can be expressed as

$$\frac{1}{k_{rate}} = \frac{1}{k^A_{association}} + \frac{1}{K^A_{eq}k_{diffusion}} + \frac{1}{K^B_{eq}k_{dissociation}} \qquad (1.37)$$

where $k^A_{association}$ is the association rate constant between an ion in solvent A and a protrusion from solvent B, K^A_{eq} is the equilibrium constant between the ion in A and the protrusion from B, $k_{diffusion}$ is the diffusion rate constant for the movement of the ion across the interfacial region, K^B_{eq} is the equilibrium constant for the formation of the ion attached to the protrusion of A extending into bulk B, and $k_{dissociation}$ is the rate constant for dissociation of the ion from the tail of A. As the ion diffuses into solvent B, the last vestiges of A form a protrusion/tail on whose tip the ion resides. The probability density of finding a protrusion of height h is denoted by $P(h)$.

$$P(h) = \frac{e^{-F(h)/k_B T}}{\int_{-\infty}^{\infty} e^{-F(h)/k_B T}\, dh} \qquad (1.38)$$

where $F(h)$ is the free energy of formation of the protrusion with $h < 0$ for B protruding into A, and $h > 0$ for A protruding into B. For small fluctuations, $F(h)$ is a quadratic function of h.

The attachment of the ion to the protrusion is treated by a "bimolecular" attachment rate, $k_A c(z) P(h)$, where $c(z)$ is the concentration of the unattached ion at the location z, where $z = ha$, a being the ionic radius. If we assume that the ion associates only to protrusion of height h_i, then we have

$$k^A_{association} = k^A P(h_i) \Delta h_i \qquad (1.39)$$

The equilibrium constants are simply given by the difference of electrochemical potential between the bulk and the interfacial position:

$$K^A_{eq} = \exp\left[-\frac{\tilde{\mu}(z) - \tilde{\mu}_A}{k_B T}\right] \qquad (1.40)$$

In this model, the liquid forming the protrusion is more likely to be the polar one, that is, the aqueous phase, and the scheme of Figure 1.12 corresponds to the transfer of an ion to the aqueous phase where the protrusion results from long-range ion–dipole interactions. For the reverse transfer, aqueous ions apply a pressure on the interface and drag water molecules, forming a protrusion as they cross the interface.

Following this seminal publication by Marcus, Kornyshev, Kuznetsov, and Urbakh (KKU) discussed the different models of ion transfer and found that three regimes of ion transfer across ITIES can be distinguished [124].

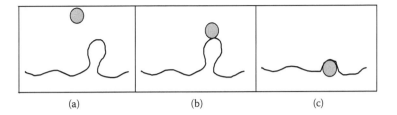

FIGURE 1.12 Schematic plot of an ion in solvent A (a) approaching a protrusion from solvent B at the A–B interface, (b) passing through the interface and (c) exiting into solvent B from another protrusion, now of B into A.

Interfacial Polaron Limit: In the absence of a barrier in the electrochemical potential of the ion, the rate constant for ion transfer is controlled by a two-dimensional diffusion of the ion and an interfacial protrusion, and is determined by an effective diffusion coefficient D_{eff}, for the ion–protrusion couple.

$$D_{eff} = \frac{k_B T}{6\eta \left[r_{ion} + \frac{4L}{3} \left(\frac{h_{max}}{\Lambda} \right)^2 \right]} \tag{1.41}$$

where L and h_{max} is the lateral size and the maximal height of the protrusion, respectively, and Λ is the half-width of the function $h_{eq}(z)$, which characterizes the equilibrium ion–interface coupling. When the relaxation of the interface is much slower than the diffusion of the ion, the kinetics of the ion transfer is controlled by slow relaxation of the interface. We may call this case interfacial polaron limit.

Bare Ion Hopping: For a high barrier in the electrochemical potential of the ion, the transfer is completely determined by the slow transition over this barrier, and the interface fluctuations just follow the ionic motion. Then, the derived equation for the rate constant reduces to the one obtained by Gurevich and Kharkats [117,118] and by Schmickler [120].

Interfacial Gating Limit: For a high barrier in the electrochemical potential of the ion, the same conditions as those discussed by Marcus prevail [123], and the rate constant is determined by the time needed for the creation of the favorable protrusion that reduces the barrier for the ionic transition.

This theory, often referred to as the KKU theory, was further developed in a series of subsequent papers [125,126].

In summary, new theories based on new concepts have evolved over the past decades, and the KKU theory provides a general framework to comprehend ion-transfer kinetics.

1.3.4 SPECTROELECTROCHEMICAL STUDIES

1.3.4.1 Voltabsorptometry and Voltfluorimetry

In 1992, Kakiuchi pioneered the development of spectroscopic techniques to monitor the diffusion layer of transferring species using voltfluorimetry [127]. The major advantage of this optical approach is to be selective to one particular transferring species even if the faradaic current includes some contributions from other charge-transfer reactions [128]. The technique was used by Ding et al. to study the transfer of tris(2,2′-bipyridine)ruthenium(II) (Ru(bpy)$_3^{2+}$) at the H_2O–DCE interface [129].

The gist of the spectroscopic approach is to use the optical property of the ITIES to reflect a light beam on the phase having the lowest refractive index. For example, at the H_2O–1,2-DCE interface, a light beam passing through the organic phase ($n = 1.442$) is reflected on the aqueous phase ($n = 1.333$), and the critical total internal reflection angle is 67.6°. In this way, it is possible to probe the absorbance through the diffusion layer when the potential is varied and charge-transfer reactions take place. The absorbance is given by

$$A_{TIR} = 2\varepsilon_i \int_0^x \frac{c_i(x,t)}{\cos\theta}\, dx \qquad (1.42)$$

where θ is the angle of incidence, ε_i the molar absorption coefficient in the phase of highest refractive index, and $c_i(x,t)$ is the concentration of the transferring ion in the case of an ion-transfer reaction at a distance x from the interface. The total concentration of the ion, i, is equal to the charge transferred through the interface, and is equal to

$$\int_0^x c_i(x,t)\, dx = \frac{1}{z_i FS} \int_0^t I(\tau)\, d\tau \qquad (1.43)$$

such that the measured absorbance provides a voltabsorptogram.

For example, for a linear sweep of potential, we obtain curves as illustrated in Figure 1.13, which, upon differentiation, provides the corresponding cyclic voltammogram [130]. Of course, the same methodology can be used to study assisted-ion-transfer reactions such as the transfer of copper(II) assisted by 6,7-dimethyl-2,3-di(2-pyridyl)quinoxaline [131], acid–base reactions such the transfer of bromophenol blue [132], and to study electron transfer. Ding et al. have in this way confirmed that electron-transfer reactions between ferricyanide–ferrocyanide in water, and 7,7,8,8-tetracyanoquinodimethane (TCNQ) in 1,2-DCE, were heterogeneous [133,134].

If the target ion is fluorescent, it is then interesting to monitor the total fluorescence generated along the beam path, as we can use very small ion concentrations. In a first approximation, one can neglect the fluorescence contribution from the evanescent wave. If the concentration of the fluorescent species is small enough

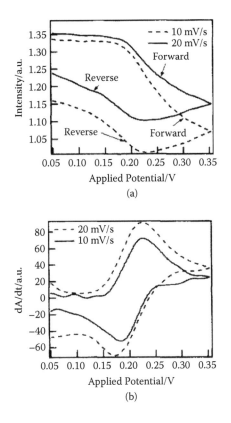

FIGURE 1.13 Transmittance data during a cyclic linear potential sweep for the transfer of $Ru(bpy)_3^{2+}$ at the water–1,2-DCE interface at 453 nm, and differential cyclic voltabsorptogram. (Ding, Z. F., R. G. Wellington, P. F. Brevet, and H. H. Girault, 1997, *J Electroanal Chem*, Vol. 420, p. 35. Used with permission.)

to neglect the variation of light intensity along the beam, then the fluorescence is simply given by

$$F(t) = 2\Phi\varepsilon_i I_0 \int_0^x c_i(x,t)\frac{dx}{\cos\theta} \qquad (1.44)$$

where Φ is the quantum yield. Of course, for fluorescence measurements one does need to operate in total internal reflection, and the factor 2 should be omitted from Equation 1.44. As earlier by differentiation, we obtain a cyclic voltfluorogram as demonstrated by Kakiuchi and Takasu for the transfer of Eosin [135]. An interesting aspect of fluorimetry is to use fluorescent ionophores, the optical properties of which vary upon complexation [136]. Of course, this spectroscopic approach can be used to monitor transfer reactions following potential step experiments to obtain kinetic data [137].

1.3.4.2 Potential Modulated Techniques

In the case of voltabsorptometry, it is possible to modulate the potential and measure reflectance variations. Indeed, reflectance is related to absorbance by

$$\frac{R}{R_{ref}} = \exp[-A_{TIR}]$$

(1.45)

such that small variations of potential ($E = E_0 + \Delta E \sin \omega t$) would generate small variations of absorbance ($A = A_0 + \Delta A \sin \omega t$) and reflectance ($R/R_{ref} = (R/R_{ref})_0 + (\Delta R/R) \sin \omega t$). In this way, Fermin et al. have measured the kinetics of methyl orange transfer by potential modulated reflectance (PMR) [138]. In 2000, Nagatani et al. studied the adsorption and transfer of fluorescent dyes by potential modulated fluorescence (PMF) spectroscopy [139]. As shown in Figure 1.14, while tris(2,2′-bipyridyl)ruthenium(II) ($Ru_{(bpy)3}^{2+}$) shows quasi-reversible ion-transfer features, the charged zinc porphyrins, namely, meso-tetrakis(N-methylpyridyl) porphyrinato zinc-(II) ($ZnTMPyP^{4+}$), and meso-tetrakis(p-sulfonatophenyl) porphyrinato zinc(II) ($ZnTPPS^{4-}$), exhibit adsorption properties at potentials close to the transfer range. The anionic $ZnTPPS^{4-}$ appears to be adsorbed at the interface at potentials after the formal transfer potential. On the other hand, the spectroelectrochemical data show that $ZnTMPyP^{4+}$ is adsorbed at the interface at potentials on either side of the formal transfer potential, that is, prior to and after the transfer. The PMF responses associated with interfacial adsorption from the aqueous side exhibit a different phase shift compared to those associated with interfacial adsorption from the organic side. The experimental results clearly demonstrate that adsorption planes at the organic and aqueous side of the interface could be distinguished. Furthermore, the PMF dependence on the light polarization of the excitation beam allows one to estimate average molecular orientation of the adsorbed species [139].

In 2001, Nagatani et al. reported a method to analyze ion adsorption–transfer kinetics using PMF [140]. The results just discussed were confirmed, and it was further shown that the PMF response for kinetically controlled adsorption is expressed as a semicircle in the complex plane in which the characteristic frequency of maximum imaginary component is proportional to the adsorption and desorption rate constants. Considering that the potential dependence of adsorption exhibits the opposite sign whether the process takes place from the aqueous or organic phase, the corresponding PMF responses appear in different quadrants of the complex plane. This work therefore confirms that the adsorption at an ITIES can take place on either side of the interface.

This method was then applied to study pyrene-sulfonate adsorption and dimerization at the H_2O–1,2-DCE interface [141]. More recently, Nagatani et al. extended their PMF spectroscopy investigation to the study of the adsorption and transfer of free-base, water-soluble porphyrins, namely, cationic meso-tetrakis (N-methylpyridyl) porphyrin (H_2TMPyP^{4+}) and anionic meso-tetrakis(4-sulfonatophenyl)porphyrin ($HTPPS^{4-}$) [142]. The PMF response indicated the presence of an adsorption process for all systems, depending on the

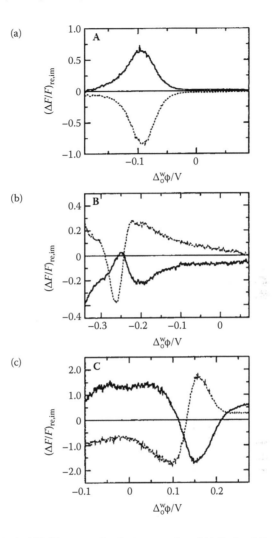

FIGURE 1.14 Typical PMF spectra for the ion transfer of (a) $Ru(bpy)_3^{2+}$, (b) $ZnTPPS^{4-}$, and (c) $ZnTMPyP^{4+}$. The solid and dashed lines denote the real and imaginary components. The concentration of dyes was $2.5 \cdot 10^{-5}$ M^{-1}, and the potential modulation was 10 mV at 6 Hz. (Reprinted from Nagatani, H., R. A. Iglesias, D. J. Fermin, P. F. Brevet, and H. H. Girault, 2000, *J Phys Chem B*, Vol. 104, p. 6869. Used with permission.)

Galvani potential difference. The adsorption on the organic side of the interface was found for cationic H_2TMPyP^{4+} at potentials more positive than its formal ion-transfer potential, that is, after its transfer to the organic phase. The emission spectrum for the interfacial species could be obtained successfully by analyzing the dependence of PMF intensity on the wavelength, and the emission maximum wavelength of the interfacial species was significantly different from the bulk species measured either in the aqueous or organic phases. The presence of adsorption on the organic side for anionic HTPPS^{4-}, and on the aqueous side for protoporphyrin IX, was also found by analyzing the PMF responses.

1.3.4.3 Photochemically Induced Ion Transfer

Considering the difficulties in measuring simple ion-transfer kinetics, photogeneration of ions close to the interface and observation of the photocurrent associated with their transfer was proposed. In this way, the interfacial polarization is kept quasi-constant, and the photocurrent does not suffer from the RC time constant of the interface. This concept was pioneered by Kotov and Kuzmin that published a series of papers on the subject [143–146]. Two kinds of photochemical systems were used: protoporphyrin (excitation 540–580 nm) quenched by quinones, and quinones (excitation 313–365 nm) quenched by tetraphenylborate. Similarly, Samec et al. have studied photoinduced ion-transfer reactions using tetra-arylborate and tetra-arylarsonium as the chromophore, thereby generating the more hydrophilic bridged tetra-aryl intermediate [147,148].

From a mass transport viewpoint, the kinetics of the ion-transfer reactions can be obtained by solving the following differential equation:

$$\frac{\partial c}{\partial t} = D\left(\frac{\partial^2 c}{\partial x^2}\right) + I_0 \varphi_c \alpha \exp[-\alpha x] - k_1 c \tag{1.46}$$

where c is the concentration of charge carriers, I_0 is the incident photon flux, φ_c is the quantum yield for the photogeneration of the charge carriers, and α is the absorption coefficient of the solution (= 2.3 ε c_0, with ε the molar absorption coefficient and c_0 the chromophore concentration). The second term of Equation 1.46 corresponds to the photogeneration term in the evanescent wave, and the third term to the charge carrier decay assumed to be first order. Kuzmin et al. have used computer simulation to extract the kinetic data. This field was nicely reviewed in 1999 by the same authors [149]. More recently, Watariguchi et al. have revisited the photoinitiated ion transfer following the irradiation of tetraphenylborate using photomodulation with an He–Cd laser at a micro-ITIES [150].

1.4   ASSISTED-ION-TRANSFER REACTIONS

1.4.1   Ion–Ionophore Reactions

Assisted-ion-transfer reactions were pioneered by Koryta [2,151] who, back in 1979, demonstrated the concept of an ion transfer assisted or facilitated by the presence of a ligand or ionophore in the organic phase, first of potassium, facilitated by di-benzo-18-crown-6, and then of sodium by nonactin, a macrotetrolide antibiotic. From a mechanistic point of view, we can classify assisted ion transfer in four different categories as proposed by Shao et al. [152] and as illustrated in Figure 1.15, where ACT stands for Aqueous Complexation followed by Transfer, TOC for Transfer to the Organic phase followed by Complexation, TIC for Transfer by Interfacial Complexation, and TID for Transfer by Interfacial Dissociation.

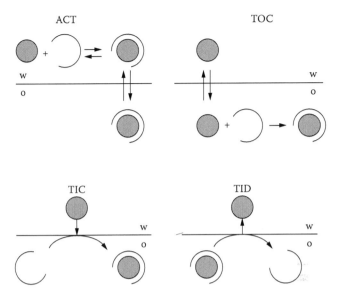

FIGURE 1.15　Assisted-ion-transfer mechanisms.

Of course, from a voltammetric viewpoint, assisted-ion-transfer reactions are characterized by a shift of apparent standard transfer potentials to lower values. From a thermodynamic viewpoint, and in the case of a 1:1 reaction between a cation and a ligand, we can consider the partition coefficient of the ion I^+, of the ligand L, and of the complex IL^+ as shown in Figure 1.16.

If all the species, that is, the cation, the ligand, and the complex can partition, the Nernst equation can be written either for the ion or the complex:

$$\Delta_o^w \phi = \Delta_o^w \phi_{I^+}^\ominus + \frac{RT}{F} \ln\left[\frac{a_{I^+}^o}{a_{I^+}^o}\right] = \Delta_o^w \phi_{IL^+}^\ominus + \frac{RT}{F} \ln\left[\frac{a_{IL^+}^o}{a_{IL^+}^o}\right] \tag{1.47}$$

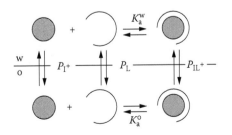

FIGURE 1.16　Thermodynamic constants for assisted-ion-transfer reactions.

from which we get

$$\Delta_o^w \phi_{IL^+}^\ominus = \Delta_o^w \phi_{I^+}^\ominus - \frac{RT}{F} \ln \left[\frac{K_a^o}{K_a^w} \right] - \frac{RT}{F} \ln P_L \qquad (1.48)$$

Comparing the general thermodynamic diagram shown in Figure 1.16 with the four main classes of reactions illustrated in Figure 1.15, we have

ACT : $pK_a^w < 0$, $pK_a^o > 0$, and $\log P_L < 0$ (hydrophilic ligand)

TOC : $pK_a^w > 0$, $pK_a^o < 0$, and $\log P_L > 0$ (lipophilic ligand)

TIC and TID: $pK_a^w > 0$, $pK_a^o < 0$, and $\log P_L > 0$ (hydrophilic ligand)

An early example of TIC transfer was that of K^+ facilitated by valinomycin [153] or di-benzo-18-crown-6 [154].

It would be too long and too tedious to list all the facilitated ion-transfer reactions that have been reported over the years. From alkali–metal ions to transition metal ions, most cations have been studied with different classes of ionophores ranging from the crown family with N, O, or S electron-donating atoms to calixarenes, not to mention all the commercial ionophores developed for ion-selective electrode applications or for solvent extraction. In the case of anions, the number of voltammetric studies reported has been much smaller [155–163], although the field of supramolecular chemistry for anion recognition is developing fast as recently reviewed [164].

Anion binding can be achieved both by neutral receptors such as urea-containing ligands, mainly through hydrogen bonds, or by positively charged ligands such as guanidinium- or polyamine-containing macrocycles, mainly through coulombic interaction.

The concept of assisted ion transfer is, of course, applicable to proton-transfer reactions assisted by the presence of an acid or base, hydrophilic or lipophilic. As pioneered by Kontturi and Murtomaki [165], voltammetry at ITIES has proved to be an excellent method to measure the $\log P$ values of protonated or deprotonated molecules. Indeed, for therapeutic molecules, the $\log P$ values, which are related the Gibbs energy of transfer as shown by Equations 1.11 and 1.12, provide an important physical parameter to assay the toxicity of a molecule. If a molecule is lipophilic, that is, $\log P > 2$, it is potentially toxic. In fact, with the concept of ionic distribution diagrams (vide infra) it is even possible to measure the $\log P$ values of the neutral associated bases. The application of voltammetry at ITIES to the study of therapeutic molecules has been one of the success stories of electrochemistry at liquid–liquid interfaces. The field has been reviewed over the years [166,167] and very recently by Gulaboski et al. [168].

1.4.2 Voltammetry for Assisted-Ion-Transfer Reaction

1.4.2.1 Successive Reactions

A pioneering series of papers to analyze the voltammetric response of assisted ion transfer of higher stoichiometry (1:n) was published by Kudo et al. mainly to treat polarographic data obtained with an ascending or dropping electrolyte system [169–171]. The theory for successive ion-transfer reactions was then thoroughly extended by Reymond et al. [172].

$$M^{z+} + L \rightleftarrows ML^{z+}$$

$$ML^{z+} + L \rightleftarrows ML_2^{z+}$$

$$ML_{j-1}^{z+} + L \rightleftarrows ML_j^{z+}$$

We can define the association constants K_a and β as

$$K_{aj} = \frac{c_{ML_j^{z+}}}{c_{ML_{j-1}^{z+}} c_L} \tag{1.49}$$

and

$$\beta_j = \frac{c_{ML_j^{z+}}}{c_{M^{z+}} (c_L)^j} = \prod_{k=0}^{j} K_{aj} \tag{1.50}$$

where $\beta_0 = K_{a0} = 1$. For each complex, we can define a Nernst equation:

$$\frac{c_{ML_j^{z+}}^o}{c_{ML_j^{z+}}^w} = \exp\left[\frac{zF}{RT} \left(\Delta_o^w \phi - \Delta_o^w \phi_{ML_j^{z+}}^{\ominus} \right) \right] \tag{1.51}$$

where the standard transfer potential of the complex is given by

$$\Delta_o^w \phi_{ML_j^{z+}}^{\ominus} = \Delta_o^w \phi_{M^{z+}}^{\ominus} - \frac{RT}{zF} \ln\left[\frac{\beta_j^o}{\beta_j^w} P_L^j \right] \tag{1.52}$$

similar to Equation 1.48 for a 1:1 stoichiometry. For each species, the current contribution could be calculated by solving the mass transport equation for each species, that is,

$$\frac{\partial c_{ML_j^{z+}}^o}{\partial t} = D_{ML_j^{z+}}^o \frac{\partial^2 c_{ML_j^{z+}}^o}{\partial x^2} + R_{ML_j^{z+}}^o \tag{1.53}$$

where $R^{\circ}_{\mathrm{ML}_j^{z+}}$ is the reaction rate of formation and dissociation. Instead of solving this set of differential equations, Matsuda has proposed [169–171] that, for each phase, a total metal concentration be defined:

$$c_{\mathrm{M}_{\mathrm{tot}}^{z+}} = c_{\mathrm{M}_i^{z+}} + c_{\mathrm{ML}_1^{z+}} + c_{\mathrm{ML}_2^{z+}} + \cdots \tag{1.54}$$

and a total ligand concentration

$$c_{\mathrm{L}_{\mathrm{tot}}} = c_{\mathrm{L}} + c_{\mathrm{ML}_1^{z+}} + 2c_{\mathrm{ML}_2^{z+}} + \cdots \tag{1.55}$$

Assuming that all the diffusion coefficients are equal in a given phase, the mass transport equations can be reduced for each phase to two Fick equations for these total quantities:

$$\frac{\partial c_{\mathrm{M}_{\mathrm{tot}}^{z+}}}{\partial t} = D\frac{\partial^2 c_{\mathrm{M}_{\mathrm{tot}}^{z+}}}{\partial x^2} \tag{1.56}$$

and

$$\frac{\partial c_{\mathrm{L}_{\mathrm{tot}}^{z+}}}{\partial t} = D\frac{\partial^2 c_{\mathrm{L}_{\mathrm{tot}}}}{\partial x^2} \tag{1.57}$$

Then, the computation of the cyclic voltammograms can be classically carried out, for example, by the method of Nicholson and Shain [173].

1.4.2.2 Half-Wave Potential for the Different Cases

For the different types of assisted-ion-transfer reactions, it is possible to express the half-wave potential as a function of experimental parameters such as the concentration of the ligand or that of the transferring ion.

TIC–TID–TOC mechanisms: Large excess of ligand $c_{\mathrm{L}}^0 \gg c_{\mathrm{M}^{z+}}^0$

$$\Delta_{\mathrm{o}}^{\mathrm{w}}\phi^{1/2}_{\mathrm{ML}_j^{z+}} = \Delta_{\mathrm{o}}^{\mathrm{w}}\phi^{\ominus\prime}_{\mathrm{M}^{z+}} - \frac{RT}{zF}\ln\left[\xi\sum_{j=0}^{m}\beta_j^\circ\left(c_{\mathrm{L}}^0\right)^j\right] \tag{1.58}$$

where ξ is the ratio of the diffusion coefficient taken all equal for the species between the two phases $\xi = \sqrt{D_{\mathrm{o}}/D_{\mathrm{w}}}$, and where $\beta_0 = 1$. In the case of a 1:1 stoichiometry, this equation reduces to

$$\Delta_{\mathrm{o}}^{\mathrm{w}}\phi^{1/2}_{\mathrm{ML}_j^{z+}} = \Delta_{\mathrm{o}}^{\mathrm{w}}\phi^{\ominus\prime}_{\mathrm{M}^{z+}} - \frac{RT}{zF}\ln\left[\xi\left(1+\beta_1^\circ c_{\mathrm{L}}^0\right)\right] \approx \Delta_{\mathrm{o}}^{\mathrm{w}}\phi^{1/2}_{\mathrm{M}^{z+}} - \frac{RT}{zF}\ln\left[\beta_1^\circ c_{\mathrm{L}}^0\right] \tag{1.59}$$

and the wave for the assisted ion transfer should shift $60/z$ mV per decade of ligand concentration. In this case, if the assisted ion transfer is fast, the current is controlled by the diffusion of the ion in the aqueous phase and that of the complex in the organic phase. This is why Equation 1.59 depends on the parameter ξ. Here, the TIC and TOC mechanisms are equivalent because we consider an excess of ligand such that the ion complexes at the interface or close to the interface.

TIC–TID–TOC mechanisms: Large excess of metal $c_{M^{z+}}^0 \gg c_L^0$

$$\Delta_o^w \phi_{ML_j^{z+}}^{1/2} = \Delta_o^w \phi_{M^{z+}}^{\ominus\prime} - \frac{RT}{zF} \ln \left[c_M^0 \sum_{j=0}^{m} j\beta_j^o \left(\frac{c_L^0}{2} \right)^{j-1} \right]$$ (1.60)

In the case of a 1:1 stoichiometry, this equation reduces to

$$\Delta_o^w \phi_{ML^{z+}}^{1/2} = \Delta_o^w \phi_{M^{z+}}^{\ominus\prime} - \frac{RT}{zF} \ln \left[c_M^0 \left(1 + \beta_1^o \right) \right] \approx \Delta_o^w \phi_{M^{z+}}^{\ominus\prime} - \frac{RT}{zF} \ln \left[c_M^0 \beta_1^o \right]$$ (1.61)

Again, the wave for the assisted ion transfer should shift $60/z$ mV per decade of ligand concentration. In this case, if the assisted ion transfer is fast, the current is controlled by the diffusion of the ligand and that of the complex, both in the organic phase. This is why Equation 1.61 does not depend on the parameter ξ. This equation is often used for assisted-proton-transfer reactions where the acidic pH is varied to measure β_1^o.

ACT mechanism

$$\Delta_o^w \phi_{ML_j^{z+}}^{1/2} = \Delta_o^w \phi_{M^{z+}}^{\ominus\prime} + \frac{RT}{zF} \ln \left[\frac{\xi \sum_{j=0}^{m} \beta_j^w \left(c_L^0 \right)^j}{\sum_{j=0}^{m} \beta_j^o \left(P_L c_L^0 \right)^j} \right]$$ (1.62)

If the terms of lowest power may be neglected, this equation reduces to

$$\Delta_o^w \phi_{ML_j^{z+}}^{1/2} = \Delta_o^w \phi_{M^{z+}}^{\ominus\prime} + \frac{RT}{zF} \ln \left[\frac{\xi \beta_m^w}{\beta_m^o P_L^m} \right]$$ (1.63)

In the case of 1:1 stoichiometry, Equation 1.63 reduces to

$$\Delta_o^w \phi_{ML_j^{z+}}^{1/2} = \Delta_o^w \phi_{M^{z+}}^{1/2} + \frac{RT}{zF} \ln \left[\frac{\beta_m^w}{\beta_m^o P_L^m} \right]$$ (1.64)

In this case, if the assisted ion transfer is fast, the current is controlled by the diffusion of the complex in the aqueous and in the organic phase. This is why Equation 1.63 depends on the parameter ξ. Figure 1.17 illustrates schematically the different diffusion regimes.

This method was recently extended by Garcia et al. for different competitive ligands [174].

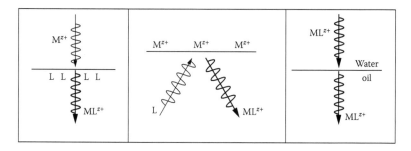

FIGURE 1.17 Diffusion regimes for TIC–TOC–TID excess ligand, TIC–TOC–TID excess cation, and ACT. The light lines represent the diffusion of the cation (left) or of the ligand (middle). The bold lines represent the diffusion of the complex.

1.4.2.3 Ion-Pair Formation at ITIES

Another interesting case is that assisted transfer by charged ligands, which is an extension of ion-pair extraction classically used in analytical chemistry. Tomazewski et al. have presented computer simulations of cyclic voltammetry experiments at liquid–liquid interfaces for the transfer of M^{z+} ions assisted by z charged ligands L^-. The main difference, when compared with neutral ligands, is that all the complexes formed have a different charge, and the flux of the ligands has to be taken into account in the definition of the current, in addition to that of the metal ion and those of the $(z-1)$ charged complexes (the complex with the highest stoichiometry being neutral) [175].

One recent example of charge-transfer reactions facilitated by counterions is the transfer of protamines using surfactant anions (dinonylnaphthalenesulfonate) as presented by Amemyia et al. [176,177] and also by Trojanek et al. [178].

1.4.3 IONIC DISTRIBUTION DIAGRAMS

On the basis of the concept of Pourbaix diagrams, Reymond et al. have proposed the concept of zone diagrams for the distribution of ionizable species such as acids or bases [179–181]. To illustrate this, let us consider first the distribution diagram for a hydrophilic AH acid in a biphasic water–organic solvent system. At a high aqueous pH, the acid is in the anionic form and can exist in the phases according to the Galvani potential difference. The Nernst equation for the distribution of the anion, ignoring the activity coefficients, is written as

$$\Delta_o^w \phi = \Delta_o^w \phi_{A^-}^\ominus - \frac{RT}{F} \ln \left(\frac{c_{A^-}^o}{c_{A^-}^w} \right) \tag{1.65}$$

Thus, the separation limit between the anionic form in water and the organic solvent ($c_{A^-}^w = c_{A^-}^o$) is a horizontal straight line. As in the Pourbaix diagrams, the

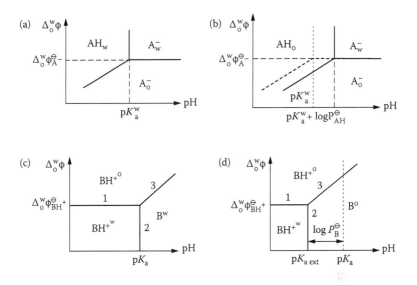

FIGURE 1.18 Distribution diagram for a hydrophilic (a) and a lipophilic (b) acid, and a hydrophilic (c) and lipophilic (d) base. (Gobry, V., S. Ulmeanu, F. Reymond, G. Bouchard, P. A. Carrupt, B. Testa, and H. H. Girault, 2001, *J Am Chem Soc*, Vol. 123, p. 10684.)

separation limit between the acid and basic forms in water is a vertical line given by pH = pK_a^w.

Finally, the line separating the neutral acid in water and the anion A_o^- in the organic phase is given by including the acidity constant in Equation 2.58 (Chapter 2) to give

$$\Delta_o^w\phi = \left[\Delta_o^w\phi_{A^-}^\ominus + \frac{RT}{F}\ln K_a^w\right] - \frac{RT}{F}\ln\left(\frac{c_{A^-}^o \cdot c_{H^+}^w}{c_{AH}^w}\right) \qquad (1.66)$$

As in the Pourbaix diagrams, we obtain a delimiting line that depends on the pH as shown in Figure 1.18. If the AH acid is lipophilic, we have to take into account the distribution of the acid in the organic phase:

$$K_a^w = \frac{a_{A^-}^w a_{H^+}^w}{a_{AH}^w} = \frac{a_{A^-}^w a_{H^+}^w}{a_{AH}^o} P_{AH}^\ominus \qquad (1.67)$$

and, neglecting the activity coefficients, the separation limit between the aqueous anion A_w^- and the neutral form in the organic solvent is described by

$$pH = pK_a^w + \log P_{AH}^\ominus \qquad (1.68)$$

This equation shows that, to extract an organic acid to an aqueous phase, one should work at a pH higher than that given by Equation 1.68. The separation limit between the two ionic forms is still the one given by the Nernst equation for the distribution of the anion. The separation limit between the anion in the water and the acid in the solvent is given by

$$\Delta_o^w \phi = \left[\Delta_o^w \phi_{A^-}^{\ominus} + \frac{RT}{F} \ln \frac{K_a^w}{P_{AH}^{\ominus}} \right] - \frac{RT}{F} \ln \left(\frac{c_{A^-}^o \cdot c_{H^+}^w}{c_{AH}^o} \right) \tag{1.69}$$

Again, this limit depends on the pH. The diagram in Figure 1.18 shows that the more lipophilic the AH acid, the smaller the stability zone of the anion A_w^-.

Similar equations can be used to draw the ionic distribution diagram of a base as illustrated in Figure 1.18. To draw ionic distribution diagrams, one should measure by voltammetry the half-wave potentials for the different ion-transfer and assisted-ion-transfer reactions.

In addition, to consider the role of the biological membranes in drug partitioning, Kontturi et al. have designed a cell with a Langmuir–Blodgett phospholipid monolayer-modified liquid–liquid interface to study the specific interactions between ionized drugs and phosphatidylcholine layers [182]. In a recent work, Jensen et al. have shown that the measurement of the ionic partition diagram in the absence and presence of ligands, for example, cholesterol in the organic phase, could be very useful in determining the interaction of either the neutral or the cationic form of a drug molecule with the ligand as illustrated in Figure 1.19 [183].

The drawing of an ionic partition diagram requires voltammetric measurements at different pH values with equilibrated solutions. This means that a new electrochemical cell has to be prepared for each pH. Different approaches have been proposed to make the measurement more automated. One approach is based

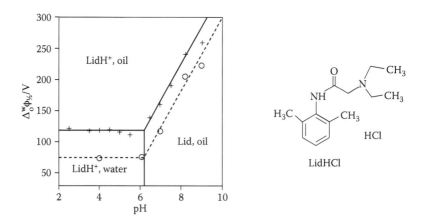

FIGURE 1.19 Ligand shift ion-partition diagram for the LidHCl–Chol system obtained at the aqueous buffer–1,2-DCE interface. (He, Q., Y. Zhang, G. Lu, R. Miller, H. Mohwald, and J. Li, 2008, *Adv Coll Interface*, Vol. 140, p. 67. Used with permission.)

on the use of a microtiter plate well equipped with a filter membrane that can be used to support the organic phase [184]. In this way, it is possible to prepare different solutions, and a voltammogram can be recorded by moving a combined aqueous reference–counter electrode from well to well. In a second approach, Lam et al. have developed a cell using a pH gradient gel classically used for the separation of proteins and peptides by isoelectric focusing (IEF) [185]. Using micromachining, an array of micro-ITIES is fabricated along the pH gradient gel such that the aqueous solution at each micro-ITIES is at a different pH as illustrated further in Figure 1.31. Again, by placing a common organic reference–counter electrode close to each micro-ITIES, a voltammogram can be recorded for different pH values.

1.4.4 Ion-Selective Electrodes

To calculate the Galvani potential difference for a biphasic system containing three species (for example, a target cation I^+, a hydrophilic anion A^- forming a salt IA in the aqueous phase, and a lipophilic anion X^- forming a salt IX in the organic phase), we should consider the three respective Nernst equations:

$$\Delta_o^w \phi = \Delta_o^w \phi_i^\ominus + \frac{RT}{z_i F} \ln\left(\frac{a_i^o}{a_i^w} \right) \tag{1.70}$$

For each species, we should consider the conservation of mass:

$$\left(c_i^w + r c_i^o = c_i^{tot} \right)$$

where $r = V^o / V^w$ is the phase ratio such that the Nernst equation now reads as

$$c_i^w + r c_i^w \exp\left[z_i F \left(\Delta_o^w \phi - \Delta_o^w \phi_i^{\ominus\prime} \right) / RT \right] = c_i^{tot} \tag{1.71}$$

Taking into account the electroneutrality condition in each phase, $(c_{I^+} - c_{A^-} - c_{X^-} = 0)$, we have only one equation to solve to calculate the resulting Galvani potential difference:

$$\frac{c_{I^+}^{tot}}{1 + r e^{F\left(\Delta_o^w \phi - \Delta_o^w \phi_{I^+}^{\ominus\prime}\right)/RT}} - \frac{c_{X^-}^{tot}}{1 + r e^{-F\left(\Delta_o^w \phi - \Delta_o^w \phi_{X^-}^{\ominus\prime}\right)/RT}} - \frac{c_{A^-}^{tot}}{1 + r e^{-F\left(\Delta_o^w \phi - \Delta_o^w \phi_{A^-}^{\ominus\prime}\right)/RT}} = 0 \tag{1.72}$$

The results obtained by numerical integration are illustrated in Figure 1.20.

When the ratio c_{IX}/c_{IA} is large, that is, when IX is in excess versus IA, the Galvani potential difference to the distribution potential of IX is represented by $\Delta_o^w \phi_{dis,IX}$. Inversely, when the ratio c_{IX}/c_{IA} is small, that is, when IA is in excess versus IX, the Galvani potential difference to the distribution potential of IA is represented by $\Delta_o^w \phi_{dis,IA}$. Between these limits, the system behaves as a Nernstian

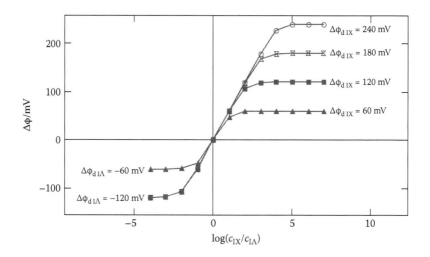

FIGURE 1.20   Galvani potential difference $\phi^w - \phi^o$ for a three-ion system; the standard transfer potential of the target ion is taken as equal to zero. The limit on either is the distribution potential given by Equation 1.13 for the salt in excess.

system, that is, a variation of 60 mV/decade of ratio of concentration of I^+. Indeed, in this case, we have $c_{I^+}^w \cong c_{A^-}^w$ and $c_{I^+}^o \cong c_{X^-}^o$, and therefore,

$$\Delta_o^w\phi = \Delta_o^w\phi_{I^+}^{\ominus} + \frac{RT}{F}\ln\left(\frac{c_{IX}^o}{c_{IA}^w}\right) \tag{1.73}$$

The further apart the distribution potential limits of IX and IA, the wider will be the Nernstian window useful for operating an ion-selective electrode.

1.4.5   ASSISTED-ION-TRANSFER KINETICS

In 1982, Samec et al. studied the kinetics of assisted alkali and alkali–earth metal cation-transfer reactions by neutral carrier and concluded that the kinetics of transfer of the monovalent ions were too fast to be measured [186]. In 1986, Kakutani et al. published a study of the kinetics of sodium transfer facilitated by di-benzo-18-crown-6 using ac-polarography [187]. They concluded that the transfer mechanism was a TIC process and that the rate constant was also high. Since then, kinetic studies of assisted-ion-transfer reactions have been mainly carried out at micro-ITIES. In 1995, Beattie et al. showed by impedance measurements that facilitated ion-transfer (FIT) reactions are somehow faster than the nonassisted ones [188,189]. In 1997, Shao and Mirkin used nanopipette voltammetry to measure the rate constant of the transfer of K^+ assisted by the presence of di-benzo-18-crown-6, and standard rate constant values of the order of 1 cm·s^{-1} were obtained [190]. A more systematic study was then published that showed the following sequence, $k_{Cs^+}^{\ominus} < k_{Li^+}^{\ominus} < k_{Rb^+}^{\ominus} < k_{Na^+}^{\ominus} < k_{K^+}^{\ominus}$, which is not in accordance with their complexation constants [191]. In 2002, Shao and his group investigated the transfer of alkali metal ions assisted by monoaza-B15C5 ionophores, and again the rate constant values were about 0.5 cm·s⁻¹ [192].

In 2004, Shao et al. published a study of FIT reactions at high driving force [193] using an SECM methodology developed earlier [194]. The FIT rate constants k_f were found to be dependent upon the reaction driving force. When the driving force for FIT was not too high, a Tafel plot indicative of a Butler–Volmer mechanism was observed. For facilitated Li⁺ transfer that can be driven over a wide potential range, the potential dependence of ln k_f showed a parabolic behavior, as in the Marcus inverted region for electron-transfer reactions.

1.5 ELECTRON TRANSFER REACTIONS

1.5.1 REDOX EQUILIBRIA

Let us consider the transfer of one electron between an oxidized species O_1 in an aqueous phase and a reduced species R_2 in the organic phase as illustrated in Figure 1.21.

$$O_1^w + R_2^o \rightleftharpoons R_1^w + O_2^o \tag{1.74}$$

At equilibrium, we have the following equality of the electrochemical potentials:

$$\tilde{\mu}_{R_1}^w + \tilde{\mu}_{O_2}^o = \tilde{\mu}_{O_1}^w + \tilde{\mu}_{R_2}^o \tag{1.75}$$

Developing this, we obtain the equivalent of the Nernst equation for this reaction of electron transfer at the interface, that is,

$$\Delta_o^w \phi = \Delta_o^w \phi_{ET}^\ominus + \frac{RT}{F} \ln \left(\frac{a_{R_1}^w a_{O_2}^o}{a_{O_1}^w a_{R_2}^o} \right) \tag{1.76}$$

with $\Delta_o^w \phi_{ET}^\ominus$ the standard redox potential for the interfacial transfer of electrons:

$$\Delta_o^w \phi_{ET}^o = \left[\tilde{\mu}_{R_1}^{\ominus,w} + \tilde{\mu}_{O_2}^{\ominus,o} - \tilde{\mu}_{O_1}^{\ominus,w} - \tilde{\mu}_{R_2}^{\ominus,o} \right] / F \tag{1.77}$$

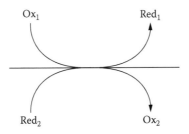

FIGURE 1.21 Heterogeneous redox reaction at a liquid–liquid interface.

It is interesting to bring into this equation the standard redox potentials. In water, the standard redox potential, $[E_{O/R}^{\ominus}]_{SHE}^{w}$, is classically defined versus the standard hydrogen electrode as [91]

$$\left[E_{O/R}^{\ominus}\right]_{SHE}^{w} = \left[\tilde{\mu}_{O}^{\ominus,w} - \tilde{\mu}_{R}^{\ominus,w} - n\mu_{H^+}^{\ominus,w} + \tfrac{n}{2}\mu_{H_2}^{\ominus}\right]/F \qquad (1.78)$$

Of course, we can define the standard redox potential of the organic couple in the organic solvent with respect to an aqueous standard hydrogen reference electrode, $[E_{O/R}^{\ominus}]_{SHE}^{o}$:

$$\left[E_{O/R}^{\ominus}\right]_{SHE}^{o} = \left[\mu_{O}^{\ominus,o} - \mu_{R}^{\ominus,o} - \mu_{H^+}^{\ominus,w} + \tfrac{1}{2}\mu_{H_2}^{\ominus}\right]/F \qquad (1.79)$$

From an experimental viewpoint, this standard redox potential is difficult to measure as it requires the determination of the liquid junction potential between the organic solution and the acid solution of the aqueous standard hydrogen electrode, which brings us back to the problem of defining an absolute Gibbs energy of transfer scale. In this case, Equation 1.77 reads

$$\Delta_o^w \phi_{ET}^{\ominus} = \left[E_{O_2/R_2}^{\ominus}\right]_{SHE}^{o} - \left[E_{O_1/R_1}^{\ominus}\right]_{SHE}^{w} \qquad (1.80)$$

To circumvent the difficulty associated with the definition of the standard redox potential in the organic phase, a classical way is to refer all the standard organic redox potentials to a reference redox couple such as ferrocinium/ferrocene. The standard redox potential of this reference couple in the organic phase is related to that in the aqueous phase by

$$\left[E_{Fc^+/Fc}^{\ominus}\right]_{SHE}^{o} = \left[E_{Fc^+/Fc}^{\ominus}\right]_{SHE}^{w} + \left(\Delta G_{tr,Fc^+}^{\ominus,w \to o} - \Delta G_{tr,Fc}^{\ominus,w \to o}\right)/F \qquad (1.81)$$

The Gibbs energy of transfer of the neutral ferrocene has been measured by a shake flask experiment and found to be equal to $\Delta G_{tr,Fc}^{\ominus} = -24.5 \pm 0.5$ kJ·mol^{-1} ($K_D \approx 2 \cdot 10^4$), and that of ferrocinium by ion-transfer voltammetry found to be equal to 0.5 kJ·mol^{-1} [195]. Considering that $[E_{Fc^+/Fc}^{\ominus\prime}]_{SHE}^{water} = 0.380$V [195], we have $[E_{Fc^+/Fc}^{\ominus}]_{SHE}^{DCE} = 0.64 \pm 0.05$V. In this way, it possible to compare the redox scale as illustrated in Figure 1.22, as we have

$$\Delta_o^w \phi_{ET}^{\ominus} = \left[E_{O_2/R_2}^{\ominus}\right]_{Fc}^{DCE} - \left[E_{O_1/R_1}^{\ominus}\right]_{SHE}^{w} + 0.64 \text{ V} \qquad (1.82)$$

In the case of the H$_2$O–NB interface, Osakai et al. have measured that the Gibbs energy of transfer of the neutral ferrocene is equal to -21.5 kJ·mol^{-1} ($K_D = 6 \cdot 10^3$), and that of ferrocinium by ion-transfer voltammetry was found to be equal to -9 kJ·mol^{-1}, which gives $[E_{Fc^+/Fc}^{\ominus}]_{SHE}^{NB} = 0.51 \pm 0.05$V [196].

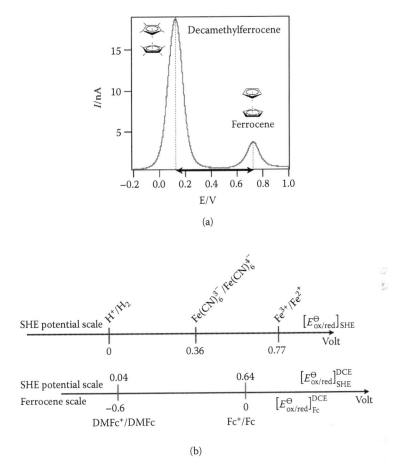

FIGURE 1.22 Determination of the standard redox potential of decamethylferrocene in 1,2-DCE versus ferrocene, and the redox scale for a biphasic system.

One of the major uses of the redox potential scale is to determine the exergonicity of redox reactions in solution. In water, for example, the oxidized form of the redox couple having the highest redox potential is the oxidant, and the reduced form of the couple having the lowest redox potential is the reductant. At ITIES, the standard Gibbs energy $\Delta G_{ET}^{\ominus}$ of the interfacial electron reaction discussed earlier (Equation 1.74) is then simply given by

$$\Delta G_{ET}^{\ominus} = nF\Delta_o^w\phi_{ET}^{\ominus} \qquad (1.83)$$

For example, we can see in Figure 1.22, that the standard redox potential of ferrocene in 1,2-DCE on the aqueous scale is higher than that of the ferri ferrocyanide redox couple (0.64 versus 0.36 V). This means that, when the Galvani potential difference between the two adjacent phases is zero (potential of zero charge), then ferricyanide cannot oxidize ferrocene. However, when we polarize the aqueous phase positively, it is like sliding the aqueous potential scale to the right, and when the polarization is larger than about 0.28 V, then the oxidation can take place.

1.5.2 EXPERIMENTAL STUDIES

Electron-transfer reactions at ITIES have been studied extensively since the seminal discovery of Samec et al. who demonstrated that it was possible to record the current associated to the oxidation of ferrocene in nitrobenzene by ferricyanide in water [197]. The study of electron-transfer reactions by voltammetry is sometimes rendered difficult by the concomitant ion-transfer reactions taking place, as shown by Cunnane et al. for different ferrocene derivatives [198]. All of these experimental difficulties such as decomposition reactions, ion pairing, and phase formation have been reviewed by Quinn and Kontturi [199]. In 2003, Osakai et al. [196] proposed that the oxidation of ferrocene by aqueous ferricyanide is not a truly heterogeneous reaction, but instead proceeds via the partitioning of neutral ferrocene to water followed by a homogeneous aqueous electron-transfer reaction, and terminated by the transfer of the ferrocinium cation back to the organic phase as illustrated in Figure 1.23. This mechanism was further supported by the work of Tatsumi and Katano [200–202].

In the case of Figure 1.23, the ferrocene concentration in the organic phase is in excess compared to that of ferricyanide in water, and this may explain the proposed mechanism. When the aqueous couple is in excess, then the electron-transfer reaction becomes more heterogeneous as the mean free path of ferrocene in water is reduced.

Over the years, many different systems have been studied. For example, in 1989, Kihara et al. studied the following reactions [203]:

$$Fe(CN)_6^{4-} + TCNQ \rightleftharpoons Fe(CN)_6^{3-} + TCNQ^-$$

$$Fe(CN)_6^{3-} + TTF \rightleftharpoons Fe(CN)_6^{3-} + TTF^+$$

$$Hydroquinone + 2 TCNQ \rightleftharpoons Benzoquinone + 2 TCNQ^-$$

These reactions were further studied at micro-ITIES by Quinn et al. who found the reactions to be reversible [204]. In the early days, most electron-transfer reactions were considered heterogeneous, but, as discussed, the locus of the ET step may occur in one of the adjacent phases. More recently, Sugihara et al. showed that the oxidation of ascorbic acid and chlorogenic acid in water by a zinc porphyrin (5,10,15,20-tetraphenylporphirinato zinc(II)) in nitrobenzene occurs on the organic side of the interface, accompanied by a back proton transfer reaction [205]. According to Osakai et al., one case of "truly" heterogeneous ET reactions was observed with a cadmium tetraphenylporphyrin in nitrobenzene and ferricyanide in water [206].

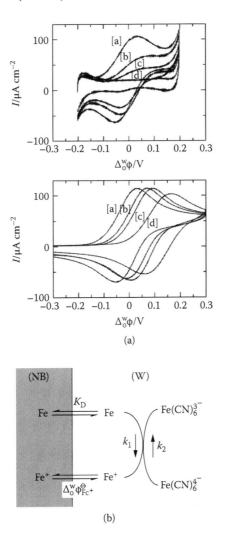

FIGURE 1.23 Cyclic voltammograms observed at the H_2O–NB interface in the presence of [a] 100, [b] 20, [c] 10, and [d] 1 mM Fc in NB and 0.5 mM $Fe(CN)_6^{3-}$ + 0.1mM $Fe(CN)_6^{4-}$ in W. Sweep rate: 0.1 V s^{-1}. (Hotta, H., S. Ichikawa, T. Sugihara, and T. Osakai, 2003, *J Phys Chem B*, Vol. 107, p. 9717. Used with permission.)

As the study of ET reactions is difficult at large interfaces, these have been studied using different experimental approaches such as SECM or liquid-film modified electrodes as presented in Section 1.6. Of course, cyclic voltammetry is a ubiquitous electrochemical method that can be applied to study electron-transfer reactions. However, the mass transport equations differ from those for classical cyclic voltammetry at a solid electrode, as we have to consider the mass transport equations for the two incoming reactants and the two outgoing products. As a result, the classical criteria for a reversible reaction such as a 60 mV peak-to-peak separation are not applicable as shown by Stewart et al. [207]. For example, the peak-to-peak separation for a one-electron transfer when the two reactants are present at the same concentration is 120 mV, as each of them carries half a charge. In such an approach, the diffusion of the reactants and products to and from the interface was considered as linear semi-infinite. Of course, when dealing with dilute concentrations, the microscopic diffusion becomes hemispherical as discussed by Osakai et al. [208], one redox couple acting as a nanoelectrode array with respect to the other. The second-order rate constant obtained from the usual kinetic measurements then involves a bimolecular-reaction effect having a certain upper limit.

1.5.3 Solvent Reorganization Energy

From a theoretical point of view, the major effort was devoted first to the calculation of the solvent reorganization energy, considering an ITIES as a planar boundary between two homogeneous dielectric media, as illustrated in Figure 1.24.

The first electrostatic model was presented by Kharkats who developed Equation 1.84 [209, 210]:

$$\lambda = \frac{1}{8\pi}\left(\frac{1}{\varepsilon_{opt}} - \frac{1}{\varepsilon_r}\right)\iiint_{\infty - V_a - V_b}\left[\mathbf{D}_f - \mathbf{D}_i\right]^2 dr^3 \tag{1.84}$$

where ε_{opt} and ε_r represent the optical and relative permittivity, respectively, $\mathbf{D}$ represents the electric displacement vector, and V_a and V_b represent the volume of the reactants. A detailed derivation of Equation 1.84 was published [211], and it can be shown that

$$\lambda = \frac{(ne)^2}{4\pi\varepsilon_0}\left(\frac{1}{\varepsilon_{opt1}} - \frac{1}{\varepsilon_{r1}}\right)\left[\frac{1}{2a}\right]$$
$$+ \frac{(ne)^2}{4\pi\varepsilon_0}\left(\frac{1}{4h}\right)\left[\frac{1}{\varepsilon_{opt1}}\left(\frac{\varepsilon_{opt1} - \varepsilon_{opt2}}{\varepsilon_{opt1} + \varepsilon_{opt2}}\right) - \frac{1}{\varepsilon_{r1}}\left(\frac{\varepsilon_{r1} - \varepsilon_{r2}}{\varepsilon_{r1} + \varepsilon_{r2}}\right)\right] \tag{1.85}$$

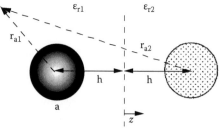

FIGURE 1.24   Geometric parameters of Equation 1.85.

where the geometric parameters are shown in Figure 1.24. Similar calculations were performed by Marcus who reached the same conclusions [212–215]. Benjamin and Kharkats extended these calculations to include the respective positions of the reactants vis-à-vis the interface [216].

The potential dependence of the rate of the interfacial ET reaction at ITIES has been the subject of debate, especially in the SECM community (see the following text). In other words, how much do the reactants feel the local electric field? Part of the answer lies in the position of the reactants vis-à-vis the interface. In 1988, Girault and Schiffrin had considered this problem and derived an equation for the ET rate constant [217]. Based on the classical encounter model for ET in solution, where the two reactants O_1 and R_2 meet at the interface, as illustrated in Figure 1.21, to form the precursor $O_1|R_2$ that reorganizes itself to form a reactant pair $\{O_1R_2\}$ in which the electron-transfer reaction can take place, thereby forming $\{R_1O_2\}$ that can relax to form the successor complex $R_1|O_2$ that finally separate into the products R_1 and O_2, they obtained

$$k_{obs} = N\delta r \int_0^{\pi/2} r\cos\alpha \int_0^\alpha \int_0^\alpha r^2 \exp[-w_p(r)/RT]\cos\zeta\, k_{ET}(r)\, d\zeta d\psi d\alpha \quad (1.86)$$

where w_p is the work required to bring the reactants reversibly to the distance r to form the precursor $O_1|R_2$ having an orientation ζ and ψ referred to the plane $x = 0$, and k_{ET} is the first-order rate constant for the electron-transfer reaction within the reorganized precursor $\{O_1R_2\}$. The angle α is defined such that $r\cos\alpha$ represents the projection of the distance between the reactants on the normal to the interface. The activation energy for the reaction, $\Delta G^{\#}$, is given classically as a function of the solvent reorganization energy and the potential dependent driving force ΔG_{ET}:

$$\Delta G^{\#} = \frac{\lambda}{4}\left(1 + \frac{\Delta G_{ET}}{\lambda}\right)^2 = \frac{\lambda}{4}\left(1 + \frac{\Delta G_{ET}^{\ominus} + w_s - w_p}{\lambda}\right)^2 \quad (1.87)$$

where w_s and w_p are the work terms associated with the formation of the successor $R_1|O_2$ and precursor $O_1|R_2$, respectively. The key question is how to define $\Delta G_{ET}^{\ominus}$. If the reaction was to happen in a bulk solution, the standard driving force is simply given by

$$\Delta G_{ET}^{\ominus} = nF\left[E_{O_2/R_2}^{\ominus}\right]_{SHE} - nF\left[E_{O_1/R_1}^{\ominus}\right]_{SHE} \quad (1.88)$$

In the case of an electrochemical reaction on a solid electrode, the standard driving force is potential-dependent and simply given by

$$\Delta G_{ET}^{\ominus} = nF\left[E - E_{O/R}^{\ominus}\right]_{SHE} \quad (1.89)$$

Of course, if all the potential drop across the interface was effective, and if the surface concentrations were equal to the bulk values, then the driving force is

$$\Delta G_{ET}^{\ominus} = nF\left(\Delta_o^w\phi - \Delta_o^w\phi_{ET}^{\ominus}\right) \tag{1.90}$$

If we consider that the reactants are located at the interface such that the local potential difference is $\Delta_{io}^{iw}\phi$, then the standard driving force is

$$\Delta G_{ET}^{\ominus} = nF\left(\Delta_{io}^{iw}\phi - \Delta_o^w\phi_{ET}^{\ominus}\right) \tag{1.91}$$

where the subscripts "iw" and "io" refer to the location of the reactants at the interface at the aqueous and organic side, respectively. The work to form a precursor from the reactants includes a noncoulombic part that can be neglected in a first approximation, and a coulombic part that be calculated as the sum of the Boltzmann contribution $z_{O_1}F\Delta_{iw}^w\phi + z_{R_2}F\Delta_{io}^o\phi$, in such a way that the difference of the work terms $w_s - w_p$ is $nF(\Delta_{iw}^w\phi - \Delta_{io}^o\phi)$. Then, we see that Equation 1.87, the driving force term, including the work term, is

$$\begin{aligned}\Delta G_{ET}^{\ominus} + w_s - w_p &= nF\left(\Delta_{io}^{iw}\phi - \Delta_o^w\phi_{ET}^{\ominus}\right) + nF\left(\Delta_{iw}^w\phi - \Delta_{io}^o\phi\right)\\ &= nF\left(\Delta_o^w\phi - \Delta_o^w\phi_{ET}^{\ominus}\right)\end{aligned} \tag{1.92}$$

In conclusion, we see that, in any case, the full potential drop acts on the activation energy term.

1.5.4 PHOTOELECTRON TRANSFER REACTIONS

The interface between two immiscible electrolyte solutions is well suited to studying artificial photosynthesis. Indeed, the products of a photoinitiated electron-transfer reaction can be separated on either side of the interface, thereby breaking the cage effect that is dominant in bulk solutions.

Back in 1988, Girault's group started to investigate photocurrent measurements at polarized ITIES using $Ru(bpy)_3^{2+}$ as a sensitizer [218,219]. This early work was followed by a series of papers of photoinduced electron reactions using porphyrin sensitizers [220–226].

First, it was shown that the presence of an electron donor or an electron acceptor in the organic phase resulted in photocurrents associated with a reductive and oxidative quenching of the water-soluble zinc tetrakis(carboxyphenyl)porphyrin (ZnTPPC), respectively, as shown in Figure 1.25 [227].

In part I of the series, it was demonstrated that photocurrent responses associated with the heterogeneous quenching of water-soluble ZnTPPC by ferrocene and DFCET are potential dependent, proportional to light intensity, and that the action spectrum followed the absorption spectrum of the porphyrin [220].

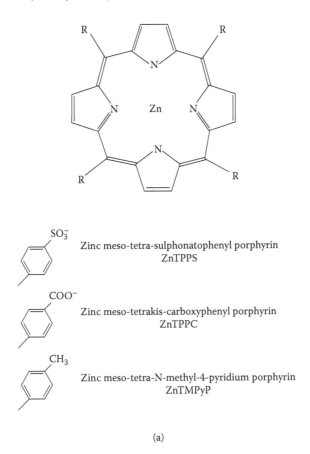

Zinc meso-tetra-sulphonatophenyl porphyrin
ZnTPPS

Zinc meso-tetrakis-carboxyphenyl porphyrin
ZnTPPC

Zinc meso-tetra-N-methyl-4-pyridium porphyrin
ZnTMPyP

(a)

FIGURE 1.25 Water-soluble porphyrins used as sensitizers at the H_2O–DCE interface. Photocurrent transient measurements at various Galvani potential differences with (a) TCNQ and (b) diferrocenylethane (DFCET) as quenchers. Illumination was provided by a 450 W Arc-Xe lamp. $\lambda < 450$ nm were cut by a Schott filter. The photocurrent is negative (electron transfer from H_2O to DCE) in the presence of TCNQ and positive for DFCET. (Fermin, D. J., Z. F. Ding, H. D. Duong, P. F. Brevet, and H. H. Girault, 1998, *Chem Commun*, p. 1125. Used with permission).

In part II, the photoresponses for the heterogeneous quenching of ZnTPPC by ferrocene derivatives were studied by intensity-modulated photocurrent spectroscopy (IMPS). The different contributions, that is, the electron injection, the recombination–product separation competition, and the attenuation due to the uncompensated resistance and interfacial capacitance (RC) time constant of the cell were deconvoluted in the frequency domain. The flux of electron injection was described as a competition between the relaxation of the porphyrin-excited state and the electron-transfer step. Experimental results confirmed that the electron-transfer rate increases with the Galvani potential difference (Butler–Volmer behavior), but the ZnTPPC coverage was potential-dependent,

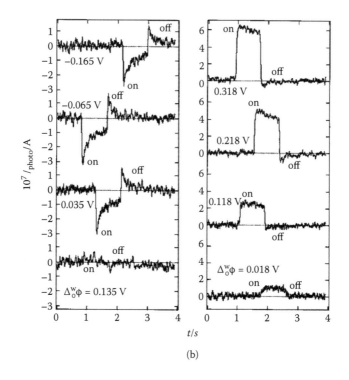

FIGURE 1.25   (Continued).

too, decreasing as the aqueous phase is polarized positively [221]. In part III, it was shown that water-soluble porphyrin ion pairs formed by mixing anionic zinc meso-tetrakis(p-sulfonatophenyl)-porphyrin (ZnTPPS$^-$) and cationic zinc tetrakis(N-methylpyridyl)porphyrin (ZnTMPP$^+$) adsorbed very strongly at the liquid–liquid interfaces. The photocurrent due to the heterogeneous quenching of the porphyrin ion pair was observed in the presence of the electron donor decamethylferrocene and the electron acceptor tetracyanoquinodimethane. No substantial photoresponses were detected in the presence of only one of the porphyrin species [222]. In part IV, the influence of light polarization on the photocurrent for the quenching of ZnTPPC specifically adsorbed at the H$_2$O–1,2-DCE by ferrocene was studied in a total internal reflection mode, and revealed a strong correlation between interfacial reactivity and molecular orientation [223]. In part V, all the developed methodologies were applied to study the molecular organization and photoreactivity of water-soluble chlorophyll (CHL) [224]. In that case, it was observed that the preferential molecular orientation was fairly independent of the applied potential and surface coverage (Figure 1.26).

In part VI, the dynamics of photoinduced heterogeneous electron transfer between a series of ferrocene derivatives and porphyrin ion pairs introduced in part IV was studied. In particular, the use of various ferrocene derivatives with

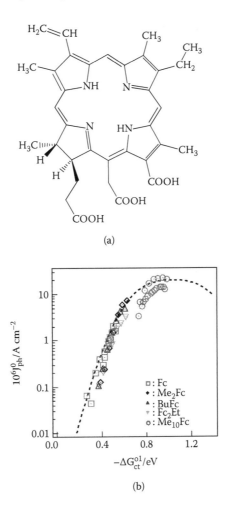

(a)

(b)

FIGURE 1.26 Water-soluble porphyrin (CHL) (Jensen, H., D. J. Fermin, and H. H. Girault, 2001, *Phys Chem Chem Phys*, Vol. 3, p. 2503) and photocurrent dependence on the formal Gibbs energy of electron transfer. (Eugster, N., D. J. Fermin, and H. H. Girault, 2002, *J Phys Chem B*, Vol. 106, p. 3428. Used with permission.)

different redox potentials, and the potentiostatic control over the Galvani potential difference across the interface, allowed the investigation of driving forces ΔG_{ET} over a range of 1 eV. The photocurrent as a function of ΔG_{ET} showed a Marcus-type behavior, and the solvent reorganization energy was estimated to be 1.05 eV, from which an average distance of 0.8 nm between the redox species could be evaluated [225]. In part VII, the self-organization of a variety of dyes, that is, Sn(IV) meso-tetra-(4-carboxyphenyl) porphyrin dichloride (SnTPPC), chlorin e-6, protoporphyrin IX (protoIX), and Fe(III) protoporphyrin IX chloride (Fe-protoIX), at the H_2O–1,2-DCE interface was studied by admittance measurements, photocurrent-potential curves, and light polarization anisotropy

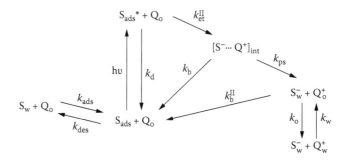

FIGURE 1.27 Schematic representation of the photoinduced heterogeneous electron-transfer reaction between a photoactive electron acceptor in the aqueous phase (S) and an electron donor in the organic phase (Q). Alternative reaction pathways are represented as dashed arrows. Values k_{ads} and k_{des} represent the rate constants related to the adsorption and desorption of the photoactive species, k_d is the rate constant associated with the decay of the excited state, and k^{II}_{et} is the bimolecular rate constant of electron transfer. Values k_b and k_{ps} are associated with back electron transfer and product separation steps, and k^{II}_b is related to a hypothetical second-order recombination. Values k_o and k_w denote the rate constants associated with the ion transfer of the photoproduct Q^+.

of the photocurrent. The results showed a clear correlation between the orientation of the transition dipole and the distribution of the peripheral carboxyl groups responsible for the hydrophilic nature of the dyes [226].

In 2003, the initial stages of the heterogeneous photoreduction of quinone species by self-assembled porphyrin ion pairs at the H_2O–1,2-DCE interface were studied by ultrafast time-resolved spectroscopy and dynamic photoelectrochemical measurements. Photoexcitation of the water-soluble ion pair formed by $ZnTPPS^{4-}$ and $ZnTMPyP^{4+}$ leads to a charge-separated state of the form $ZnTPPS^{3-}$–$ZnTMPyP^{3+}$ within 40 ps that can inject electrons to acceptors in the organic phase in the microsecond time scale, resulting in photocurrent responses [228].

In 2005, Samec et al. presented a thorough kinetic model to analyze photocurrent responses according to the reaction scheme shown in Figure 1.27 [229]. At low photon fluxes, the effect of diffusion is negligible, and the expression for the current density across the interface simplifies to the steady photocurrent given by

$$j^{ss}_{ph} = F\sigma_s\Gamma_s\Phi\left(\frac{k^{II}_{et}c^o_Q}{k^{II}_{et}c^o_Q+k_d}\right)\left(\frac{k_{ps}}{k_{ps}+k_d}\right) \tag{1.93}$$

At higher light intensities, the decay of the photocurrent can be effectively attributed to the depletion of the organic reactant at the interface. The solution of the diffusion problem allows rationalization of the photocurrent dependence on the illumination over a wide range of light intensities. Another interesting system studied by Nagatani et al. consists of zinc tetraphenylporphyrin as organic synthe- sizer at the polarized H_2O–1,2-DCE interface. Photocurrent transient responses using tris(tetraoctylammonium)tungstophosphate $((TOcA)_3PW_{12}O_{40})$ as organic supporting electrolyte exhibited a pH dependence showing that photoreduction of hydrogen ions probably took place [230].

1.5.5 Proton-Coupled Electron-Transfer Reactions

The liquid–liquid interface is particularly well suited to study proton-coupled electron-transfer reactions. Indeed, the protons can be provided by the aqueous phase, and the electrons by an organic donor. Kihara and his group were the first to illustrate this point by voltammetric studies. In 1997, they reported the oxi- dation of ascorbic acid when teterachlorobenzoquinone, dibromobenzoquinone, and Meldola's Blue were used as oxidizing agents in organic solution [231]. Then they studied the reduction of flavin mononucleotide (FMN) in the presence of oxygen and decamethylferrocene [231], and the oxidation of NADH by chloranil and toluquinone [232]. The major work is the study of oxygen reduction by tet- rahydroquinone showing that, depending on the interfacial polarization, the final product is either water or hydrogen peroxide [233].

Oxygen reduction at ITIES has recently been revisited. Using decamethylferro- cene as the electron donor, Su et al. have shown that the potential control of the pro- ton pumping at the interface can be used to control the overall reaction [234]. Also, Samec et al. have shown that cobalt porphyrins and free-base porphyrins can be used to reduce oxygen at the interface in the presence of lipophilic donors [235].

1.6 EXPERIMENTAL METHODS

1.6.1 Micro-ITIES

1.6.1.1 Micro- and Nanopipettes

In 1986, Taylor and Girault reported the transfer of tetraethylammonium ion at an H_2O–DCE interface supported at the tip of a glass micropipette [236]. Micropipette-supported ITIES offer different type of mass transport regimes, as discussed by Stewart et al. [237]. For example, in the case of an ion-transfer reac- tion from outside to inside the pipette, the ingress takes place under steady-state conditions, as for a redox reaction on a microdisc electrode, but the egress occurs under linear diffusion conditions. This results in an asymmetric voltammogram as displayed in Figure 1.28a. For an ion transfer from inside to outside, the egress occurs under linear diffusion, and the return peak can hardly be observed, as displayed in Figure 1.28b. This asymmetry of the diffusion fields can be used to determine which ions limit the potential window [238].

On the other hand, in the case of assisted-ion-transfer reactions in which the ligand is inside the pipette, the mass transport regime is completely analogous to a microdisc electrode. Since these early days, many groups have developed the pipette technique to a high level. In particular, Shao and Mirkin have made

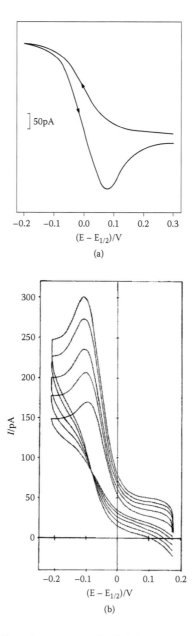

FIGURE 1.28 (a) Cyclic voltammogram for the ingress transfer of: TEA⁺ from the organic phase in a 15-μm-radius pipette. (b) Sweep rate dependence for the egress transfer of TEA⁺ from the aqueous phase. (Inverse voltage scale. Negative potential = water positive versus organic.) (Stewart, A. A., G. Taylor, H. H. Girault, and J. McAleer, 1990, *J Electroanal Chem*, Vol. 296, p. 491. Used with permission.)

extensive use of this technique to study ion-transfer and assisted-ion-transfer reactions. In 1997, they used nanopipettes to measure the rate of potassium transfer assisted by di-benzo-18-crown-6 and could measure rates up to 10 cm·s⁻¹ [190]. One year later, they demonstrated that, with controlled silanization of either the inner or the outer wall of the micropipette, it was possible to control both the position and the curvature of the interface as shown in Figure 1.29 [239].

In 1998, Shao et al. developed a dual micropipette to be used in a generator–collector mode, thereby circumventing the restriction due to the potential window [240]. This concept was further developed in 2000 using theta-pipettes as illustrated, and it is shown that these pipettes could be used in the air as a gas sensor [241]. Nanopipettes were recently used to investigate the water–octanol interface that is very difficult to investigate with macroscopic interfaces [242].

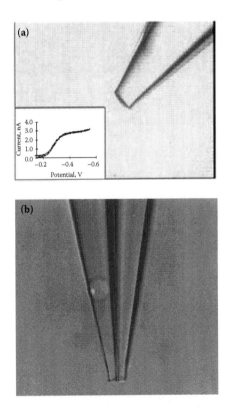

FIGURE 1.29   (a) Video micrographs of a 15.5-μm-radius micropipette filled with an aqueous KCl solution and immersed in a DCE solution of DB18C6. No external pressure was applied to the pipette, and the micro-ITIES is flat. The insets show corresponding steady-state voltammograms of facilitated transfer of potassium (From Shao, Y. H. and M. V. Mirkin, 1998, *Anal. Chem.*, Vol. 70, p. 3155. Used with permission). (b) Photomicrograph of a theta-pipette filled with an aqueous solution. A thin (≤1-μm-thick) glass wall separates two barrels, one of which is blocked by an air bubble. Both orifices are about 4.5 μm radius. (Liu, B., Y. H. Shao, and M. V. Mirkin, 2000, *Anal Chem*, Vol. 72, p. 510. Used with permission.)

A very recent development of ITIES in micropipettes is the electrochemical attosyringe by Laforge et al. [243] who demonstrated electrochemical control of the fluid motion inside the pipette to sample and dispense attoliter-to-picoliter volumes, for example, in an immobilized biological cell. The movement of the interface supported at the tip of a micropipette was recently observed by Dale and Unwin using confocal fluorescence microscopy [244]. This motion was found to be reversible upon cycling.

In summary, micropipettes have been useful not only for kinetic measurements of ion-transfer or assisted-ion-transfer reactions but the asymmetry of the diffusion fields can be used to determine which charge-transfer reactions take place.

1.6.1.2 Microhole-Supported ITIES

In 1989, Campbell and Girault reported ion-transfer reactions at ITIES supported in microholes drilled in a thin polyethylene terephthalate film by UV laser photoablation [245]. The main advantage of this system is that it is completely analogous from a mass transport aspect to a microdisc electrode when studying ion-transfer reactions. For example, it allows one to reduce the amount of supporting electrolyte as demonstrated by Osborne et al. who studied an ITIES between pure water and an organic electrolyte solution in order to measure the Gibbs energy of transfer of lipophilic ions [246]. Another advantage is the possibility to produce micro-ITIES arrays for analytical purposes. For example, Lee et al. have developed the so-called ionodes illustrated in Figure 1.30 that are micro-ITIES arrays for the amperometric detection of ionic species [247,248], for example, as a detector for ion chromatography [249]. These arrays have been characterized both by simulation and experimentally by Wilke et al. [250] and Murtomaki et al. [251]. Also, Osborne and Girault have demonstrated how a microhole could be integrated in the design of a biosensor, namely, a creatinine enzyme assay with creatinine deiminase [252] or a urea sensor using urease to produce ammonium ions [253]. Mass transport at microhole-supported ITIES have been simulated to understand the influence of the thickness of the film and to study how the position of the interface with respect to the film determines the amperometric response [254]. The localization of the interface with respect to the hole and its influence on the mass regimes was further discussed by Ohde et al. [255] and Peulon et al. [256].

Microholes are particularly useful in sustaining stable interfaces and for working with small-solution volumes as shown by Quinn et al. for the water–NPOE interface [257] or Silva et al. for NPOE gels [258]. These microholes are also useful to study water–ionic liquid interfaces [259].

More recently, microholes have been integrated in a system to measure the pH dependence of the lipophilicity of drug molecules [185] as illustrated in Figure 1.31.

1.6.2 Scanning Electrochemical Microscopy (SECM)

In 1995, Bard and his group published a series of papers on the application of SECM to the study of electron-transfer reactions at ITIES [260–262]. Here, the classic scanning microtip electrode was replaced by a micropipette filled with an

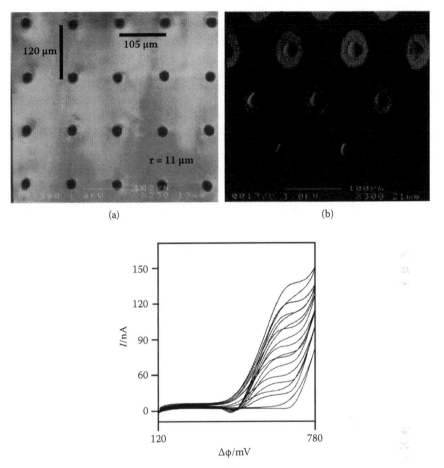

(a)   (b)

(c)

FIGURE 1.30   (a) Scanning electron photomicrograph of a microhole array in a PET film covered with a PVC–NPOE gel cast at 70°C. (b) The PVC gel can be seen filling the holes. (c) Steady-state voltammogram for choline transfer with concentrations of 0, 0.1, 0.2, 0.3…0.9, 1.0 mM at the microhole array illustrated on the left. (Lee, H. J., P. D. Beattie, B. J. Seddon, M. D. Osborne, and H. H. Girault, 1997, *J Electroanal Chem*, Vol. 440, p. 73. Used with permission.)

aqueous solution of ferricyanide and was used to image a metallic substrate in an organic phase [262]. In the second work, a classic metallic microtip was approached to a macroscopic ITIES to run an uphill electron-transfer reaction [262]. Tsionski et al. then made kinetic measurements by varying the ITIES polarization using salts of different concentrations as illustrated in Figure 1.32 [263]. The observed change in the ET rate with the interfacial potential drop could not be attributed to

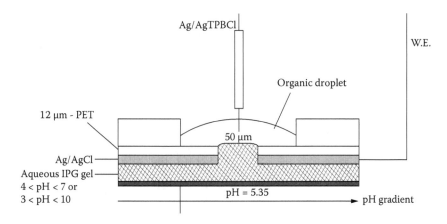

FIGURE 1.31 Schematic presentation of the setup for the ITIES measurement at a single microhole at pH 5.35.

concentration effects, and was considered as the true potential dependence indicative of a Butler–Volmer mechanism. This methodology was further simulated by Unwin et al. [264]. In 2001, Ding et al. showed that, at high driving forces, the measured experimental rate constants decreased with increasing driving force, deviating from predictions based on classical Butler–Volmer kinetics, which is consistent with the inverted region predicted by the Marcus theory. At low driving forces, the potential dependence of the forward and reverse ET rate constants were found to follow the Butler–Volmer theory [265]. Such behavior was confirmed by Barker et al. [266], and more recently by Li et al. for fast electron transfer reactions [267].

Of course, it is now possible to control the polarization of the ITIES either by using a drop on a solid electrode as pioneered by Zhang et al. [268], or by using a 5-electrode potentiostat.

As discussed earlier (see Section 1.4.5), SECM has been extremely useful in studies of assisted-ion-transfer reactions following the seminal work of Shao and Mirkin to study the transfer of K^+ assisted by the presence of di-benzo-18-crown-6 in the organic phase [269].

Since these early days, SECM has been widely used to study charge-transfer reactions, and excellent reviews have been published recently [270–272]. Among the recent work, one should mention the possibility of generating electrochemiluminescence at ITIES using SECM [273,274], or of exploring cell metabolism and detect Ag^+ toxicity in living cells [275].

1.6.3 SOLID-SUPPORTED ITIES

1.6.3.1 Organic Electrolyte Layer on Electrodes

In 1998, Shi and Anson reported an original method to study heterogeneous electron transfer at ITIES [276]. The gist of this approach is to coat a pyrolytic graphite electrode with a thin layer of nitrobenzene and to immerse it in an aqueous

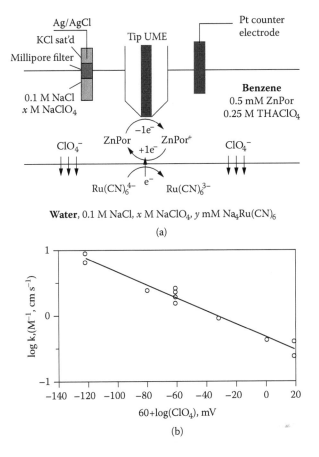

FIGURE 1.32 (a) Schematic diagram of the application of SECM in the feedback mode measurement of the kinetics of ET between ZnPor$^+$ in benzene and Ru(CN)$_6^{4-}$ in water. Electroneutrality maintained by distribution of perchlorate ions across the interface. (b) Dependence of the effective heterogeneous rate constant on potential drop across the ITIES at 5 mM concentrations of Ru(CN)$_6^{4-}$. The value $\Delta_o^w\phi$ is expressed in terms of log[ClO^{4-}] . Potential dependence of an effective bimolecular rate constant k) k f/[Ru(CN)$_6^{4-}$] (M^{-1} cm·s^{-1}). (Tsionsky, M., A. J. Bard, and M. V. Mirkin, 1996, *J Phys Chem-US*, Vol. 100, p. 17881. Used with permission.)

solution. Redox reactants dissolved in the nitrobenzene film included decamethylferrocene and zinc–cobalt tetraphenylporphyrin, while the redox reactants in the aqueous phase included Fe(CN)$_6^{3-/4-}$, Ru(CN)$_6^{4-}$, Mo(CN)$_8^{4-}$, and IrCl$_6^{2-}$. Rate constants for cross-phase electron transfer could be easily obtained. The system developed by Shi and Anson comprises two polarizable interfaces in series, namely, the graphite–NB electrochemical interface and the H$_2$O–NB ITIES. Because of the thin layer of NB, catalytic currents can be observed as shown in Figure 1.33.

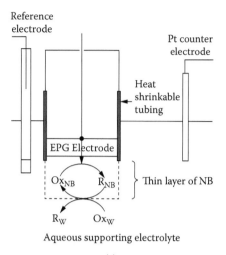

FIGURE 1.33   Left: Schematic diagram (not to scale) of the electrochemical cell and the EPG electrode coated with a thin layer of NB. The diameter of the EPG electrode was 0.64 cm, and the thickness of the NB layer was typically 20–30 μm. Right: Voltammetric observation of electron transfer from DMFc in a thin layer of NB to $Fe(CN)_6^{3-}$ in water. (A) Cyclic voltammogram for 0.47 mM $Fe(CN)_6^{3-}$ at an uncoated EPG electrode: supporting electrolyte, 0.1 M $NaClO_4$ + 0.1 M NaCl; scan rate, 5 mV·s^{-1} throughout. (B) Repeat of A after the electrode surface was covered with 1 μL of NB. (C) Cyclic voltammogram with the electrode covered with 1 μL of NB containing 0.55 mM DMFc+ (generated by oxidation of DMFc at the initial potential). The aqueous solution contained only supporting electrolyte (0.1 M $NaClO_4$ + 0.1 M NaCl). (D) Solid line: Repeat of C with 1.7 mM $Fe(CN)_6^{3-}$ present in the aqueous phase. The plotted points are the steady-state currents recorded as the potential was stepped to more negative (O) and to more positive (b) values. (Shi, C. N., and F. C. Anson, 1999, *J Phys Chem B*, Vol. 103, p. 6283. Used with permission.)

In subsequent papers, the same authors developed the technique further. In particular, they showed that it was very well suited for the study of metalloporphyrins [277]. However, when using different solvents such as 4-methylbenzonitrile, chloroform, or benzene, they also showed that the coupling between ion-transfer and electron-transfer reactions can render the quantitative analysis difficult [278–280].

Finally, Chung and Anson also showed that thin organic films are very useful in the study of proton-coupled redox reactions such as the reduction of tetrachloro-1,4-benzoquinone in nitrobenzene or benzonitrile [281].

1.6.3.2   Thin Aqueous Layer on Electrodes

Following the layer-by-layer deposition method pioneered by Decher et al. [282], Cheng and Corn have demonstrated [283] that it was possible to form an aqueous thin layer on a solid electrode by modifying the electrode surface with charged groups, such as by adsorption of mercapto-undecanoic acid on gold, and then

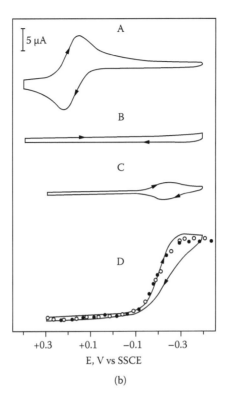

(b)

FIGURE 1.33 (Continued).

by successive dipping in alternate solutions of polycations and polyanions. The aqueous-film-coated electrode can then be immersed in an organic solvent, and the ITIES then formed can be studied. Slevin et al. have characterized these films by capacitance measurements [284], and Hoffmannova et al. by quartz microbalance EQCM [285]. These films can then be used on an open-cell-structure solid such as reticulated carbon to form 3d ITIES as demonstrated by Tan et al. [286].

A major interesting characteristic of these films is their ability to be loaded with charged nanoparticles as shown by Santos et al. [287], which can, in certain cases, mediate electron transfer through the film, as shown by Fermin et al. [288]. Quantum dots have also been incorporated in aqueous thin films to form photoanodes or photocathodes [289,290].

An alternative method is to deposit an aqueous drop on a solid electrode such as a silver–silver chloride electrode or a platinum electrode and immerse it in an organic solvent [291]. This method provides a fast and convenient method to measure standard transfer potentials, for example, of ionized drug molecules. It differs from the three-phase junction discussed in the following text, in the sense that the electrode is in contact with one phase only, namely, the aqueous one. In this case, we have two interfaces in series.

1.6.3.3   Membrane-Supported ITIES

In 1991, Kihara et al. pioneered the concept of voltammetry for organic liquid membranes with two ITIES in series, that is, for a system: aqueous electrolyte (AE1)–organic electrolyte–aqueous electrolyte (AE2) [292,293]. Using a 6-electrode cell, they have clearly shown that, as we polarize a liquid membrane, AE2–AE1, the two polarized ITIES in series, are coupled by the equation of continuity of current, and the overall potential window is the sum of the two individual potential windows in series. In 1998, Beriet and Girault used this approach to selectively transfer metal ions for metal recovery [294].

In 2002, Ulmeanu et al. used a thin organic layer that is supported by a porous hydrophobic membrane such as porous Teflon or polyvinylidenedifluoride (PVDF), or sandwiched between two aqueous dialysis membranes [295]. With this setup, they showed that the transfer of highly hydrophilic ions at one interface can be studied by limiting the mass transfer of the other ion-transfer reaction at the other interface. They have also shown that cyclic voltammetry for coupled ion-transfer reactions at the two interfaces in series is analogous to cyclic voltammetry for electron-transfer reactions studied by Stewart et al. [207], as the diffusion equations of the reactants and products are analogous, and as the overall Nernst equation for the coupled ion transfer equal to the two individual Nernst equations for ion distribution is also analogous to the Nernst equation for the heterogeneous ET.

1.6.4   Three-Phase Junctions

In 1997, Marken and the Compton group published a series of papers with a very interesting method to study liquid–liquid interfaces [296–301]. The gist of this approach is to modify the solid electrode, such as a graphite electrode, with organic droplets containing redox molecules, or even to use a redox liquid such N,N,N',N'-tetraoctylphenylenediamine (TOPD). For a charge-transfer reaction taking place at the solid–organic liquid interface, another charge transfer reaction must occur at the liquid–liquid interface. In the absence of a redox couple in the aqueous phase, the electroneutrality balance is guaranteed by ion-transfer reactions, such as anion insertion during an oxidation at the electrode, as illustrated in Figure 1.34. The early work was nicely reviewed by Compton et al. in 2003 [302]. Since then, many papers have been published in this field of so-called three-phase junctions.

One advantage of these droplet-covered electrodes is that it possible to study the Gibbs energy of transfer of aqueous ions into the organic droplet that can be ion free. In 2000, Scholz et al. demonstrated this approach to measuring the Gibbs energy of transfer of aqueous ions [303]. Indeed, in the case illustrated in Figure 1.34, the electrode potential versus a standard hydrogen electrode in the aqueous phase is given by

$$E_{\text{SHE}} = \left[E_{\text{ox/red}}^{\ominus} \right]_{\text{SHE}}^{\text{o}} + \frac{RT}{nF} \ln\left(\frac{a_{\text{O}}}{a_{\text{R}}} \right) - \Delta_{\text{o}}^{\text{w}}\phi \qquad (1.94)$$

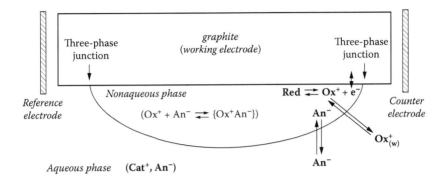

FIGURE 1.34 Electrode assembly with a droplet of an aprotic solvent containing a neutral redox probe (Red) attached to a graphite electrode, which is immersed in an aqueous electrolyte solution.

where $[E_{\text{ox/red}}^{\ominus}]_{\text{SHE}}^{\text{o}}$ represents the standard redox potential of the organic couple with respect to an aqueous standard hydrogen electrode as defined by Equation 1.79, and where $\Delta_{\text{o}}^{\text{w}}\phi$ represents the Galvani potential difference between the two solutions. If the organic phase contains decamethylferrocene, then, upon oxidation, an anion must penetrate the droplet, and in this case, the Galvani potential difference can be considered in a first approximation to be given by the Nernst equation (1.73) as illustrated in Figure 1.20.

This methodology has been applied to many solvent systems both for anions [304–306] and cations [307], for chiral ions [308], for amino acids [309], and even to determine the log$P$ of therapeutic molecules [310,311]. An interesting review of this methodology was published in 2005 by Scholz and Gulaboski [312]. More recently, L'Her and his group reported the transfer of very hydrophilic ions from water to nitrobenzene using lutetium bisphthalocyanines, as these molecular sandwich complexes can be reduced as well as oxidized, and because the products of the reactions have a very low affinity for water [313]. The theoretical aspects of charge transfer at three-phase junctions have also been addressed by Aoki et al. [314,315], who in particular have demonstrated some strong convection effects inside the droplet onset by a Marangoni flow associated to gradient in interfacial tension [316].

Another interesting aspect of three-phase junctions is electrowetting. Indeed, Monroe et al. have shown that electrowetting phenomena at ITIES can be controlled by small voltages [317,318] compared to the higher voltages used in commercial electrowetting-based devices such as the zoom lens in some mobile phones.

In summary, many novel systems have been developed over the past two decades to study charge-transfer reactions. Micro-ITIES and microdroplet electrodes have circumvented most of the drawbacks of classical flat liquid–liquid interfaces, namely, the large iR drop and the large capacitance. Also, both systems allow a reduction in the amount of supporting electrolyte needed.

1.7 PHOSPHOLIPID-FUNCTIONALIZED ITIES

1.7.1 Ion Transfer through an Adsorbed Phospholipid Monolayer

Back in 1982, Koryta pioneered the concept of ion-transfer reactions across a phospholipid layer adsorbed as a monolayer at ITIES [319]. In 1984, Girault and Schiffrin had shown that the adsorption of phospholipids at ITIES was both potential and pH dependent [320]. Indeed, neutral phospholipids do adsorb very strongly, forming a monolayer that can desorb when becoming charged, and ion-transfer reactions are consequently studied through a compact monolayer. Soon after, Cunnane et al. studied ion-transfer reactions across a phospholipid layer and related the kinetics of the ion-transfer process to a pore formation mechanism, calculating the energy required to create a pore in a compact monolayer [321]. In 1992, Kakiuchi et al. reported a couple of studies of ion-transfer kinetics across a phospholipid monolayer. The first study compared six L-α-phosphatidylcholines (PC), namely, dilauroyl (DLPC), dimiristoyl (DMPC), dipalmitoyl (DPPC), distearoyl (DSPC), diarachidoyl (DAPC), and dibehenoyl-phosphatidyl-choline (DBPC). Those monolayers, in a liquid-condensed state, were observed to reduce the rate of ion transfer for both tetramethylammonium ion (TMA$^+$) and tetraethylammonium ion (TEA$^+$). In contrast, those monolayers in a liquid-expanded state were observed to accelerate the transfer of both ions [322]. The authors concluded that a PC layer in the liquid-condensed state exerts a hydrodynamic friction on transferring ions, whereas a PC layer in the liquid-expanded state itself is transparent to the ion transfer. In the second study, they showed that a dilauroylphosphatidyl-ethanolamine (DLPE) layer can hinder the transfer of TEA$^+$ but not that of ClO$_4^-$ [323]. More recently, Monzon and Yudi confirmed this aspect using distearoyl phosphatidic acid (DSPA) layers, and showed that the layer is tightly compact and the transfer of TEA$^+$ by permeation does not take place. Instead of this, TEA$^+$ ions were found to adsorb at the polar head groups of DSPA, and that these adsorbed cations were acting as nucleation centers of DSPA molecules when the DSPA amount was low [324].

Very interestingly, the influence of a phospholipid, dipalmitoyl phosphatidylcholine layer at an H$_2$O–NB interface on the transfer of TMA$^+$ and a polyammonium antifungus agent, poly [(dimethylimino)(2-oxo-1,2-ethanediyl) imino-1,6-hexanediylimino(1-oxo-1,2-ethanediyl)(dimethylimino)-1,6-hexanediyl] ion, across the interface was studied by Katano et al. using normal pulse voltammetry [325]. In accordance with what has been described, when the phospholipid was adsorbed to form a monolayer at the H$_2$O–NB interface by addition to the organic phase, the half-wave potential for the transfer of TMA$^+$ did not change, but the limiting current was significantly decreased, indicating a retarding effect of the layer on the ion transfer. On the other hand, in the current versus potential curves for the transfer of the polyammonium ion, no significant change in either the half-wave potential or the limiting current was observed upon adding the phospholipid, indicating that the polyammonium ion can easily

TABLE 1.3
Association constant for facilitated ion transfer between phosphatidylcholine and differentials

Ions	$\Delta\phi_{desorb}$	$\Delta\phi^{\ominus}$	$\log K_{ass}$
Li^+	0.13	0.58	8.6
Na^+	0.13	0.58	8.6
K^+	0.13	0.54	7.9
Cs^+	0.10	0.39	5.9
NH_4^+	0.08	0.45	7.3
$MeNH_3^+$	0.03	0.36	6.6
$Me_2NH_2^+$	−0.05	0.27	6.4
Me_3NH^+	−0.10	0.17	5.6
Arg^+	0.15	0.64	9.3

permeate through the phospholipid layer. The results suggest a new application of the voltammetric technique to the study of cell-membrane permeability to poly-ionic bioactive compounds.

In a breakthrough paper, Yoshida has shown that phospholipids were acting as an ionophore to facilitate the transfer of alkali metal ions and also arginine cations but not for TEA$^+$, Cl$^-$, HCO$_3^-$, SO$_4^{2-}$, and picrate [326]. This facilitated transfer phenomenon has since been widely investigated in a series of papers that all confirmed this ionophoric property of the phospholipids [327]. In particular, Yoshida et al. established a list of association constants (Table 1.3) [328].

1.7.2 Ion Adsorption on a Phospholipid Monolayer

To analyze ion adsorption at the phospholipid monolayer at an ITIES, one needs to set up a thermodynamic model. In 2003, Samec et al. proposed a simple model comprising the adsorption of the zwitterionic form L$^\pm$, the formation of the cat-ionic complex RL$^+$ with an aqueous cation R$^+$, and the desorption of the complex in the organic phase and its dissociation in the organic phase [329].

$$L^{\pm o} \rightleftarrows L^{\pm ad}$$

$$R^{+w} + L^{\pm ad} \rightleftarrows RL^{+ad}$$

$$RL^{+ad} \rightleftarrows RL^{+o}$$

$$RL^{+o} \rightleftarrows R^{+o} + L^{+o}$$

The Gibbs adsorption equation includes the contributions from ions of the aque-ous (RX) and organic (SY) electrolytes, as well as from the zwitterionic (L$^\pm$) and cationic adsorbed complex (RL^{+ad}), and can be expressed as not only a function of the surface concentration of the different charged species but also as a function of the variation of the experimental parameters such as the electrical polarization, the concentration of phospholipids in the organic phase, and the concentrations of the respective salts.

$$
\begin{aligned}
-d\gamma &= \Gamma_{R^+}d\tilde{\mu}_{R^+} + \Gamma_{X^-}d\tilde{\mu}_{X^-} + \Gamma_{L^\pm}d\mu_{L^\pm} + \Gamma_{RL^+}d\tilde{\mu}_{RL^+} + \Gamma_{S^+}d\tilde{\mu}_{S^+} + \Gamma_{Y^-}d\tilde{\mu}_{Y^-} \\
&= QdE + (\Gamma_{L^\pm} + \Gamma_{RL^+})d\mu_{L^\pm} + (\Gamma_{R^+} + \Gamma_{RL^+})d\mu_{RX} + \Gamma_{Y^-}d\mu_{SY}
\end{aligned}
\tag{1.95}
$$

where Γ represents the surface excess concentration with respect to both solvents, and dE is the variation of the applied potential with respect to an aqueous reference electrode reversible to the aqueous anion (X^-, e.g., Cl^-) and an organic reference electrode reversible to the organic cation (S^+, e.g., TBA^+):

$$
dE = -(d\tilde{\mu}_{S^+} + d\tilde{\mu}_{X^-})/F
\tag{1.96}
$$

The thermodynamic charge is defined as a difference of surface excesses:

$$
Q = -\left(\frac{\partial\gamma}{\partial E}\right)_{\mu_i} = F(\Gamma_{L^\pm} + \Gamma_{RL^+} - \Gamma_{X^-}) = -F(\Gamma_{S^+} - \Gamma_{Y^-})
\tag{1.97}
$$

It represents the total charge that is necessary to supply so as to maintain the state of the interface when the interfacial area is increased by a unit amount. Additionally, the interfacial capacitance can also be defined as the curvature of the electrocapillary curve, or can be obtained directly from impedance measurements:

$$
C = \left(\frac{\partial Q}{\partial E}\right)_{\mu_i} = -\left(\frac{\partial^2\gamma}{\partial E^2}\right)_{\mu_i}
\tag{1.98}
$$

The total surface concentration of phospholipids can be measured by the concentration dependence of the surface tension measurements:

$$
\Gamma_{L^\pm} + \Gamma_{RL^+} = -\left(\frac{\partial\gamma}{\partial\mu_{L^\pm}}\right)_{E,\mu_{RX},\mu_{SY}}
\tag{1.99}
$$

The charge defined by Equation 1.97 cannot be read directly as the slope of the electrocapillary curve, especially when the monolayer desorbs upon complexation. Indeed, due to the charge-transfer reaction taking place during the desorption of the complex, it is impossible to vary the potential E while keeping constant the chemical potential of the phospholipid $L^\pm$. Hence, the differentiation of the interfacial tension with respect to the potential does not give the surface charge density but rather the sum of two terms:

$$\frac{d\gamma}{dE} = Q + F\left(\Gamma_{L^\pm} + \Gamma_{RL^+}\right)\frac{d\mu_{L^\pm}}{dE} \tag{1.100}$$

Analogously, the second differentiation with respect to the potential does not provide the double-layer capacity.

The adsorption of the zwitterionic lipid and cationic complex can be considered as a particular case of the adsorption of a molecule in two different states (e.g., two different orientations), which can be described by two Frumkin isotherms [66,67]. Assuming that the ratio of the interfacial area occupied by either the zwitterionic or cationic form to the area occupied by the solvent molecules that either form replaces is unity, the adsorption of the zwitterionic form can then be described by the Frumkin isotherm

$$B_\pm c_{L^{\pm o}} = \frac{\theta_\pm}{1 - \theta_\pm - \theta_+}\exp(-2a_{\pm\pm}\theta_\pm - 2a_{\pm+}\theta_+) \tag{1.101}$$

where θ represents the respective surface coverage, a is the attraction constant, and B is the adsorption coefficient. For the cationic form, we have

$$B_+ c_{R^{+w}}c_{L^{\pm o}} = B'_+ c_{RL^{+o}} = \frac{\theta_+}{1 - \theta_\pm - \theta_+}\exp(-2a_{++}\theta_+ - 2a_{\pm+}\theta_\pm) \tag{1.102}$$

Taking into account the assisted cation transfer for which the half-wave potential is given by

$$E_{1/2} = E^{\ominus\prime} - \frac{RT}{F}\ln\left[\frac{K_{ass}c_{RX}^b D_{RL^+}^{1/2}}{D_{L^\pm}^{1/2}}\right] \tag{1.103}$$

we obtain an expression for the interfacial tension:

$$\gamma = \gamma_0 + RT\Gamma_{max}[\ln(1 - \theta_\pm - \theta_+) + a_\pm\theta_\pm^2 + +a_+\theta_+^2 + 2a_{\pm+}\theta_\pm\theta_+] \tag{1.104}$$

where γ_0 is the interfacial tension in the absence of phospholipid. Figure 1.35 shows how this model can fit the electrocapillary curve for DPPC adsorption in the presence of Li^+ and Ca^{2+} in the aqueous phase.

In summary, it is clear that ITIES provide a unique support to study phospholipid adsorption and the interaction of the phosphatidyl moiety with aqueous cation. It has been observed many times that a compact monolayer can hinder the transfer of some cations such TEA$^+$, but does seem to be an effective barrier to the transfer of anions. Finally, charge-transfer studies combined to electrocapillary data have clearly shown that phosphatidylcholine acts as a strong ionophore for alkali metal cations and peptides.

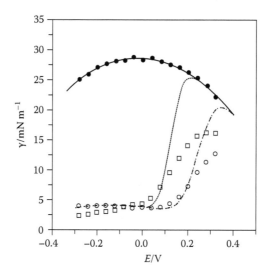

FIGURE 1.35 Experimental (points) and theoretical (lines) interfacial tension versus potential in the absence (•) and presence (o, □) of 10 μmol dm^{-3} DPPC in the organic phase. Aqueous phase: 0.1 M LiCl +0.05 M Tris (pH 8.9) (•,o), 0.1 M LiCl +0.05 M Tris (pH 8.9) +5 mM CaCl$_2$ (□). Parameters of the theoretical model: $Epzc,1 = 0$ V, $Epzc,2 = -0.2$ V, $B_{10}c^0 = 50$, $B_{20}c^0 = 5$, $a1 = a3 = -0.5$, $a2 = -2$, $\Gamma_m = 2.5 \cdot 10^{-6}$ mol·m^{-2}, $C_0 = C_2 = 0.1$ F·m^{-2}, $C_1 = 0.01$ F·m^{-2} and (a) $z = 1$, $E_{1/2} = 0.2$ V (—··—) or (b) $z = 2$, $E_{1/2} = 0.2$ V (Samec, Z. A. Trojanek, and H. H. Girault, 2003, *Electrochem Commun.* Vol. 5, p. 98. Used with permission.)

1.8   NANOPARTICLES AT ITIES

1.8.1   NANOPARTICLE SYNTHESIS AT ITIES

One property of liquid–liquid interfaces in general, and of ITIES in particular, is their ability to adsorb nanoparticles and, in certain cases, to form metal-like films. Pioneering measurements were carried out by Guainazzi et al. [330] who demonstrated that a direct current applied across the interface between Cu^{2+} ion in water and a vanadium complex in 1,2- DCE causes deposition of a copper layer at the liquid–liquid boundary. Another seminal work is that of Efrima et al. who showed in 1988 the formation of silver metal-like films at the H$_2$O–DCE interface [331]. Since then, many publications have addressed this fascinating topic, and the field was excellently reviewed recently by Boerker et al. [332].

More specifically at ITIES, a lot of effort has been dedicated to nucleation and growth of nanoparticles for electrocatalytic studies. In 1998, Schiffrin deposited gold particles at an ITIES by electrochemical reduction of tetraoctylammonium tetrachloroaurate in 1,2-DCE using ferrocyanide in water as the electron donor. Their growth was monitored in situ by transmission UV-VIS spectroscopy, and the spectra have been qualitatively analyzed using Mie's theory [331]. The nucleation mechanism was later addressed by Johans et al. [334,335] using dibutyl-ferrocene as electron donor to reduce PdCl$_4^{2-}$.

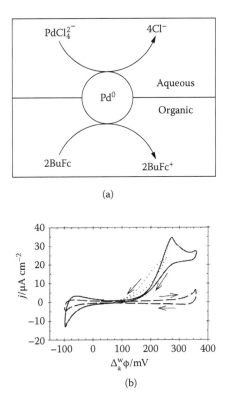

(a)

(b)

FIGURE 1.36 (a) nucleation and growth scheme. (b) cyclic voltammograms obtained at a sweep rate of 25 mV s^{-1}. Dashed line represents (---) 0.5 mM BuFc (1,2-DCE) + 5 mM TBATPFB (1,2-DCE) in contact with 100 mM LiCl + 100 mM Li$_2$SO$_4$ (H$_2$O). Solid line represents (—) as before but with the addition of 1.0 mM (NH$_4$)$_2$PdCl$_4$ to the aqueous phase. Dotted line (⋯) represents a voltammetric scan reversed at a potential less positive than the peak potential featuring a clear nucleation loop. (Johans, C., R. Lahtinen, K. Kontturi, and D. J. Schiffrin, 2000, *J Electroanal Chem*, Vol. 488, p. 99. Used with permission.)

More recently, Samec et al. have studied the nucleation of Pt NPs at an ITIES [336] and showed by repeated potential-step experiments on newly prepared interface that the initial rate of the Pt deposition can attain a broad range of values that span over two orders of magnitude and even approach zero. This may be due to the random rate of nuclei formation with a critical size required for a stable growth to occur (Figure 1.36).

Dryfe et al. extended this study by depositing Pd and Pt nanoparticles on ITIES supported on porous alumina so as to control the mass transport conditions and prevent nanoparticle aggregation [337–343]. Cunnane and his group have shown that nanoparticle deposition could be carried out concomitantly with a polymerization process so as to obtain well-dispersed NPs [344–346].

(a)

(b)

FIGURE 1.37 SEM images of Au:Ag alloy nanostructures. (Agrawal, V. V., G. U. Kulkarni, and C. N. R. Rao, 2008, *J Colloid Interface Sci*, Vol. 318, p. 501. Used with permission.)

Recently, Rao and his group have shown that beautiful fractal structures could be formed at ITIES when using surfactants to control the growth process (Figure 1.37) [347].

1.8.2 NANOPARTICLE ADSORPTION AT ITIES

In 2004, Su et al. reported the voltage-induced assembly of mercaptosuccinic-acid-stabilized Au nanoparticles of about 1.5 nm diameter at the H_2O–DCE interface. Admittance measurements and quasi-elastic laser scattering (QELS) were used to show that the surface concentration of the nanoparticle is reversibly controlled by the interfacial polarization. No evidence of irreversible aggregation of the particles at the interface was observed, and the electrocapillary curves provide an estimate of the maximum particle surface density corresponding to 67% of a square closed-pack arrangement [348]. Similarly, Su et al. studied the voltage-induced assembly and photoreactivity of cadmium selenide (CdSe) nanoparticles protected by mercaptosuccinic acid at the polarizable H_2O–DCE interface [349]. In 2007, Bresme and Faraudo studied the behavior of ellipsoidal nanoparticles adsorbed at liquid–liquid interfaces and under the influence of an external field. They showed that the interplay of particle geometry, particle size, and field strength results in discontinuous orientational transitions [350].

Nanoparticle adsorption has also been monitored by spectroscopic techniques such as SHG to study aggregation at the water–heptanone interfaces reported by Galetto et al. [351]. In 2007, Abid et al. showed, also by SHG, that core-shell gold–silver nanoparticles could be reversibly adsorbed at the polarized H_2O–DCE, proving the excellent stability against aggregation of core-shell nanoparticles [352]. In the same year, Yamamoto and Watarai used TIR light scattering microscopy and TIR surface-enhanced Raman scattering (SERS) to study the adsorption and domain formation of dodecanethiol (DT)-bound silver nanoparticles (SNPs) at the cyclohexane–water interface [353]. Very interestingly, they could observe fluorescence exaltation of dyes in the vicinity of the SNPs. Such a phenomenon was also observed by Cohanoschi et al. who further observed surface plasmon enhancement at a metal-like film at the xylene–water interface using a pseudo-Kretschmann geometry [354].

The theoretical aspects of nanoparticle adsorption at ITIES was recently reviewed by Flatte et al. [355]. In particular, they discuss the effects that drive or hamper the localization at the interface, namely, competitive wetting, solvation of the charged nanoparticles, shift in the external electric field, polarizability drive, and line tension. A simple model is presented to account for these different contributions, and the results are shown in Figure 1.38.

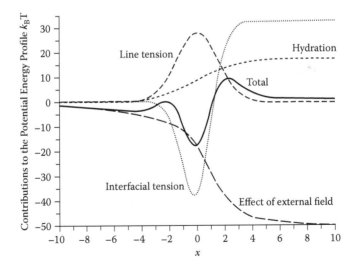

FIGURE 1.38 Different contributions to the nanoparticle energy profile at ITIES, compared to the total profile. Parameters—potential drop across the interface is −250 mV. (Flatte, M. E., A. A. Kornyshev, and M. Urbakh, 2008, *J Phys-Condens Mat*, Vol. 20, p. 073102. Used with permission.)

The main conclusions are that classically charged nanoparticles cannot be spontaneously adsorbed at the H_2O–1,2-DCE interface without applying an external electric field, but small potential difference of 0.1 V are sufficient. Also, the variation of surface coverage with applied electric field can be extremely strong for highly charged particles.

1.8.3 ELECTROCATALYSIS BY NANOPARTICLE-FUNCTIONALIZED ITIES

There are a few examples of electrocatalysis at ITIES using adsorbed nanoparticles. In 2000, Lathinen et al. electrogenerated interfacial palladium particles as reductive photocatalysts for the heterogeneous photoreduction of tetracyanoquinodimethane (TCNQ) by hydrophilic porphyrin species [356]. In 2002, Jensen et al. showed that TiO_2 nanoparticles could be adsorbed at ITIES, either as positively charged particles at low pH or as negatively charged nanoparticles at high pH. In the former case, photocurrents were measured in the presence of electron donors in the organic phase, whereas in the latter case the photocurrents were obtained in the presence of acceptors such as TCNQ [357]. These photocurrents were extended to the visible region by dye sensitization [358]. More recently, Trojanek et al. observed the reduction of oxygen catalyzed by Pt nanoparticles [359].

Considering the importance of nanoparticles in catalysis, we expect this aspect of electrochemistry at ITIES to develop further in the future.

1.9 THERMOELECTRIC EFFECTS AT ITIES

In 1988, Girault showed that it was possible to build an ionic thermocouple using a water–organic phase–water system, with the two interfaces being at different temperatures, thereby demonstrating an ionic Seebeck effect [360]. This methodology was used by Osakai et al. to measure ionic entropy of transfer [361]. More recently, a new technique called thermal modulation voltammetry (TMV) was developed to study ion-transfer reactions [362]. The gist of the method is to heat a microinterface supported on a thin polymer foil with a 325 nm laser that is not absorbed by the aqueous phase but only by the nitrobenzene phase. The authors could then measure the entropy of transfer by measuring the shift of the half-wave potential with the temperature.

1.10 CONCLUSION

The field of electrochemistry at liquid–liquid interfaces has progressed in many directions in the past two decades. Of course, molecular dynamic calculations are getting more accurate, and new information on interfacial structure, such as molecular, orientation is available. Spectroscopic studies have also contributed to a better understanding of interfacial structure.

Perhaps the bulk of the work has been to adapt the classical voltammetric methodologies to the study of charge-transfer reactions. New tools such as SECM have been used to measure kinetic data both for ion- and electron-transfer reactions. One should not forget the investigation of new solvent systems, including the newly developed hydrophobic ionic liquids.

The trend to functionalize the interface, for example, with dyes or quantum dots for photoelectrochemistry, with metallic nanoparticles for electrocatalysis, or with phospholipids for biomimetic studies, will continue as it is very promising. Devices combining electrowetting phenomena [317,318] and polarized liquid–liquid interfaces functionalized with nanoparticles are being developed.

Another promising and fundamental aspect is the study of proton-coupled electron-transfer reactions, for example, for the study of oxygen and carbon dioxide reduction, and that of hydrogen photosynthesis.

All in all, the field of electrochemistry at liquid–liquid interfaces has grown from an exotic curiosity to an established field in which charge-transfer reactions are reasonably well characterized, and it would not be surprising that, after many years of fundamental research, new applications, especially in the field of energy related chemistry, will emerge.

ACKNOWLEDGMENTS

The review was partly written while I was on sabbatical leave at the University of Kyoto and at Peking University. I thank Takashi Kakiuchi and Yuan Hua Shao for their hospitality and for the many fruitful discussions about electrochemistry at liquid–liquid interfaces. Many thanks also to Bin Su and Michael Urbakh for their useful comments on the manuscript.

REFERENCES

1. Girault, H. H. J. and D. J. Schiffrin, Electrochemistry of liquid-liquid interfaces, *Electroanal Chem*, Vol. 15, (1989): p. 1.
2. Koryta, J., Electrochemical polarization phenomena at the interface of 2 immiscible electrolyte-solutions, *Electrochim Acta*, Vol. 24, (1979): p. 293.
3. Linse, P., Monte-Carlo simulation of liquid-liquid benzene-water interface, *J Chem Phys*, Vol. 86, (1987): p. 4177.
4. Theoretical study of the water/1,2-dichloroethane interface: Structure, dynamics and conformational equilibria at the liquid-liquid interface Benjamin, I., *J Chem Phys*, Vol. 97, 1992, p. 1432.
5. Benjamin, I., Chemical reactions and solvation at liquid interfaces: A microscopic perspective, *Chem Rev*, Vol. 96, (1996): p. 1449.
6. Benjamin, I., Molecular structure and dynamics at liquid-liquid interfaces, *Annu Rev Phys Chem*, Vol. 48, (1997): p. 407.
7. Jedlovszky, P., A. Vincze, and G. Horvai, New insight into the orientational order of water molecules at the water/1,2-dichloroethane interface: A Monte Carlo simulation study, *J Chem Phys*, Vol. 117, (2002): p. 2271.
8. Partay, L. B., G. Hantal, P. Jedlovszky, A. Vincze, and G. Horvai, A new method for determining the interfacial molecules and characterizing the surface roughness in computer simulations. Application to the liquid-vapor interface of water, *J Comput Chem*, Vol. 29, (2008): p. 945.

9. Partay, L. B., G. Horvai, and P. Jedlovszky, Molecular level structure of the liquid/liquid interface. Molecular dynamics simulation and ITIM analysis of the water-CCl4 system, *Phys Chem Chem Phys*, Vol. 10, (2008): p. 4754.

10. Benjamin, I., Hydrogen bond dynamics at water/organic liquid interfaces, *J Phys Chem B*, Vol. 109, (2005): p. 13711.

11. Jedlovszky, P., A. Vincze, and G. Horvai, Properties of water/apolar interfaces as seen from Monte Carlo simulations, *J Mol Liq*, Vol. 109, (2004): p. 99.

12. Jorge, M., M. Natalia, and D. S. Cordeiro, Intrinsic structure and dynamics of the water/nitrobenzene interface, *J Phys Chem C*, Vol. 111, (2007): p. 17612.

13. Michael, D. and I. Benjamin, Molecular dynamics simulation of the water | nitrobenzene interface, *J Electroanal Chem*, Vol. 450, (1998): p. 335.

14. Schweighofer, K. J. and I. Benjamin, Electric field effects on the structure and dynamics at a liquid|liquid interface, *J Electroanal Chem*, Vol. 391, (1995): p. 1.

15. Evidence for a diffuse interfacial region at the dichloroethane/water interface, Walker, D. S., M. Brown, C. L. McFearin, and G. L. Richmond, *J Phys Chem B*, Vol. 108, 2004, p. 2111.

16. Girault, H. H. and D. J. Schiffrin, Thermodynamic surface excess of water and ionic solvation at the interface between two immiscible liquids, *J Electroanal Chem*, Vol. 150, (1983): p. 43.

17. Walker, D. S. and G. L. Richmond, Depth profiling of water molecules at the liquid-liquid interface using a combined surface vibrational spectroscopy and molecular dynamics approach, *J Am Chem Soc*, Vol. 129, (2007): p. 9446.

18. Walker, D. S. and G. L. Richmond, Interfacial depth profiling of the orientation and bonding of water molecules across liquid-liquid interfaces, *J Phys Chem C*, Vol. 112, (2008): p. 201.

19. Fernandes, P. A., M. N. D. S. Cordeiro, and J. A. N. F. Gomes, Molecular dynamics simulation of the water/2-heptanone liquid-liquid interface, *J Phys Chem B*, Vol. 103, (1999): p. 6290.

20. Wang, H. B., E. Carlson, D. Henderson, and R. L. Rowley, Molecular dynamics simulation of the liquid-liquid interface for immiscible and partially miscible mixtures, *Mol Simulat*, Vol. 29, (2003): p. 777.

21. Jorge, M., M. Natalia, and D. S. Cordeiro, Molecular dynamics study of the interface between water and 2-nitrophenyl octyl ether, *J Phys Chem B*, Vol. 112, (2008): p. 2415.

22. Benjamin, I., Mechanism and dynamics of ion transfer across a liquid-liquid interface, *Science*, Vol. 261, (1993): p. 1558.

23. Schweighofer, K. J. and I. Benjamin, Transfer of small ions across the water/1,2-dichloroethane interface, *J Phys Chem*, Vol. 99, (1995): p. 9974.

24. Schweighofer, K. and I. Benjamin, Transfer of a tetramethylammonium ion across the water-nitrobenzene interface: Potential of mean force and nonequilibrium dynamics, *J Phys Chem A*, Vol. 103, (1999): p. 10274.

25. Kharkats, Y. I. and J. Ulstrup, The electrostatic Gibbs energy of finite-size ions near a planar boundary between 2 dielectric media, *J Electroanal Chem*, Vol. 308, (1991): p. 17.

26. dos Santos, D. J. V. A. and J. A. N. F. Gomes, Molecular dynamics study of the calcium ion transfer across the water/nitrobenzene interface, *Chemphyschem*, Vol. 3, (2002): p. 946.

27. Wick, C. D. and L. X. Dang, Molecular dynamics study of ion transfer and distribution at the interface of water and 1,2-dichlorethane, *J Phys Chem C*, Vol. 112, (2008): p. 647.

28. Jorge, M., R. Gulaboski, C. M. Pereira, and M. N. D. S. Cordeiro, Molecular dynamics study of 2-nitrophenyl octyl ether and nitrobenzene, *J Phys Chem B*, Vol. 110, (2006): p. 12530.

29. Chevrot, G., R. Schurhammer, and G. Wipff, Molecular dynamics study of dicarbollide anions in nitrobenzene solution and at its aqueous interface. Synergistic effect in the Eu(III) assisted extraction, *Phys Chem Chem Phys*, Vol. 9, (2007): p. 5928.

30. Chevrot, G., R. Schurhammer, and G. Wipff, Surfactant behavior of "ellipsoidal" dicarbollide anions: A molecular dynamics study, *J Phys Chem B*, Vol. 110, (2006): p. 9488.

31. Sieffert, N. and G. Wipff, The [BMI][Tf2N] ionic liquid/water binary system: A molecular dynamics study of phase separation and of the liquid-liquid interface, *J Phys Chem B*, Vol. 110, (2006): p. 13076.

32. Chaumont, A., E. Engler, and G. Wipff, Uranyl and strontium salt solvation in room-temperature ionic liquids. A molecular dynamics investigation, *Inorg Chem*, Vol. 42, (2003): p. 5348.

33. Michael, D. and I. Benjamin, Proposed experimental prove of the liquid-liquid interface structure: Molecular dynamics of charge transfer at the water/octanol interface, *J Phys Chem*, Vol. 99, (1995): p. 16810.

34. Ishizaka, S., S. Habuchi, H. B. Kim, and N. Kitamura, Excitation energy transfer from sulforhodamine 101 to Acid Blue 1 at a liquid/liquid interface: Experimental approach to estimate interfacial roughness, *Anal Chem*, Vol. 71, (1999): p. 3382.

35. Ishizaka, S., H. B. Kim, and N. Kitamura, Time-resolved total internal reflection fluorometry study on polarity at a liquid/liquid interface, *Anal Chem*, Vol. 73, (2001): p. 2421.

36. Kornyshev, A. A. and M. Urbakh, Direct energy transfer at electrified liquid-liquid interfaces: a way to study interface morphology on mesoscopic scales, *Electrochem Commun*, Vol. 6, (2004): p. 703.

37. Wang, H. F., E. Borguet, and K. B. Eisenthal, Generalized interface polarity scale based on second harmonic spectroscopy, *J Phys Chem B*, Vol. 102, (1998): p. 4927.

38. Uber pyridinium-N-phenol-betaine und ihre verwendung zur charakterisierung der polaritat von losungmittel, Dimroth, K., F. Bohlmann, C. Reichard, and T. Siepmann, *Liebigs Ann Chem*, Vol. 661, 1963, p. 1.

39. Steel, W. H. and R. A. Walker, Measuring dipolar width across liquid-liquid interfaces with 'molecular rulers', *Nature*, Vol. 424, (2003): p. 296.

40. Zhao, X. L., S. Subrahmanyan, and K. B. Eisenthal, Determination of pKa at the air-water interface by 2nd harmonic generation, *Chem Phys Lett*, Vol. 171, (1990): p. 558.

41. TamburelloLuca, A. A., P. Hebert, R. Antoine, P. F. Brevet, and H. H. Girault, Optical surface second harmonic generation study of the two acid/base equilibria of eosin B at the air/water interface, *Langmuir*, Vol. 13, (1997): p. 4428.

42. Higgins, D. A. and R. M. Corn, Second harmonic generation studies of adsorption at a liquid-liquid electrochemical interface, *J Phys Chem-Us*, Vol. 97, (1993): p. 489.

43. Naujok, R. R., D. A. Higgins, D. G. Hanken, and R. M. Corn, Optical second harmonic generation measurements of molecular adsorption and orientation at the liquid-liquid electrochemical interface, *J Chem Soc Faraday T*, Vol. 91, (1995): p. 1411.

44. Naujok, R. R., D. A. Higgins, D. G. Hanken, and R. M. Corn, Optical second harmonic generation measurements of molecular adsorption and orientation at the liquid-liquid electrochemical interface (Vol 91, page 1411, 1995), *J Chem Soc Faraday T*, Vol. 91, (1995): p. 2353.

45. Pant, D., M. Le Guennec, B. Illien, and H. H. Girault, The pH dependent adsorption of Coumarin 343 at the water/dichloroethane interface, *Phys Chem Chem Phys*, Vol. 6, (2004): p. 3140.

46. Benjamin, I., Static and dynamic electronic spectroscopy at liquid interfaces, *Chem Rev*, Vol. 106, (2006): p. 1212.

47. Kakiuchi, T. and N. Tsujioka, Electrochemical polarizability of the interface between an aqueous electrolyte solution and a room-temperature molten salt, *J Electroanal Chem*, Vol. 599, (2007): p. 209.

48. Samec, Z., Electrical double layer at the interface between 2 immiscible electrolyte solutions, *Chem Rev*, Vol. 88, (1988): p. 617.

49. Verwey, E. J. W., K. F. Niessen, and N. V. Philips, The electrical double layer at the interface of two liquids, *Philos Mag*, Vol. 28, (1939): p. 435.

50. Gros, M., S. Gromb, and C. Gavach, Double layer and ion adsorption at the interface between two non-miscible electrolyte solutions. 2. Electrocapillary behavior of some water-nitrobenzene systems., *J Electroanal Chem*, Vol. 89, (1978): p. 29.

51. Gavach, C., P. Seta, and B. D'Epenoux, Double layer and ion adsorption at the interface between two non miscible solutions. Part1 . Interfacial tension measurements for the water-nitrobenzene tetraalkylammonium bromide systems, *J Electroanal Chem*, Vol. 83, (1977): p. 225.

52. Parsons, R. and F. G. R. Zobel, Interphase between mercury and aqueous sodium dihydrogenphosphate, *J Electroanal Chem*, Vol. 9, (1965): p. 333.

53. Pereira, C. M., W. Schmickler, A. F. Silva, and M. J. Sousa, On the capacity of liquid-liquid interfaces, *Chem Phys Lett*, Vol. 268, (1997): p. 13.

54. Pereira, C. M., F. Silva, M. J. Sousa, K. Kontturi, and L. Murtomaki, Capacitance and ionic association at the electrified oil I water interface: the effect of the oil phase composition, *J Electroanal Chem*, Vol. 509, (2001): p. 148.

55. Torrie, G. M. and J. P. Valleau, Double layer structure at the interface between two immiscible electrolyte solutions, *J Electroanal Chem*, Vol. 206, (1986): p. 69.

56. Huber, T., O. Pecina, and W. Schmickler, The influence of the ions on the capacity of liquid I liquid interfaces, *J Electroanal Chem*, Vol. 467, (1999): p. 203.

57. Pereira, C. M., W. Schmickler, F. Silva, and M. J. Sousa, Ion association at liquid I liquid interfaces, *J Electroanal Chem*, Vol. 436, (1997): p. 9.

58. Pecina, O. and J. P. Badiali, Influence of a modulated surface on the properties of liquid-liquid interfaces, *Phys Rev E*, Vol. 58, (1998): p. 6041.

59. Pecina, O. and J. P. Badiali, Specific adsorption at modulated liquid I liquid interfaces, *J Electroanal Chem*, Vol. 475, (1999): p. 46.

60. Daikhin, L. I., A. A. Kornyshev, and M. Urbakh, Effect of capillary waves on the double layer capacitance of the interface between two immiscible electrolytes, *Electrochim Acta*, Vol. 45, (1999): p. 685.

61. Daikhin, L. I., A. A. Kornyshev, and M. Urbakh, Capillary waves at soft electrified interfaces, *J Electroanal Chem*, Vol. 483, (2000): p. 68.

62. Daikhin, L. I., A. A. Kornyshev, and M. Urbakh, Ion penetration into an 'unfriendly medium' and the double layer capacitance of the interface between two immiscible electrolytes, *J Electroanal Chem*, Vol. 500, (2001): p. 461.

63. Daikhin, L. I. and M. Urbakh, Double layer capacitance and a microscopic structure of electrified liquid-liquid interfaces, *J Electroanal Chem*, Vol. 560, (2003): p. 59.

64. Monroe, C. W., M. Urbakh, and A. A. Kornyshev, Understanding the anatomy of capacitance at interfaces between two immiscible electrolytic solutions, *J Electroanal Chem*, Vol. 582, (2005): p. 28.

65. Markin, V. S., A. G. Volkov, and M. I. Volkova-Gugeshashvili, Structure of non-polarizable water/nitrobenzene interface: Potential distribution, ion adsorption, and interfacial tension, *J Phys Chem B*, Vol. 109, (2005): p. 16444.

66. Markin, V. S., M. I. Volkova-Gugeshashvili, and A. G. Volkov, Adsorption at liquid interfaces: The generalized Langmuir isotherm and interfacial structure, *J Phys Chem B*, Vol. 110, (2006): p. 11415.

67. Volkova-Gugeshashvili, M. I., A. G. Volkov, and V. S. Markin, Adsorption at liquid interfaces: The generalized Frumkin isotherm and interfacial structure, *Russ J Electrochem*, Vol. 42, (2006): p. 1073.

68. Luo, G., S. Malkova, J. Yoon, D. G. Schultz, B. Lin, M. Meron, I. Benjamin, P. Vanysek, and M. L. Schlossman, Ion distributions at the nitrobenzene-water interface electrified by a common ion, *J Electroanal Chem*, Vol. 593, (2006): p. 142.

69. Luo, G. M., S. Malkova, S. V. Pingali, D. G. Schultz, B. H. Lin, M. Meron, I. Benjamin, P. Vanysek, and M. L. Schlossman, Structure of the interface between two polar liquids: Nitrobenzene and water, *J Phys Chem B*, Vol. 110, (2006): p. 4527.

70. Luo, G. M., S. Malkova, J. Yoon, D. G. Schultz, B. H. Lin, M. Meron, I. Benjamin, P. Vanysek, and M. L. Schlossman, Ion distributions near a liquid-liquid interface, *Science*, Vol. 311, (2006): p. 216.

71. Onuki, A., Surface tension of electrolytes: Hydrophilic and hydrophobic ions near an interface, *J Chem Phys*, Vol. 128, (2008): p. 224704.

72. Guastalla, J., *Proc. 2nd Int. Congr. Surface Activity 3, 112*, (1957): p.

73. Watanabe, A., M. Matsumoto, H. Tamai, and R. Gotoh, Electrocapillary phenomena at oil-water interfaces. Part 1. Electrocapillary curves of oil-water systems containing surface active agents., *Kolloid Z Z Polym*, Vol. 220, (1967): p. 152.

74. Watanabe, A., M. Matsumoto, H. Tamai, and R. Gotoh, Electrocapillary phenomena at oil-water interfaces. Part 2. Counter ions binding at oil-water interfaces, *Kolloid Z Z Polym*, Vol. 221, (1967): p. 47.

75. Dupeyrat, M. and E. Nakache, Electrocapillarity and electroadsorption, *J Colloid Interf Sci*, Vol. 73, (1980): p. 332.

76. Kakiuchi, T., M. Kobayashi, and M. Senda, The effect of the electrical potential difference on the adsorption of the hexadecyltrimethylammonium ion at the polarized water-nitrobenzene interface, *B Chem Soc Jpn*, Vol. 60, (1987): p. 3109.

77. Kakiuchi, T., Potential-dependent adsorption and partitioning of ionic components at a liquid | liquid interface, *J Electroanal Chem*, Vol. 496, (2001): p. 137.

78. Su, B., N. Eugster, and H. H. Girault, Simulations of the adsorption of ionic species at polarisable liquid | liquid interfaces, *J Electroanal Chem*, Vol. 577, (2005): p. 187.

79. Arai, K., S. Fukuyama, F. Kusu, and K. Takamura, Role of a surfactant in the electrical potential oscillation across a liquid membrane, *Electrochim Acta*, Vol. 40, (1995): p. 2913.

80. Kihara, S. and K. Maeda, Membrane oscillations and ion transport, *Prog Surf Sci*, Vol. 47, (1994): p. 1.

81. Maeda, K., S. Nagami, Y. Yoshida, H. Ohde, and S. Kihara, Voltammetric elucidation of the process of self-sustained potential oscillation observed with a liquid membrane system composed of water containing cetyltrimethylammonium chloride | nitrobenzene containing picric acid | pure water, *J Electroanal Chem*, Vol. 496, (2001): p. 124.

82. Kakiuchi, T., Electrochemical instability of the liquid | liquid interface in the presence of ionic surfactant adsorption, *J Electroanal Chem*, Vol. 536, (2002): p. 63.

83. Kakiuchi, T., M. Chiba, N. Sezaki, and M. Nakagawa, Cyclic voltammetry of the transfer of anionic surfactant across the liquid-liquid interface manifests electrochemical instability, *Electrochem Commun*, Vol. 4, (2002): p. 701.

84. Kakiuchi, T., N. Nishi, T. Kasahara, and M. Chiba, Regular irregularity in the transfer of anionic surfactant across the liquid/liquid interface, *Chemphyschem*, Vol. 4, (2003): p. 179.

85. Kakiuchi, T., Electrochemical instability in facilitated transfer of alkaline-earth metal ions across the nitrobenzene water interface, *J Electroanal Chem*, Vol. 569, (2004): p. 287.

86. Kasahara, T., N. Nishi, M. Yamamoto, and T. Kakiuchi, Electrochemical instability in the transfer of cationic surfactant across the 1,2-dichloroethane/water interface, *Langmuir*, Vol. 20, (2004): p. 875.

87. Kitazumi, Y. and T. Kakiuchi, Emergence of the electrochemical instability in transfer of decylammonium ion across the 1,2-dichloroethane I water interface formed at the tip of a micropipette, *J Phys-Condens Mat*, Vol. 19, (2007): p.

88. Daikhin, L. I. and M. Urbakh, What is the origin of irregular current oscillations in the transfer of ionic surfactants across liquid/liquid interfaces?, *J Chem Phys*, Vol. 128, (2008): p. 014706.

89. Ishimatsu, R., F. Shigematsu, T. Hakuto, N. Nishi, and T. Kakiuchi, Structure of the electrical double layer on the aqueous solution side of the polarized interface between water and a room-temperature ionic liquid, tetrahexylammonium bis(trifluoromethylsulfonyl)imide, *Langmuir*, Vol. 23, (2007): p. 925.

90. Yasui, Y., Y. Kitazumi, R. Ishimatsu, N. Nishi, and T. Kakiuchi, Ultraslow Response of Interfacial Tension to the Change in the Phase-Boundary Potential at the Interface between Water and a Room-Temperature Ionic Liquid, Trioctylmethylammonium bis(nonafluorobutanesulfonyl)amide, *J Phys Chem B*, Vol. 113, (2009): p. 3273.

91. Girault, H. H., *Analytical and Physical Electrochemistry*, EPFL Press, Lausanne, Switzerland, 2004.

92. Markin, V. S. and A. G. Volkov, Distribution potential in small liquid-liquid systems, *J Phys Chem B*, Vol. 108, (2004): p. 13807.

93. Wandlowski, T., V. Marecek, and Z. Samec, Kinetic analysis of the picrate ion transfer across the interface between two immiscible electrolyte solutions from impedance measurements at equilibrium potential, *J Electroanal Chem*, Vol. 242, (1988): p. 291.

94. Wandlowski, T., V. Maecek, K. Holub, and Z. Samec, Ion transfer across liquid-liquid phase boundaries. Electrochemical kinetics by Faradaic impedance, *J Phys Chem-Us*, Vol. 93, (1989): p. 8204.

95. Wandlowski, T., V. Marecek, Z. Samec, and R. Fuoco, Effect of the temperature on the ion transfer across the interface between two immiscible electrolyte solutions. Ion transfer dynamics, *J Electroanal Chem*, Vol. 331, (1992): p. 765.

96. Shao, Y. and H. H. Girault, Kinetics of the transfer of acetylcholine across the water+sucrose/1-2-dichloroethane interface. A comparison between ion transport and ion transfer, *J Electroanal Chem*, Vol. 282, (1990): p. 59.

97. Shao, Y., J. A. Campbell, and H. H. Girault, Kinetics of the transfer of acetylcholine across the water nitobenzene-tetrachloromethane interface: The Gibbs energy of transfer dependence of the standard rate constant, *J Electroanal Chem*, Vol. 300, (1991): p. 415.

98. Samec, Z., V. Marecek, and D. Homolka, Charge transfer between two immiscible electrolyte solutions. Part 9. Kinetics of the transfer of choline and acetylcholine cations across the water-nitrobenzene interface, *J Electroanal Chem*, Vol. 158, (1983): p. 25.

99. Gavach, C., B. D'Epenoux, and F. Henry, Transfer of tetra-n-alkyl ammonium ions from water to nitrobenzene. Chronopotentiometric determination of the kinetic parameters, *J Electroanal Chem*, Vol. 64, (1975): p. 107.

100. Kakiuchi, T., J. Noguchi, and M. Senda, Kinetics of the transfer of monovalent anion across the nitrobenzene-water interface, *J Electroanal Chem*, Vol. 327, (1992): p. 63.

101. Kakiuchi, T., J. Noguchi, and M. Senda, Double layer effect on the transfer of some monovalent ions across the polarised oil-water interface, *J Electroanal Chem*, Vol. 336, (1992): p. 137.

102. Samec, Z., T. Kakiuchi, and M. Senda, Double layer effects on the Cs+ ion transfer kinetics at the water-nitrobenzene interface, *Electrochim Acta*, Vol. 40, (1995): p. 2971.

103. Murtomaki, L., K. Kontturi, and D. J. Schiffrin, Some remarks on the double layer correction to the kinetics of ion transfer at the interface of immiscible electrolytes, *J Electroanal Chem*, Vol. 474, (1999): p. 89.

104. Kakiuchi, T. and Y. Teranishi, Effect of the viscosoty of the aqueous phase on the rate of ion transfer across the nitrobenzene-water interface, *J Electroanal Chem*, Vol. 396, (1995): p. 401.

105. Solomon, T., Linear Gibbs energy relationships and the activated transport model for ion transfer across liquid-liquid systems, *J Electroanal Chem*, Vol. 313, (1991): p. 29.

106. Brummer, S. B. and G. J. Hills, Kinetics of ionic conductance. Part 1. Energies of activation and constant volume principle, *T Faraday Soc*, Vol. 57, (1961): p. 1816.

107. Brummer, S. B. and G. J. Hills, Kinetics of ionic conductance. Part 2. Temperature and pressure coefficients of conductance, *T Faraday Soc*, Vol. 57, (1961): p. 1823.

108. Barreira, F. and G. J. Hills, Kinetics of ionic migration. Part 3. Pressure and temperature coefficients of conductance in nitrobenzene, *T Faraday Soc*, Vol. 64, (1968): p. 1359.

109. Cai, C. X., Y. H. Tong, and M. V. Mirkin, Probing rapid ion transfer across a nanoscopic liquid-liquid interface, *J Phys Chem B*, Vol. 108, (2004): p. 17872.

110. Rodgers, P. J. and S. Amemiya, Cyclic voltammetry at micropipet electrodes for the study of ion-transfer kinetics at liquid/liquid interfaces, *Anal Chem*, Vol. 79, (2007): p. 9276.

111. D'Epenoux, B., P. Seta, G. Amblard, and C. Gavach, Transfer mechanism of tetraalkylammonium ions across a water-nitrobenzene interface and the structure of the double layer, *J Electroanal Chem*, Vol. 99, (1979): p. 77.

112. Kontturi, K., J. A. Manzanares, and L. Murtomaki, Effect of concentration polarization on the current-voltage characteristics of ion transfer across ITIES, *Electrochim Acta*, Vol. 40, (1995): p. 2979.

113. Kontturi, K., J. A. Manzanares, L. Murtomaki, and D. J. Schiffrin, Rate constant for ion transfer in inhomogeneous media at the interface of immiscible electrolytes, *J Phys Chem B*, Vol. 101, (1997): p. 10801.

114. Kakiuchi, T., Current potential characteristic of ion transfer across the interface between 2 immiscible electrolyte solutions based on the Nernst-Planck equation, *J Electroanal Chem*, Vol. 322, (1992): p. 55.

115. Kakiuchi, T., DC and AC responses of ion transfer across an oil-water interface with a Goldmann type current-potential characteritics, *J Electroanal Chem*, Vol. 344, (1993): p. 1.

116. Samec, Z., Y. I. Kharkats, and Y. Y. Gurevich, Stochastic approach to the ion transfer kinetics across the interface between two immiscible electrolyte solutions. Comparison with experimental data, *J Electroanal Chem*, Vol. 204, (1986): p. 257.

117. Gurevich, Y. Y. and Y. I. Kharkats, Theory of ion tranfer across interfaces between 2 media, *Sov Electrochem*, Vol. 22, (1986): p. 463.
118. Gurevich, Y. Y. and Y. I. Kharkats, Ion transfer through a phase boundary: A stochastic approach, *J Electroanal Chem*, Vol. 200, (1986): p. 3.
119. Indenbom, A. V. and O. Dvorak, Electrochemical kinetics of facilitated transfer of Na+ ion through the polymer gel-water interface, *Biol Membrany*, Vol. 8, (1991): p. 1314.
120. Schmickler, W., A model for ion transfer through liquid | liquid interfaces, *J Electroanal Chem*, Vol. 426, (1997): p. 5.
121. Aoki, K., Theory of ion-transfer kinetics at a viscous immiscible liquid/liquid interface by means of the Langevin equation, *Electrochim Acta*, Vol. 41, (1996): p. 2321.
122. Ferrigno, R. and H. H. Girault, Hydrodynamic approach to ion transfer reactions across an ITIES, *J Electroanal Chem*, Vol. 496, (2001): p. 131.
123. Marcus, R. A., On the theory of ion transfer rates across the interface of two immiscible liquids, *J Chem Phys*, Vol. 113, (2000): p. 1618.
124. Kornyshev, A. A., A. M. Kuznetsov, and M. Urbakh, Coupled ion-interface dynamics and ion transfer across the interface of two immiscible liquids, *J Chem Phys*, Vol. 117, (2002): p. 6766.
125. Kornyshev, A. A., A. M. Kuznetsov, and M. Urbakh, Surface polaron effect on the ion transfer across the interface of two immiscible electrolytes, *Russ J Electrochem*, Vol. 39, (2003): p. 119.
126. Verdes, C. G., M. Urbakh, and A. A. Kornyshev, Surface tension and ion transfer across the interface of two immiscible electrolytes, *Electrochem Commun*, Vol. 6, (2004): p. 693.
127. Kakiuchi, T., Y. Takasu, and M. Senda, Voltage scan fluorometry of Rose Bengal ion at the 1,2-dichloroethane-water interface, *Anal Chem*, Vol. 64, (1992): p. 3096.
128. Kakiuchi, T. and Y. Takasu, Ion selectivity of voltage scan fluorometry at the 1,2-dichloroethane-water interface, *J Electroanal Chem*, Vol. 365, (1994): p. 293.
129. Ding, Z. F., R. G. Wellington, P. F. Brevet, and H. H. Girault, Spectroelectrochemical studies of Ru(bpy)3(2+) at the water/1,2-dichloroethane interface, *J Phys Chem*, Vol. 100, (1996): p. 10658.
130. Ding, Z. F., R. G. Wellington, P. F. Brevet, and H. H. Girault, Differential cyclic voltabsorptometry and chronoabsorptometry studies of ion transfer reactions at the water | 1,2-dichloroethane interface, *J Electroanal Chem*, Vol. 420, (1997): p. 35.
131. Tomaszewski, L., Z. F. Ding, D. J. Fermin, H. M. Cacote, C. M. Pereira, F. Silva, and H. H. Girault, Spectroelectrochemical study of the copper(II) transfer assisted by 6,7-dimethyl-2,3-di(2-pyridyl)quinoxaline at the water vertical bar 1,2-dichloroethane interface, *J Electroanal Chem*, Vol. 453, (1998): p. 171.
132. Cacote, M. H. M., C. M. Pereira, and F. Silva, Monitoring bromophenol blue transfer across water/1,2-DCE interface, *Electroanal*, Vol. 14, (2002): p. 935.
133. Ding, Z. F., P. F. Brevet, and H. H. Girault, Heterogeneous electron transfer at the polarised water/1,2-dichloroethane interface studied by in situ UV-VIS spectroscopy and differential cyclic voltabsorptometry, *Chem Commun*, (1997): p. 2059.
134. Ding, Z. F., D. J. Fermin, P. F. Brevet, and H. H. Girault, Spectroelectrochemical approaches to heterogeneous electron transfer reactions at the polarised water /1,2-dichloroethane interfaces, *J Electroanal Chem*, Vol. 458, (1998): p. 139.
135. Kakiuchi, T. and Y. Takasu, Differential cyclic voltfluorometry and chronofluorometry of the transfer of fluorescent ions across the 1,2-dichloroethane-water interface, *Anal Chem*, Vol. 66, (1994): p. 1853.

136. Kakiuchi, T., K. Ono, Y. Takasu, J. Bourson, and B. Valeur, Voltage-controlled fluorometry of the transfer of nonfluorescent ions across the 1,2-dichloroethane/water interface using fluorescent ionophores, *Anal Chem*, Vol. 70, (1998): p. 4152.

137. Kakiuchi, T. and Y. Takasu, Potential step chronofluorometric response of fluorescent ion transfer across a liquid-liquid interface, *J Electroanal Chem*, Vol. 381, (1995): p. 5.

138. Fermin, D. J., Z. Ding, P. F. Brevet, and H. H. Girault, Potential-modulated reflectance spectroscopy of the methyl orange transfer across the water | 1,2-dichloroethane interface, *J Electroanal Chem*, Vol. 447, (1998): p. 125.

139. Nagatani, H., R. A. Iglesias, D. J. Fermin, P. F. Brevet, and H. H. Girault, Adsorption behavior of charged zinc porphyrins at the water/1,2-dichloroethane interface studied by potential modulated fluorescence spectroscopy, *J Phys Chem B*, Vol. 104, (2000): p. 6869.

140. Nagatani, H., D. J. Fermin, and H. H. Girault, A kinetic model for adsorption and transfer of ionic species at polarized liquid | liquid interfaces as studied by potential modulated fluorescence spectroscopy, *J Phys Chem B*, Vol. 105, (2001): p. 9463.

141. Nakatani, K., H. Nagatani, D. J. Fermin, and H. H. Girault, Transfer and adsorption of 1-pyrene sulfonate at the water|1,2-dichloroethane interface studied by potential modulated fluorescence spectroscopy, *J Electroanal Chem*, Vol. 518, (2002): p. 1.

142. Nagatani, H., T. Ozeki, and T. Osakai, Direct spectroelectrochemical observation of interfacial species at the polarized water/1,2-dichloroethane interface by ac potential modulation technique, *J Electroanal Chem*, Vol. 588, (2006): p. 99.

143. Kotov, N. A. and M. G. Kuzmin, A photoelectrochemical effect at the interface of immiscible electrolyte solutions, *J Electroanal Chem*, Vol. 285, (1990): p. 223.

144. Kotov, N. A. and M. G. Kuzmin, Nature of the processes of charge carrier generation at ITIES by the photoexcitation of porphyrins, *J Electroanal Chem*, Vol. 338, (1992): p. 99.

145. Kotov, N. A. and M. G. Kuzmin, Computer analysis of photoinduced charge transfer at the ITIES in protoporphyrin-quinone systems, *J Electroanal Chem*, Vol. 341, (1992): p. 47.

146. Kotov, N. A. and M. G. Kuzmin, Computer analysis of photoinduced charge transfer at the ITIES and potential of the optimisation technique, *J Electroanal Chem*, Vol. 327, (1992): p. 47.

147. Samec, Z., A. R. Brown, L. J. Yellowlees, H. H. Girault, and K. Base, Photochemical ion transfer across the interface between two immiscible electrolyte solutions, *J Electroanal Chem*, Vol. 259, (1989): p. 309.

148. Samec, Z., A. R. Brown, L. J. Yellowlees, and H. H. Girault, Photochemical transfer of tetra-aryl ions across the interface between two immiscible electrolyte solutions, *J Electroanal Chem*, Vol. 288, (1990): p. 245.

149. Kuzmin, M. G., I. V. Soboleva, and N. A. Kotov, Kinetics of photoinduced charge transfer at microscopic and macroscopic interfaces, *Anal Sci*, Vol. 15, (1999): p. 3.

150. Watariguchi, S., E. Ikeda, and T. Hinoue, Ultraviolet laser photo-modulation voltammetry of tetraphenylborate at a liquid/liquid interface, *Anal Sci*, Vol. 21, (2005): p. 1233.

151. Homolka, D., L. Q. Hung, A. Hofmanova, M. W. Khalil, J. Koryta, V. Marecek, Z. Samec, S. K. Sen, P. Vanysek, J. Weber, M. Brezina, M. Janda, and I. Stibor, Faradaic ion transfer across the the interface of two immiscible electrolyte solutions. Chronopotentiometry and cyclic voltammetry, *Anal Chem*, Vol. 52, (1980): p. 1606.

152. Shao, Y., M. D. Osborne, and H. H. Girault, Assisted ion transfer at micro-ITIES supported at the tip of micropipettes, *J Electroanal Chem*, Vol. 318, (1991): p. 101.

153. Vanysek, P., W. Ruth, and J. Koryta, Valinomycin mediated transfer of potassium across the water-nitrobenzene interface. A study by voltammetry at the interface between two immiscible electrolyte solutions, *J Electroanal Chem*, Vol. 148, (1983): p. 117.

154. Campbell, J. A., A. A. Stewart, and H. H. Girault, Determination of the kinetics of facilitated ion transfer reactions across the micro-interface between two immiscible electrolyte solutions, *J Chem Soc Farad T 1*, Vol. 85, (1989): p. 843.

155. Shioya, T., S. Nishizawa, and N. Teramae, Anion recognition at the liquid-liquid interface. Sulfate transfer across the 1,2-dichloroethane-water interface facilitated by hydrogen-bonding ionophores, *J Am Chem Soc*, Vol. 120, (1998): p. 11534.

156. Nishizawa, S., T. Yokobori, R. Kato, T. Shioya, and N. Teramae, Chloride transfer across the liquid-liquid interface facilitated by a mono-thiourea as a hydrogen-bonding ionophore, *B Chem Soc Jpn*, Vol. 74, (2001): p. 2343.

157. Nishizawa, S., T. Yokobori, T. Shioya, and N. Teramae, Facilitated transfer of hydrophilic anions across the nitrobenzene-water interface by a hydrogen-bonding ionophore: Applicability for multianalyte detection, *Chem Lett*, (2001): p. 1058.

158. Shao, Y. H., B. Linton, A. D. Hamilton, and S. G. Weber, Electrochemical studies on molecular recognition of anions: complex formation between xylylenyl bis-iminoimidazolinium and dicarboxylates in nitrobenzene and water, *J Electroanal Chem*, Vol. 441, (1998): p. 33.

159. Qian, Q. S., G. S. Wilson, K. Bowman-James, and H. H. Girault, MicroITIES detection of nitrate by facilitated ion transfer, *Anal Chem*, Vol. 73, (2001): p. 497.

160. Qian, Q. S., G. S. Wilson, and K. Bowman-James, MicroITIES detection of adenosine phosphates, *Electroanal*, Vol. 16, (2004): p. 1343.

161. Katano, H., Y. Murayama, and H. Tatsumi, Voltammetric study of the transfer of fluoride ion at the nitrobenzene | water interface assisted by tetraphenylantimony, *Anal Sci*, Vol. 20, (2004): p. 553.

162. Dryfe, R., S. Hill, A. Davis, J. Joos, and E. Roberts, Electrochemical quantification of high-affinity halide binding by a steroid-based receptor, *Org Biomol Chem*, Vol. 2, (2004): p. 2716.

163. Cui, R., Q. Li, D. E. Gross, X. Meng, B. Li, M. Marquez, R. Yang, J. L. Sessler, and Y. Shao, Anion Transfer at a Micro-Water/1,2-Dichloroethane Interface Facilitated by beta-Octafluoro-meso-octamethylcalix[4]pyrrole, *J Am Chem Soc*, Vol. 130, (2008): p. 14364.

164. Gale, P. A., Structural and molecular recognition studies with acyclic anion receptors, *Accounts Chem Res*, Vol. 39, (2006): p. 465.

165. Kontturi, K. and L. Murtomaki, Electrochemical determination of partition coefficients of drugs, *J Pharm Sci*, Vol. 81, (1992): p. 970.

166. Reymond, F.,Transfer mechanism and lipophilicity of ionisable drugs, in" *Liquid interfaces in chemical, biological and pharmaceutical applications*," Volkov, A. G. Eds, Marcel Dekker, New York, 2001, p. 729.

167. Reymond, F., V. Gobry, G. Bouchard, and H. H. Girault, Elecrochemical aspects of drug partitioning, in "*Pharmacokinetic optimisation in drug research: Biological, physicochemical and computational strategies*," Testa, B., H. van de Waterbeemd, G. Folkers, and R. Guy Eds, Helvetica Chimica Acta Verlag, 2001, p. 327.

168. Gulaboski, R., F. Borges, C. M. Pereira, M. N. D. S. Cordeiro, J. Garrido, and A. F. Silva, Voltammetric insights in the transfer of ionizable drugs across biomimetic membranes—Recent achievements, *Comb Chem High T Scr*, Vol. 10, (2007): p. 514.

169. Kudo, Y., H. Imamizo, K. Kanamori, S. Katsuta, Y. Takeda, and H. Matsuda, On the facilitating effect of neutral macrocyclic ligands on the ion transfer across the interface between aqueous and organic solutions Part III. Competitive facilitated ion-transfer, *J Electroanal Chem*, Vol. 509, (2001): p. 128.

170. Matsuda, H., Y. Yamada, K. Kanamori, Y. Kudo, and Y. Takeda, On the facilitating effect of neutral macrocyclic ligands on ion transfer across the interface between aqueous and organic solutions. Part 1. Theoretical equation of ion transfer polarographic current-potential curves and its experimental verification, *B Chem Soc Jpn*, Vol. 64, (1991): p. 1497.

171. Kudo, Y., Y. Takeda, and H. Matsuda, On the facilitating effect of neutral macrocyclic ligands on ion transfer across the interface between aqueous and organic solutions. Part 2. Alkali-metal ion complexes with hydrophilic crown ethers, *J Electroanal Chem*, Vol. 396, (1995): p. 333.

172. Reymond, F., P. A. Carrupt, and H. H. Girault, Facilitated ion transfer reactions across oil vertical bar water interfaces. Part I. Algebraic development and calculation of cyclic voltammetry experiments for successive complex formation, *J Electroanal Chem*, Vol. 449, (1998): p. 49.

173. Nicholson, R. S. and I. Shain, Theory of stationary electode polarography. Single scan and cyclic methods applied to reversible, irreversible and kinetic systems, *Anal Chem*, Vol. 36, (1964): p. 706.

174. Garcia, J. I., R. A. Iglesias, and S. A. Dassie, Facilitated ion transfer reactions across oil I water interfaces based on different competitive ligands, *J Electroanal Chem*, Vol. 580, (2005): p. 255.

175. Tomaszewski, L., F. Reymond, P. F. Brevet, and H. H. Girault, Facilitated ion transfer across oil vertical bar water interfaces. Part III. Algebraic development and calculation of cyclic voltammetry experiments for the formation of a neutral complex, *J Electroanal Chem*, Vol. 483, (2000): p. 135.

176. Amemiya, S., X. T. Yang, and T. L. Wazenegger, Voltammetry of the phase transfer of polypeptide protamines across polarized liquid/liquid interfaces, *J Am Chem Soc*, Vol. 125, (2003): p. 11832.

177. Yuan, Y. and S. Amemiya, Facilitated protamine transfer at polarized water/1,2-dichloroethane interfaces studied by cyclic voltammetry and chronoamperometry at micropipet electrodes, *Anal Chem*, Vol. 76, (2004): p. 6877.

178. Trojanek, A., J. Langmaier, E. Samcova, and Z. Samec, Counterion binding to protamine polyion at a polarised liquid-liquid interface, *J Electroanal Chem*, Vol. 603, (2007): p. 235.

179. Reymond, F., G. Steyaert, P. A. Carrupt, B. Testa, and H. Girault, Ionic partition diagrams: A Potential pH representation, *J Am Chem Soc*, Vol. 118, (1996): p. 11951.

180. Reymond, F., V. Chopineaux-Courtois, G. Steyaert, G. Bouchard, P. A. Carrupt, B. Testa, and H. H. Girault, Ionic partition diagrams of ionisable drugs: pH-lipophilicity profiles, transfer mechanisms and charge effects on solvation, *J Electroanal Chem*, Vol. 462, (1999): p. 235.

181. Gobry, V., S. Ulmeanu, F. Reymond, G. Bouchard, P. A. Carrupt, B. Testa, and H. H. Girault, Generalization of ionic partition diagrams to lipophilic compounds and to biphasic systems with variable phase volume ratios, *J Am Chem Soc*, Vol. 123, (2001): p. 10684.

182. Malkia, A., P. Liljeroth, A. Konturi, and K. Kontturi, Electrochemistry at lipid monolayer-modified liquid-liquid interfaces as an improvement to drug partitioning studies, *J Phys Chem B*, Vol. 105, (2001): p. 10884.

183. Deryabina, M. A., S. H. Hansen, and H. Jensen, Molecular interactions in lipophilic environments studied by electrochemistry at interfaces between immiscible electrolyte solutions, *Anal Chem*, Vol. 80, (2008): p. 203.

184. Ulmeanu, S. M., H. Jensen, G. Bouchard, P. A. Carrupt, and H. H. Girault, Water-oil partition profiling of ionized drug molecules using cyclic voltammetry and a 96-well microfilter plate system, *Pharmaceut Res*, Vol. 20, (2003): p. 1317.

185. Lam, H. T., C. M. Pereira, C. Roussel, P. A. Carrupt, and H. H. Girault, Immobilized pH gradient gel cell to study the pH dependence of drug lipophilicity, *Anal Chem*, Vol. 78, (2006): p. 1503.

186. Samec, Z., D. Homolka, and V. Marecek, Charge transfer between two immiscible electrolyte solutions. Part 8. Transfer of alkali and alkaline earth metal cations across the water-nitrobenzene interface facilitated by synthetic neutral ion carriers, *J Electroanal Chem*, Vol. 135, (1982): p. 265.

187. Kakutani, T., Y. Nishiwaki, T. Osakai, and M. Senda, On the mechanism of transfer of sodium ion across the nitrobenzene-water interface facilitated by di-benzo-18-crown-6, *B Chem Soc Jpn*, Vol. 59, (1986): p. 781.

188. Beattie, P. D., A. Delay, and H. H. Girault, Investigation of the kinetics of assisted potassium ion transfer by di-benzo-18-crown-6 at the micro-ITIES by means of steady state voltammetry, *J Electroanal Chem*, Vol. 380, (1995): p. 167.

189. Beattie, P. D., A. Delay, and H. H. Girault, Investigation of the kinetics of ion and assisted ion transfer by the technique of AC-impedanceof the micro-ITIES, *Electrochim Acta*, Vol. 40, (1995): p. 2961.

190. Shao, Y. H. and M. V. Mirkin, Fast kinetic measurements with nanometer-sized pipets. Transfer of potassium ion from water into dichloroethane facilitated by dibenzo-18-crown-6, *J Am Chem Soc*, Vol. 119, (1997): p. 8103.

191. Yuan, Y. and Y. H. Shao, Systematic investigation of alkali metal ion transfer across the micro- and nano-water/1,2-dichloroethane interfaces facilitated by dibenzo-18-crown-6, *J Phys Chem B*, Vol. 106, (2002): p. 7809.

192. Zhan, D. P., Y. Yuan, Y. J. Xiao, B. L. Wu, and Y. H. Shao, Alkali metal ions transfer across a water/1,2-dichloroethane interface facilitated by a novel monoaza-B15C5 derivative, *Electrochim Acta*, Vol. 47, (2002): p. 4477.

193. Li, F., Y. Chen, P. Sun, M. Q. Zhang, Z. Gao, D. P. Zhan, and Y. H. Shao, Investigation of facilitated ion-transfer reactions at high driving force by scanning electrochemical Microscopy, *J Phys Chem B*, Vol. 108, (2004): p. 3295.

194. Sun, P., Z. Q. Zhang, Z. Gao, and Y. H. Shao, Probing fast facilitated ion transfer across an externally polarized liquid-liquid interface by scanning electrochemical microscopy, *Angew Chem Int Edit*, Vol. 41, (2002): p. 3445.

195. Fermin, D. J. and R. Lahtinen,Dynamic aspects of heterogeneous electron-transfer reactions at liquid-liquid interfaces, in" *Liquid interfaces in chemical, biological, and pharmaceutical applications,*" Volkov, A. G. Eds, Marcel Dekker, New York, 2001, p. 179.

196. Hotta, H., S. Ichikawa, T. Sugihara, and T. Osakai, Clarification of the mechanism of interfacial electron-transfer reaction between ferrocene and hexacyanoferrate(III) by digital simulation of cyclic voltammograms, *J Phys Chem B*, Vol. 107, (2003): p. 9717.

197. Samec, Z., V. Marecek, J. Weber, and D. Homolka, Charge transfer between two immiscible electrolyte solutions.Part 7. Convolution potential sweep votlammetry of Cs+ ion transfer and electron transfer between ferrocene and hexacyanoferrate(II) ion across the water-nitrobenzene interface, *J Electroanal Chem*, Vol. 126, (1981): p. 105.

198. Cunnane, V. J., G. Geblewicz, and D. J. Schiffrin, Electron and ion transfer potentials of ferrocene and derivatives at liquid-liquid interfaces, *Electrochim Acta*, Vol. 40, (1995): p. 3005.

199. Quinn, B. and K. Kontturi, Aspects of electron transfer at ITIES, *J Electroanal Chem*, Vol. 483, (2000): p. 124.

200. Tatsumi, H. and H. Katano, Voltammetric study of interfacial electron transfer between bis(cyclopentadienyl)iron in organic solvents and hexacyanoferrate in water, *Anal Sci*, Vol. 20, (2004): p. 1613.

201. Tatsumi, H. and H. Katano, Voltammetric study of the interfacial electron transfer between bis(cyclopentadienyl)iron in 1,2-dichloroethane and in nitrobenzene and hexacyanoferrate in water, *J Electroanal Chem*, Vol. 592, (2006): p. 121.

202. Tatsumi, H. and H. Katano, Cyclic voltammetry of the electron transfer reaction between bis(cyclopentadienyl)iron in 1,2-dichloroethane and hexacyanoferrate in water, *Anal Sci*, Vol. 23, (2007): p. 589.

203. Kihara, S., M. Suzuki, K. Maeda, K. Ogura, M. Matsui, and Z. Yoshida, The electron transfer at a liquid-liquid interface studied by current scann polarograph at the electrolyte dropping electrode, *J Electroanal Chem*, Vol. 271, (1989): p. 107.

204. Quinn, B., R. Lahtinen, L. Murtomaki, and K. Kontturi, Electron transfer at micro liquid-liquid interfaces, *Electrochim Acta*, Vol. 44, (1998): p. 47.

205. Sugihara, T., T. Kinoshita, S. Aoyagi, Y. Tsujino, and T. Osakai, A mechanistic study of the oxidation of natural antioxidants at the oil/water interface using scanning electrochemical microscopy, *J Electroanal Chem*, Vol. 612, (2006): p. 241.

206. Osakai, T., S. Ichikawa, H. Hotta, and H. Nagatani, A true electron-transfer reaction between 5,10,15,20-tetraphenylporphyrinato cadmium(II) and the hexacyanoferrate couple at the nitrobenzene/water interface, *Anal Sci*, Vol. 20, (2004): p. 1567.

207. Stewart, A. A., J. A. Campbell, H. H. Girault, and M. Eddowes, Cyclic voltammetry for electron transfer reactions at liquid-liquid interfaces, *Ber Bunsen Phys Chem*, Vol. 94, (1990): p. 83.

208. Osakai, T., H. Hotta, T. Sugihara, and K. Nakatani, Diffusion-controlled rate constant of electron transfer at the oil | water interface, *J Electroanal Chem*, Vol. 571, (2004): p. 201.

209. Kharkats, Y. I., *Elektrokhimiya*, Vol. 12, (1976): p. 1370.

210. Kharkats, Y. I., *Elektrokhimiya*, Vol. 15, (1979): p. 409.

211. Girault, H. H., Solvent reorganisation energy for heterogeneous electron transfer reactions at liquid-liquid interfaces, *J Electroanal Chem*, Vol. 388, (1995): p. 93.

212. Marcus, R. A., Theory of electon transfer rates across liquid-liquid interfaces, *J Phys Chem-Us*, Vol. 94, (1990): p. 4152.

213. Marcus, R. A., Theory of Electron-Transfer Rates across Liquid-Liquid Interfaces. Correction 94(1990) 4153, *J. Phys. Chem.*, Vol. 94, (1990): p. 7742.

214. Marcus, R. A., Theory of electon transfer rates across liquid-liquid interfaces. 2. Relationships and application, *J Phys Chem*, Vol. 95, (1991): p. 2010.

215. Marcus, R. A., Theory of electon transfer rates across liquid-liquid interfaces(Vol 94, p 4154, 1990), *J Phys Chem*, Vol. 99, (1995): p. 5742.

216. Benjamin, I. and Y. I. Kharkats, Reorganization free energy for electron transfer reactions at liquid/liquid interfaces, *Electrochim Acta*, Vol. 44, (1998): p. 133.

217. Girault, H. H. J. and D. J. Schiffrin, Electron transfer reactions at the interface between two immiscible electrolyte solutions, *J Electroanal Chem*, Vol. 244, (1988): p. 15.

218. Thomson, F. L., L. J. Yellowlees, and H. H. Girault, Photocurrent measurements at the interface betwee two immiscible electrolyte solutions, *J Chem Soc Chem Comm*, (1988): p. 1547.

219. Brown, A. R., L. J. Yellowlees, and H. H. Girault, Photo-initiated electron transfer reactions across the interface between two immiscible electrolyte solutions, *J Chem Soc Faraday T*, Vol. 89, (1993): p. 207.
220. Fermin, D. J., Z. F. Ding, H. D. Duong, P. F. Brevet, and H. H. Girault, Photoinduced electron transfer at liquid/liquid interfaces. I. Photocurrent measurements associated with heterogeneous quenching of zinc porphyrins, *J Phys Chem B*, Vol. 102, (1998): p. 10334.
221. Fermin, D. J., H. D. Duong, Z. F. Ding, P. F. Brevet, and H. H. Girault, Photoinduced electron transfer at liquid/liquid interfaces - Part II. A study of the electron transfer and recombination dynamics by intensity modulated photocurrent spectroscopy (IMPS), *Phys Chem Chem Phys*, Vol. 1, (1999): p. 1461.
222. Fermin, D. J., H. D. Duong, Z. F. Ding, P. F. Brevet, and H. H. Girault, Photoinduced electron transfer at liquid/liquid interfaces. Part III. Photoelectrochemical responses involving porphyrin ion pairs, *J Am Chem Soc*, Vol. 121, (1999): p. 10203.
223. Jensen, H., J. J. Kakkassery, H. Nagatani, D. J. Fermin, and H. H. Girault, Photoinduced electron transfer at liquid I liquid interfaces. Part IV. Orientation and reactivity of zinc tetra(4-carboxyphenyl) porphyrin self-assembled at the water I 1,2-dichloroethane junction, *J Am Chem Soc*, Vol. 122, (2000): p. 10943.
224. Jensen, H., D. J. Fermin, and H. H. Girault, Photoinduced electron transfer at liquid/liquid interfaces. Part V. Organisation of water-soluble chlorophyll at the water/1,2-dichloroethane interface, *Phys Chem Chem Phys*, Vol. 3, (2001): p. 2503.
225. Eugster, N., D. J. Fermin, and H. H. Girault, Photoinduced electron transfer at liquid/liquid interfaces. Part VI. On the thermodynamic driving force dependence of the phenomenological electron-transfer rate constant, *J Phys Chem B*, Vol. 106, (2002): p. 3428.
226. Eugster, N., H. Jensen, D. J. Fermin, and H. H. Girault, Photoinduced electron transfer at liquid I liquid interfaces. Part VII. Correlation between self-organisation and structure of water-soluble photoactive species, *J Electroanal Chem*, Vol. 560, (2003): p. 143.
227. Fermin, D. J., Z. F. Ding, H. D. Duong, P. F. Brevet, and H. H. Girault, Photocurrent responses associated with heterogeneous electron transfer at liquid/liquid interfaces, *Chem Commun*, (1998): p. 1125.
228. Eugster, N., D. J. Fermin, and H. H. Girault, Photoinduced electron transfer at liquid I liquid interfaces: Dynamics of the heterogeneous photoreduction of quinones by self-assembled porphyrin ion pairs, *J Am Chem Soc*, Vol. 125, (2003): p. 4862.
229. Samec, Z., N. Eugster, D. J. Fermin, and H. H. Girault, A generalised model for dynamic photocurrent responses at dye-sensitised liquid/liquid interfaces, *J Electroanal Chem*, Vol. 577, (2005): p. 323.
230. Nagatani, H., S. Dejima, H. Hotta, T. Ozeki, and T. Osakai, Photoinduced electron transfer of 5,10,15,20-tetraphenylporphyrinato zinc(II) at the polarized water/1,2-dichloroethane interface, *Anal Sci*, Vol. 20, (2004): p. 1575.
231. Suzuki, M., S. Umetani, M. Matsui, and S. Kihara, Oxidation of ascorbate and ascorbic acid at the aqueous I organic solution interface, *J Electroanal Chem*, Vol. 420, (1997): p. 119.
232. Ohde, H., K. Maeda, Y. Yoshida, and S. Kihara, Redox reactions between NADH and quinone derivatives at a liquid/liquid interface, *Electrochim Acta*, Vol. 44, (1998): p. 23.
233. Ohde, H., K. Maeda, Y. Yoshida, and S. Kihara, Redox reactions between molecular oxygen and tetrachlorohydroquinone at the water I 1,2-dichloroethane interface, *J Electroanal Chem*, Vol. 483, (2000): p. 108.

234. Su, B., R. P. Nia, F. Li, M. Hojeij, M. Prudent, C. Corminboeuf, Z. Samec, and H. H. Girault, H2O2 generation by decamethylferrocene at a liquid | liquid interface, *Angew Chem Int Edit*, Vol. 47, (2008): p. 4675.

235. Trojainek, A., V. Marecek, H. Janchenova, and Z. Samec, Molecular electrocatalysis of the oxygen reduction at a polarised interface between two immiscible electrolyte solutions by Co(II) tetraphenylporphyrin, *Electrochem Commun*, Vol. 9, (2007): p. 2185.

236. Taylor, G. and H. H. J. Girault, Ion transfer reactions across a liquid-liquid interface supported on a micropipette tip, *J Electroanal Chem*, Vol. 208, (1986): p. 179.

237. Stewart, A. A., G. Taylor, H. H. Girault, and J. McAleer, Voltammetry at micro-ITIES supported at the tip of a micropipette. 1. Linear sweep voltammetry, *J Electroanal Chem*, Vol. 296, (1990): p. 491.

238. Stewart, A. A., Y. Shao, C. M. Pereira, and H. H. Girault, Micropipette as a tool for the determination of the ionic species limiting the potential window at liquid-liquid interfaces, *J Electroanal Chem*, Vol. 305, (1991): p. 135.

239. Shao, Y. H. and M. V. Mirkin, Voltammetry at micropipet electrodes, *Anal Chem*, Vol. 70, (1998): p. 3155.

240. Shao, Y. H., B. Liu, and M. V. Mirkin, Studying ionic reactions by a new generation/collection technique, *J Am Chem Soc*, Vol. 120, (1998): p. 12700.

241. Liu, B., Y. H. Shao, and M. V. Mirkin, Dual-pipet techniques for probing ionic reactions, *Anal Chem*, Vol. 72, (2000): p. 510.

242. Jing, P., M. Zhang, H. Hu, X. Xu, Z. Liang, B. Li, L. Shen, S. Xie, C. M. Pereira, and Y. Shao, Ion-transfer reactions at the nanoscopic water/n-octanol interface, *Angew Chem Int Edit*, Vol. 45, (2006): p. 6861.

243. Laforge, F. O., J. Carpino, S. A. Rotenberg, and M. V. Mirkin, Electrochemical attosyringe, *P Natl Acad Sci Usa*, Vol. 104, (2007): p. 11895.

244. Dale, S. E. C. and P. R. Unwin, Polarised liquid/liquid micro-interfaces move during charge transfer, *Electrochemistry Communications*, Vol. 10, (2008): p. 723.

245. Campbell, J. A. and H. H. Girault, Steady state current for ion transfer reactions at a micro liquid-liquid interface, *J Electroanal Chem*, Vol. 266, (1989): p. 465.

246. Osborne, M. C., Y. Shao, C. M. Pereira, and H. H. Girault, Micro-hole interface for the amperometric determination of ionic species in aqueous solutions, *J Electroanal Chem*, Vol. 364, (1994): p. 155.

247. Lee, H. J., C. Beriet, and H. H. Girault, Amperometric detection of alkali metal ions on micro-fabricated composite polymer membranes, *J Electroanal Chem*, Vol. 453, (1998): p. 211.

248. Lee, H. J., P. D. Beattie, B. J. Seddon, M. D. Osborne, and H. H. Girault, Amperometric ion sensors based on laser-patterned composite polymer membranes, *J Electroanal Chem*, Vol. 440, (1997): p. 73.

249. Lee, H. J. and H. H. Girault, Amperometric ion detector for ion chromatography, *Anal Chem*, Vol. 70, (1998): p. 4280.

250. Wilke, S., M. D. Osborne, and H. H. Girault, Electrochemical characterisation of liquid | liquid microinterface arrays, *J Electroanal Chem*, Vol. 436, (1997): p. 53.

251. Murtomaki, L. and K. Kontturi, Electrochemical characteristics of the microhole ITIES, *J Electroanal Chem*, Vol. 449, (1998): p. 225.

252. Osborne, M. D. and H. H. Girault, The micro water/1,2-dichloroethane interface as a transducer for creatinine assay, *Mikrochim Acta*, Vol. 117, (1995): p. 175.

253. Osborne, M. D. and H. H. Girault, The liquid-liquid micro-interface for the amperometric detection of urea, *Electroanal*, Vol. 7, (1995): p. 714.

254. Josserand, J., J. Morandini, H. J. Lee, R. Ferrigno, and H. H. Girault, Finite element simulation of ion transfer reactions at a single micro-liquid I liquid interface supported on a thin polymer film, *J Electroanal Chem*, Vol. 468, (1999): p. 42.

255. Ohde, H., A. Uehara, Y. Yoshida, K. Maeda, and S. Kihara, Some factors in the voltammetric measurement of ion transfer at the micro aqueous I organic solution interface, *J Electroanal Chem*, Vol. 496, (2001): p. 110.

256. Peulon, S., V. Guillou, and M. L'Her, Liquid I liquid microinterface. Localization of the phase boundary by voltammetry and chronoamperometry; influence of the microchannel dimensions on diffusion, *J Electroanal Chem*, Vol. 514, (2001): p. 94.

257. Quinn, B., R. Lahtinen, and K. Kontturi, Ion transfer at a micro water nitrophenyl octyl ether interface, *J Electroanal Chem*, Vol. 436, (1997): p. 285.

258. Silva, F., M. J. Sousa, and C. M. Pereira, Electrochemical study of aqueous-organic gel micro-interfaces, *Electrochim Acta*, Vol. 42, (1997): p. 3095.

259. Kakiuchi, T., Ionic-Liquid I Water two-phase systems, *Anal Chem*, Vol. 79, (2007): p. 6442.

260. Wei, C., A. J. Bard, and M. V. Mirkin, Scanning electrochemical microscopy. 31. Application of SECM to the study of charge transfer processes at the liquid-liquid interfaces, *J Phys Chem*, Vol. 99, (1995): p. 16033.

261. Solomon, T. and A. J. Bard, Reverse (uphill) electron transfer at the liquid-liquid interface, *J Phys Chem*, Vol. 99, (1995): p. 17487.

262. Solomon, T. and A. J. Bard, Scnning electrochemical microscopy. 30. Application of glass micropipet tips and electron transfer at the interface between two immiscible electrolyte solutions for SECM imaging, *Anal Chem*, Vol. 67, (1995): p. 2787.

263. Tsionsky, M., A. J. Bard, and M. V. Mirkin, Scanning electrochemical microscopy .34. Potential dependence of the electron-transfer rate and film formation at the liquid/liquid interface, *J Phys Chem*, Vol. 100, (1996): p. 17881.

264. Barker, A. L., P. R. Unwin, S. Amemiya, J. F. Zhou, and A. J. Bard, Scanning electrochemistry microscopy (SECM) in the study of electron transfer kinetics at liquid/liquid interfaces: Beyond the constant composition approximation, *J Phys Chem B*, Vol. 103, (1999): p. 7260.

265. Ding, Z. F., B. M. Quinn, and A. J. Bard, Kinetics of heterogeneous electron transfer at liquid/liquid interfaces as studied by SECM, *J Phys Chem B*, Vol. 105, (2001): p. 6367.

266. Barker, A. L., P. R. Unwin, and J. Zhang, Measurement of the forward and back rate constants for electron transfer at the interface between two immiscible electrolyte solutions using scanning electrochemical microscopy (SECM): Theory and experiment, *Electrochem Commun*, Vol. 3, (2001): p. 372.

267. Li, F., A. L. Whitworth, and P. R. Unwin, Measurement of rapid electron transfer across a liquid/liquid interface from 7,7,8,8-tetracyanoquinodimethane radical anion in 1,2-dichloroethane to aqueous tris(2,2-bipyridyl)-ruthenium (III), *J Electroanal Chem*, Vol. 602, (2007): p. 70.

268. Zhang, Z. Q., Y. Yuan, P. Sun, B. Su, J. D. Guo, Y. H. Shao, and H. H. Girault, Study of electron-transfer reactions across an externally polarized water/1,2-dichloroethane interface by scanning electrochemical microscopy, *J Phys Chem B*, Vol. 106, (2002): p. 6713.

269. Shao, Y. H. and M. V. Mirkin, Scanning electrochemical microscopy (SECM) of facilitated ion transfer at the liquid/liquid interface, *J Electroanal Chem*, Vol. 439, (1997): p. 137.

270. Wittstock, G., M. Burchardt, S. E. Pust, Y. Shen, and C. Zhao, Scanning electrochemical microscopy for direct imaging of reaction rates, *Angew Chem Int Edit*, Vol. 46, (2007): p. 1584.

271. Sun, P., F. O. Laforge, and M. V. Mirkin, Scanning electrochemical microscopy in the 21st century, *Phys Chem Chem Phys*, Vol. 9, (2007): p. 802.

272. Lu, X., Q. Wang, and X. Liu, Review: Recent applications of scanning electrochemical microscopy to the study of charge transfer kinetics, *Anal Chim Acta*, Vol. 601, (2007): p. 10.

273. Zu, Y. B., F. R. F. Fan, and A. J. Bard, Inverted region electron transfer demonstrated by electrogenerated chemiluminescence at the liquid/liquid interface, *J Phys Chem B*, Vol. 103, (1999): p. 6272.

274. Kanoufi, F., C. Cannes, Y. B. Zu, and A. J. Bard, Scanning electrochemical microscopy. 43. Investigation of oxalate oxidation and electrogenerated chemiluminescence across the liquid-liquid interface, *J Phys Chem B*, Vol. 105, (2001): p. 8951.

275. Zhan, D., X. Li, W. Zhan, F. F. Fan, and A. J. Bard, Scanning electrochemical microscopy. 58. Application of a micropipet-supported ITIES tip to detect Ag+ and study its effect on fibroblast cells, *Anal Chem*, Vol. 79, (2007): p. 5225.

276. Shi, C. N. and F. C. Anson, Simple electrochemical procedure for measuring the rates of electron transfer across liquid/liquid interfaces formed by coating graphite electrodes with thin layers of nitrobenzene, *J Phys Chem B*, Vol. 102, (1998): p. 9850.

277. Shi, C. N. and F. C. Anson, A simple method for examining the electrochemistry of metalloporphyrins and other hydrophobic reactants in thin layers of organic solvents interposed between graphite electrodes and aqueous solutions, *Anal Chem*, Vol. 70, (1998): p. 3114.

278. Shi, C. N. and F. C. Anson, Electron transfer between reactants located on opposite sides of liquid/liquid interfaces, *J Phys Chem B*, Vol. 103, (1999): p. 6283.

279. Shi, C. N. and F. C. Anson, Rates of electron-transfer across Liquid/Liquid interfaces. Effects of changes in driving force and reaction reversibility, *J Phys Chem B*, Vol. 105, (2001): p. 8963.

280. Shi, C. N. and F. C. Anson, Selecting experimental conditions for measurement of rates of electron-transfer at liquid-liquid interfaces by thin-layer electrochemistry, *J Phys Chem B*, Vol. 105, (2001): p. 1047.

281. Chung, T. D. and F. C. Anson, Electrochemical monitoring of proton transfer across liquid/liquid interfaces on the surface of graphite electrodes, *Anal Chem*, Vol. 73, (2001): p. 337.

282. Decher, G., J. D. Hong, and J. Schmitt, Buildup of ultrathin multilayer films by a self-assembly process.3. Consecutively alternating adsorption of anionic and cationic polyelectrolytes on charged surfaces, *Thin Solid Films*, Vol. 210, (1992): p. 831.

283. Cheng, Y. F. and R. M. Corn, Ultrathin polypeptide multilayer films for the fabrication of model liquid/liquid electrochemical interfaces, *J Phys Chem B*, Vol. 103, (1999): p. 8726.

284. Slevin, C., A. Malkia, P. Liljeroth, M. Toiminen, and K. Kontturi, Electrochemical characterization of polyelectrolyte multilayers deposited at liquid-liquid interfaces, *Langmuir*, Vol. 19, (2003): p. 1287.

285. Hoffmannova, H., D. Fermin, and P. Krtil, Growth and electrochemical activity of the poly-L-lysine|poly-L-glutamic acid thin layer films: an EQCM and electrochemical study, *J Electroanal Chem*, Vol. 562, (2004): p. 261.

286. Tan, S., M. Hojeij, B. Su, G. Meriguet, N. Eugster, and H. H. Girault, 3D-ITIES supported on porous reticulated vitreous carbon, *J Electroanal Chem*, Vol. 604, (2007): p. 65.

287. Santos, H. A., M. Chirea, V. Garcia-Morales, F. Silva, J. A. Manzanares, and K. Kontturi, Electrochemical study of interfacial composite nanostructures: Polyelectrolyte/gold nanoparticle multilayers assembled on phospholipid/dextran sulfate monolayers at a liquid-liquid interface, *J Phys Chem B*, Vol. 109, (2005): p. 20105.

288. Zhao, J. J., C. R. Bradbury, S. Huclova, I. Potapova, M. Carrara, and D. J. Fermin, Nanoparticle-mediated electron transfer across ultrathin self-assembled films, *J Phys Chem B*, Vol. 109, (2005): p. 22985.

289. Hojeij, M., N. Eugster, B. Su, and H. H. Girault, CdSe sensitized thin aqueous films: Probing the potential distribution inside multilayer assemblies, *Langmuir*, Vol. 22, (2006): p. 10652.

290. Hojeij, M., B. Su, S. Tan, G. Meriguet, and H. H. Girault, Nanoporous photocathode and photoanode made by multilayer assembly of quantum dots, *ACS Nano*, Vol. 2, (2008): p. 984.

291. Ulmeanu, S., H. Lee, D. Fermin, H. Girault, and Y. Shao, Voltammetry at a liquid-liquid interface supported on a metallic electrode, *Electrochem Commun*, Vol. 3, (2001): p. 219.

292. Shirai, O., S. Kihara, M. Suzuki, K. Ogura, and M. Matsui, Voltammetry for the ion transfer through a membrane, *Anal Sci*, Vol. 7 suppl, (1991): p. 607.

293. Shirai, O., S. Kihara, Y. Yoshida, and M. Matsui, Ion transfer through a liquid membrane or a bilayer lipid membrane in the presence of sufficient electrolytes, *J Electroanal Chem*, Vol. 389, (1995): p. 61.

294. Beriet, C. and H. H. Girault, Electrochemical studies of ion transfer at micro-machined supported liquid membranes, *J Electroanal Chem*, Vol. 444, (1998): p. 219.

295. Ulmeanu, S. M., H. Jensen, Z. Samec, G. Bouchard, P. A. Carrupt, and H. H. Girault, Cyclic voltammetry of highly hydrophilic ions at a supported liquid membrane, *J Electroanal Chem*, Vol. 530, (2002): p. 10.

296. Ball, J. C., F. Marken, F. L. Qiu, J. D. Wadhawan, A. N. Blythe, U. Schroder, R. G. Compton, S. D. Bull, and S. G. Davies, Voltammetry of electroactive oil droplets. Part II: Comparison of experimental and simulation data for coupled ion and electron insertion processes and evidence for microscale convection, *Electroanal*, Vol. 12, (2000): p. 1017.

297. Marken, F., R. G. Compton, C. H. Goeting, J. S. Foord, S. D. Bull, and S. G. Davies, Anion detection by electro-insertion into N,N,N',N'-tetrahexylphenylenediamine (THPD) microdroplets studied by voltammetry, EQCM, and SEM techniques, *Electroanal*, Vol. 10, (1998): p. 821.

298. Marken, F., R. D. Webster, S. D. Bull, and S. G. Davies, Redox processes in microdroplets studied by voltammetry, microscopy, and ESR spectroscopy: oxidation of N,N,N',N'-tetrahexylphenylene diamine deposited on solid electrode surfaces and immersed in aqueous electrolyte solution, *J Electroanal Chem*, Vol. 437, (1997): p. 209.

299. Qiu, F. L., J. C. Ball, F. Marken, R. G. Compton, and A. C. Fisher, Voltammetry of electroactive oil droplets. Part I: Numerical modelling for three mechanistic models using the dual reciprocity finite element method, *Electroanal*, Vol. 12, (2000): p. 1012.

300. Wadhawan, J. D., R. G. Compton, F. Marken, S. D. Bull, and S. G. Davies, Photoelectrochemically driven processes at the N,N,N',N'-tetrahexylphenylenediamine microdroplet/electrode/aqueous electrolyte triple interface, *J Solid State Electrochem*, Vol. 5, (2001): p. 301.

301. Wadhawan, J. D., U. Schroder, A. Neudeck, S. J. Wilkins, R. G. Compton, F. Marken, C. S. Consorti, R. F. de Souza, and J. Dupont, Ionic liquid modified electrodes. Unusual partitioning and diffusion effects of Fe(CN)(6)(4-/3-) in droplet and thin layer deposits of 1-methyl-3-(2, 6-(S)-dimethylocten-2-yl)-imidazolium tetrafluoroborate, *Journal of Electroanalytical Chemistry*, Vol. 493, (2000): p. 75.

302. Banks, C. E., T. J. Davies, R. G. Evans, A. Hignett, A. J. Wain, N. S. Lawrence, J. D. Wadhawan, F. Marken, and R. G. Compton, Electrochemistry of immobilised redox droplets: Concepts and applications, *Phys Chem Chem Phys*, Vol. 5, (2003): p. 4053.

303. Scholz, F., S. Komorsky-Lovric, and M. Lovric, A new access to Gibbs energies of transfer of ions across liquid | liquid interfaces and a new method to study electrochemical processes at well-defined three-phase junctions, *Electrochem Commun*, Vol. 2, (2000): p. 112.

304. Komorsky-Lovric, S., M. Lovric, and F. Scholz, Cyclic voltammetry of decamethylferrocene at the organic liquid | aqueous solution | graphite three-phase junction, *J Electroanal Chem*, Vol. 508, (2001): p. 129.

305. Gulaboski, R., V. Mirceski, and F. Scholz, An electrochemical method for determination of the standard Gibbs energy of anion transfer between water and n-octanol, *Electrochem Commun*, Vol. 4, (2002): p. 277.

306. Gulaboski, R., K. Riedl, and F. Scholz, Standard Gibbs energies of transfer of halogenate and pseudohalogenate ions, halogen substituted acetates, and cycloalkyl carboxylate anions at the water | nitrobenzene interface, *Phys Chem Chem Phys*, Vol. 5, (2003): p. 1284.

307. Scholz, F., R. Gulaboski, and K. Caban, The determination of standard Gibbs energies of transfer of cations across the nitrobenzene | water interface using a three-phase electrode, *Electrochem Commun*, Vol. 5, (2003): p. 929.

308. Scholz, F., R. Gulaboski, V. Mirceski, and P. Langer, Quantification of the chiral recognition in electrochemically driven ion transfer across the interface water/chiral liquid, *Electrochem Commun*, Vol. 4, (2002): p. 659.

309. Gulaboski, R., V. Mirceski, and F. Scholz, Determination of the standard Gibbs energies of transfer of cations and anions of amino acids and small peptides across the water nitrobenzene interface, *Amino Acids*, Vol. 24, (2003): p. 149.

310. Bouchard, G., A. Galland, P. A. Carrupt, R. Gulaboski, V. Mirceski, F. Scholz, and H. H. Girault, Standard partition coefficients of anionic drugs in the n-octanol/water system determined by voltammetry at three-phase electrodes, *Phys Chem Chem Phys*, Vol. 5, (2003): p. 3748.

311. Gulaboski, R., A. Galland, G. Bouchard, K. Caban, A. Kretschmer, P. A. Carrupt, Z. Stojek, H. H. Girault, and F. Scholz, A comparison of the solvation properties of 2-nitrophenyloctyl ether, nitrobenzene, and n-octanol as assessed by ion transfer experiments, *J Phys Chem B*, Vol. 108, (2004): p. 4565.

312. Scholz, F. and R. Gulaboski, Determining the Gibbs energy of ion transfer across water-organic liquid interfaces with three-phase electrodes, *Chemphyschem*, Vol. 6, (2005): p. 16.

313. Charreteur, K., F. Quentel, C. Elleouet, and M. L'Her, Transfer of highly hydrophilic ions from water to nitrobenzene, studied by three-phase and thin-film modified electrodes, *Anal Chem*, Vol. 80, (2008): p. 5065.

314. Tasakorn, P., J. Y. Chen, and K. Aoki, Voltammetry of a single oil droplet on a large electrode, *J Electroanal Chem*, Vol. 533, (2002): p. 119.

315. Aoki, K., P. Tasakorn, and J. Y. Chen, Electrode reactions at sub-micron oil | water | electrode interfaces, *J Electroanal Chem*, Vol. 542, (2003): p. 51.

316. Aoki, K., M. Satoh, J. Chen, and T. Nishiumi, Convection caused by three-phase boundary reactions, *J Electroanal Chem*, Vol. 595, (2006): p. 103.

317. Monroe, C. W., L. I. Daikhin, M. Urbakh, and A. A. Kornyshev, Electrowetting with electrolytes, *Phys Rev Lett*, Vol. 97, (2006): p.

318. Girault, H. H., Electrowetting: Shake, rattle and roll, *Nat Mater*, Vol. 5, (2006): p. 851.

319. Koryta, J., L. Q. Hung, and A. Hofmanova, Biomembrane transport processes at the interface of two immiscible electrolyte solutions with an adsorbed phospholipid monolayer, *Stud Biophys*, Vol. 90, (1982): p. 25.

320. Girault, H. H. J. and D. J. Schiffrin, Adsorption of phosphatidylcholine and phosphati-dylethanolamine at the polarised wtaer/1,2-dichloroethane interface, *J Electroanal Chem*, Vol. 179, (1984): p. 277.

321. Cunnane, V. J., D. J. Schiffrin, M. Fleischmann, G. Geblewicz, and D. Williams, The kinetics of ionic transfer across adsorbed phospholipid layers, *J Electroanal Chem*, Vol. 243, (1988): p. 455.

322. Kakiuchi, T., M. Kotani, J. Noguchi, M. Nakanishi, and M. Senda, Phase transition and ion permeability of phosphatidylcholine monolayers at the polarised oil-water interface, *J Colloid Interf Sci*, Vol. 149, (1992): p. 279.

323. Kakiuchi, T., T. Kondo, M. Kotani, and M. Senda, Ion permeability of dilauroylphos-phatidylethanolamine monolayer at the polarised nitrobenzene-water interface, *Langmuir*, Vol. 8, (1992): p. 169.

324. Monzon, L. M. A. and L. M. Yudi, Cation adsorption at a distearoylphosphatidic acid layer adsorbed at a liquid/liquid interface, *Electrochim Acta*, Vol. 52, (2007): p. 6873.

325. Katano, H., Y. Murayama, H. Tatsumi, T. Hibi, T. Ikeda, I. Kameoka, and T. Tsukatani, Voltammetric behavior of the transfer of mono- and polyammonium ions across a phospholipid monolayer at the nitrobenzene/water interface, *Anal Sci*, Vol. 21, (2005): p. 1529.

326. Yoshida, Y., Ion transfer reactions Across the Aqueous|Organic Solution Interface in the Presence of Phospholipid Layer Adsorbed on the Interface, *Anal Sci*, Vol. vol.17 supplement, (2001): p. 3.

327. Gulaboski, R., C. M. Pereira, M. N. D. S. Cordeiro, I. Bogeski, E. Ferreira, D. Ribeiro, M. Chirea, and A. F. Silva, Electrochemical study of ion transfer of acetylcholine across the interface of water and a lipid-modified 1,2-dichloroethane, *J Phys Chem B*, Vol. 109, (2005): p. 12549.

328. Yoshida, Y., K. Maeda, and O. Shirai, The complex formation of ions with a phospho-lipid monolayer adsorbed at an aqueous | 1,2-dichloroethane interface, *J Electroanal Chem*, Vol. 578, (2005): p. 17.

329. Samec, Z., A. Trojanek, and H. H. Girault, Thermodynamic analysis of the cation binding to a phosphatidylcholine monolayer at a polarised interface between two immiscible electrolyte solutions, *Electrochem Commun*, Vol. 5, (2003): p. 98.

330. Guainazzi, M., G. Silvestri, and G. Serravalle, Electrochemical metallization at liquid-lquid interfaces of non-miscible electrolytic solutions, *J Chem Soc Chem Comm*, (1975): p. 200.

331. Yogev, D. and S. Efrima, Novel silver metal liquid-like films, *J Phys Chem-Us*, Vol. 92, (1988): p. 5754.

332. Boeker, A., J. He, T. Emrick, and T. P. Russell, Self-assembly of nanoparticles at interfaces, *Soft Matter*, Vol. 3, (2007): p. 1231.

333. Cheng, Y. F. and D. J. Schiffrin, Electrodeposition of metallic gold clusters at the water/1,2-dichloroethane interface, *J Chem Soc Faraday T*, Vol. 92, (1996): p. 3865.

334. Johans, C., K. Kontturi, and D. J. Schiffrin, Nucleation at liquid | liquid interfaces: galvanostatic study, *J Electroanal Chem*, Vol. 526, (2002): p. 29.

335. Johans, C., R. Lahtinen, K. Kontturi, and D. J. Schiffrin, Nucleation at liquid | liquid interfaces: electrodeposition without electrodes, *J Electroanal Chem*, Vol. 488, (2000): p. 99.

336. Trojanek, A., J. Langmaier, and Z. Samec, Random nucleation and growth of Pt nanoparticles at the polarised interface between two immiscible electrolyte solu-tions, *J Electroanal Chem*, Vol. 599, (2007): p. 160.

337. Platt, M., R. A. W. Dryfe, and E. P. L. Roberts, Controlled deposition of nanoparticles at the liquid-liquid interface, *Chem Commun*, (2002): p. 2324.

338. Dryfe, R. A. W., A. O. Simm, and B. Kralj, Electroless deposition of palladium at bare and templated liquid/liquid interfaces, *J Am Chem Soc*, Vol. 125, (2003): p. 13014.

339. Platt, M., R. A. W. Dryfe, and E. P. L. Roberts, Electrodeposition of palladium nanoparticles at the liquid-liquid interface using porous alumina templates, *Electrochim Acta*, Vol. 48, (2003): p. 3037.

340. Platt, M., R. A. W. Dryfe, and E. P. L. Roberts, Structural and electrochemical characterisation of Pt and Pd nanoparticles electrodeposited at the liquid/liquid interface, *Electrochim Acta*, Vol. 49, (2004): p. 3937.

341. Platt, M. and R. A. W. Dryfe, Structural and electrochemical characterisation of Pt and Pd nanoparticles electrodeposited at the liquid/liquid interface: Part 2, *Phys Chem Chem Phys*, Vol. 7, (2005): p. 1807.

342. Dryfe, R. A. W., Modifying the liquid/liquid interface: pores, particles and deposition, *Phys Chem Chem Phys*, Vol. 8, (2006): p. 1869.

343. Platt, M. and R. A. W. Dryfe, Electrodeposition at the liquid/liquid interface: The chronoamperometric response as a function of applied potential difference, *J Electroanal Chem*, Vol. 599, (2007): p. 323.

344. Johans, C., J. Clohessy, S. Fantini, K. Kontturi, and V. J. Cunnane, Electrosynthesis of polyphenylpyrrole coated silver particles at a liquid-liquid interface, *Electrochem Commun*, Vol. 4, (2002): p. 227.

345. Knake, R., A. W. Fahmi, S. A. M. Tofail, J. Clohessy, M. Mihov, and V. J. CUNNANE, Electrochemical nucleation of gold nanoparticles in a polymer film at a liquid-liquid interface, *Langmuir*, Vol. 21, (2005): p. 1001.

346. Lepkova, K., J. Clohessy, and V. J. Cunnane, The pH-controlled synthesis of a gold nanoparticle/polymer matrix via electrodeposition at a liquid-liquid interface, *J Phys-Condens Mat*, Vol. 19, (2007): p. 375106.

347. Agrawal, V. V., G. U. Kulkarni, and C. N. R. Rao, Surfactant-promoted formation of fractal and dendritic nanostructures of gold and silver at the organic-aqueous interface, *J Colloid Interf Sci*, Vol. 318, (2008): p. 501.

348. Su, B., J. P. Abid, D. J. Fermin, H. H. Girault, H. Hoffmannova, P. Krtil, and Z. Samec, Reversible voltage-induced assembly of Au nanoparticles at liquid vertical bar liquid interfaces, *J Am Chem Soc*, Vol. 126, (2004): p. 915.

349. Su, B., D. Fermin, J. Abid, N. Eugster, and H. Girault, Adsorption and photoreactivity of CdSe nanoparticles at liquid | liquid interfaces, *J Electroanal Chem*, Vol. 583, (2005): p. 241.

350. Bresme, F. and J. Faraudo, Orientational transitions of anisotropic nanoparticles at liquid-liquid interfaces, *J Phys-Condens Mat*, Vol. 19, (2007): p.

351. Galletto, P., H. H. Girault, C. Gomis-Bas, D. J. Schiffrin, R. Antoine, M. Broyer, and P. F. Brevet, Second harmonic generation response by gold nanoparticles at the polarized water/2-octanone interface: from dispersed to aggregated particles, *J Phys-Condens Mat*, Vol. 19, (2007): p.

352. Abid, J. P., M. Abid, C. Bauer, H. H. Girault, and P. F. Brevet, Controlled reversible adsorption of core—Shell metallic nanoparticles at the polarized water/1,2-dichloroethane interface investigated by optical second-harmonic generation, *J Phys Chem C*, Vol. 111, (2007): p. 8849.

353. Yamamoto, S. and H. Watarai, Surface-enhanced Raman spectroscopy of dodecanethiol-bound silver nanoparticles at the liquid/liquid interface, *Langmuir*, Vol. 22, (2006): p. 6562.

354. Cohanoschi, I., A. Thibert, C. Toro, S. Zou, and F. E. Hernandez, Surface plasmon enhancement at a liquid-metal-liquid interface, *Plasmonics*, Vol. 2, (2007): p. 89.

355. Flatte, M. E., A. A. Kornyshev, and M. Urbakh, Understanding voltage-induced localization of nanoparticles at a liquid-liquid interface, *J Phys-Condens Mat*, Vol. 20, (2008): p. 073102.

356. Lahtinen, R. M., D. J. Fermin, H. Jensen, K. Kontturi, and H. H. Girault, Two-phase photocatalysis mediated by electrochemically generated Pd nanoparticles, *Electrochem Commun*, Vol. 2, (2000): p. 230.

357. Jensen, H., D. J. Fermin, J. E. Moser, and H. H. Girault, Organization and reactivity of nanoparticles at molecular interfaces. Part 1. Photoelectrochemical responses involving TiO2 nanoparticles assembled at polarizable water | 1,2-dichloroethane junctions, *J Phys Chem B*, Vol. 106, (2002): p. 10908.

358. Fermin, D. J., H. Jensen, J. E. Moser, and H. H. Girault, Organisation and reactivity of nanoparticles at molecular interfaces. Part II. Dye sensitisation of TiO2 nanoparticles assembled at the water vertical bar 1,2-dichloroethane interface, *Chemphyschem*, Vol. 4, (2003): p. 85.

359. Trojanek, A., J. Langmaier, and Z. Samec, Electrocatalysis of the oxygen reduction at a polarised interface between two immiscible electrolyte solutions by electrochemically generated Pt particles, *Electrochem Commun*, Vol. 8, (2006): p. 475.

360. Girault, H. H., The water-oil-water thermocouple and the ionic Seebeck effect, *J Chem Soc Farad T 1*, Vol. 84, (1988): p. 2147.

361. Osakai, T., H. Ogawa, T. Ozeki, and H. H. Girault, Determination of the entropy of ion transfer between two immiscible liquids using the wateroilwater thermocouple, *J Phys Chem B*, Vol. 107, (2003): p. 9829.

362. Hinoue, T., E. Ikeda, S. Watariguchi, and Y. Kibune, Thermal modulation voltammetry with laser heating at an aqueous | nitrobenzene solution microinterface: Determination of the standard entropy changes of transfer for tetraalkylammonium ions, *Anal Chem*, Vol. 79, (2007): p. 291.

2 Reduction of Platinum under Superdry Conditions

An Electrochemical Approach

Philippe Hapiot and Jacques Simonet

CONTENTS

2.1   INTRODUCTION

2.1.1   Does the Perfectly Inert Electrode Really Exist?

The search for the perfect working electrode in terms of stability, ease of use, cost, and wide potential range remains nowadays one of the goals of most electrochemists in the fields of preparative organic and inorganic electrochemistry, fuel cells, and electroanalysis [1,2]. In fact, any valuable cathodic and anodic reaction is generally associated with a specific set of experimental conditions such as the nature of the solvent, choice of the electrolyte, and, more particularly for electrochemical processes, reactivity of the electrode material that could be tuned through intrinsic modifications of the electrode interface [1,2]. Until now, mercury, platinum, and glassy carbon have been the most used electrode materials [3]. After research by electrochemists determined its environmental impact, mercury has been banned [4] both in industry and also from the laboratories. However, are platinum [5] and all forms of carbon [6], generally considered as stable materials under cathodic and anodic polarizations, the best choices? Many substrates that are reported as inert materials are not always very stable [7]. As we will detail in this chapter, this is clearly the case for platinum, and certainly for other noble metals.

2.1.2   REPORTED REACTIVITY OF COMMON INTERFACES UNDER ELECTRON TRANSFER

2.1.2.1. Addition of Electrogenerated Species onto Surfaces

Many free radicals, essentially produced after the bond cleavages of some electrogenerated intermediates, lead to activated, alkyl, or aryl radicals, which could be grafted onto conducting surfaces. This type of reaction is found in essentially organic, strongly activated substrates and surfaces possessing efficient catalytic capacities because the condition of immobilization on the surface depends on the nonreducibility of the radical. One of the most useful substrates belongs to the family of aryl diazonium cations, and many studies have demonstrated their properties [7,8]. More generally, during many conventional reactions at solid cathodes, formation of free radicals may react on surfaces and cause inhibiting films (by the addition of free alkyl radicals onto metallic polarized surfaces or carbons).

2.1.2.2   Insertion of Salts into the Electrode Material

In electrochemistry, the electrolyte plays a crucial role. Generally, the salt is not expected to affect the interfacial reaction within the range in which it is totally inactive. In most cases, the supporting electrolyte is composed of nonredox reactive ions that are especially chosen for this purpose to avoid the interference of their own reactivities (if any) with the considered (or expected) electrochemical reaction. However, in many studies, strong interactions between the electroactive organic substrates, cations, and anions have been noted and were therefore considered, respectively, as acceptors and donors [9]. Antecedent and subsequent reactions brought on by salts could influence product distribution. The specific choice of an electrolyte could be used to direct the global electrode reaction toward a precise direction (stereochemistry, polymerization, protonation, etc.). For example, the cathodic reduction of acrylonitrile in the presence of alkali metal salts yields propionitrile, whereas the use of quaternary ammonium salts orients the reaction to dimer formation (synthesis of adiponitrile) [10]. Different models of electronic double layers were proposed to explain such determining effects of the interface [11].

As a classical material used in electrochemistry, mercury is known to react with electrogenerated alkali metals (sodium, potassium) to form amalgams. This property impedes the use of alkali metal salts with aprotic organic solvents (dimethylformamide, acetonitrile, propylene carbonate) at potentials <-1.8 V versus SCE. Surprisingly, ammonium salts (cation tetraalkylammonium) have been said to react as well, but at much more negative potentials. The structure of the so-called tetramethylammonium amalgam has been characterized [12]. The formation of a material analogous to an amalgam was obtained with tetraalkylammonium of longer chains [13]. It was proposed that it could correspond to the dilution of a hypothetical $NR_4^\cdot$ radical inside the mercury bulk ($n \gg 1$):

$$n\ Hg + R_4N^+ + e^- \rightarrow R_4N(Hg)_n$$

This hypothesis was supported by Southworth et al. [14] who reported, under electrolysis of aliphatic ammonium salts on a mercury pool, the formation of a silvered grey solid. Such electrogenerated materials were later used by Horner and Neumann [15] as efficient reducing reagents capable of preventing contact of the reagent with air. More recently, it was established by Garcia and coworkers [16] that such quaternary ammonium amalgams were in fact Zintl ion salts.

Quite similarly and in the same spirit, Svetlicic and coworkers showed that mercury is not the lone electrode material that could electrochemically react in this way with quaternary ammonium salts [17]. Indeed, some posttransition metals also react with quaternary ammonium salts in a similar way [18]. In particular, lead, tin, antimony, and bismuth were reported to produce thin organic layers under electrochemical reduction. From coulometric measurements, the stoichiometries of these electroformed materials were established. Under the experimental conditions chosen by Kariv-Miller, the cathodic material M was deeply changed: each ammonium is associated with a homopolyanion derived from reduction of the metal M according to the following general equation:

$$5\ M^+ + R_4N^+ + e^- \rightarrow R_4N^+(M_5)^-$$

The potentials required for the formation of these organometallic layers and their stability in aprotic solvents depend on the size of the concerned cations and on the nature of the metal used as electrode. Thus, in the presence of dimethylpyrrolidinium cations, the stability of the layer obtained with posttransition metals increases as follows: $Sb > Sn > Bi > Pb > Hg$. All these phases were found to exhibit a certain reducing power toward weak π-acceptors (aliphatic ketones, weakly activated phenyl rings) and organic compounds capable of giving an irreversible scission under electron transfer (such as alkyl halides). Quite similarly, the cathodic insertion of metallic ions (Li^+) is well known in rechargeable battery systems [19,20], and that of tetraalkylammonium cations into graphite (mainly Me_4N^+, which occurs without exfoliation of the material) must certainly be cited as well [21–24]. Several insertion stages were reported with the maximum insertion achieved at somewhat reducing potentials and found to correspond to C_6Li and $C_{12}NMe_4$ as limits. Those species were used as efficient reducing reagents [24] but were reported to be sensitive to dioxygen.

2.1.2.3 Chemical Formation of Homopolyanions with Meta-Metals: Zintl Phases, State of the Art

Related to the subject of this chapter are obviously the well-known Zintl phases. The generation of homopolyanions M_m^{n-} (Figure 2.1) was first reported by Zintl et al. [25] at the beginning of the 20th century. Their synthesis is based on the reduction of meta-metal salts (M = lead, tin, antimony, bismuth, and many others) by alkaline and alkaline-earth metals in liquid ammonia in chemically inert vessels [26–28]. It was found that the stability of complexes with Na^+ or Cs^+ is fairly weak. Their stability (case of posttransition metals) was found to be strongly increased by the use of cryptates [29,30]. Nevertheless, it should be emphasized

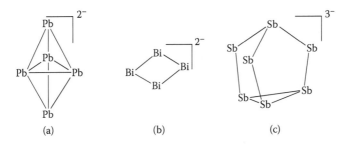

FIGURE 2.1 Structures of some Zintl homopolyanions established by means of x-ray diffraction.

that the electrochemical method (cathodic dissolution of alloys used as cathode material) permits, at room temperature, the synthesis of a large palette of Zintl-type complexes, and allows the insertion of a large variety of cations [31,32]. After Zintl, the chemistry of heteropolyanions was developed. It is remarkable that species obtained by means of electrochemistry are those of weaker entropy (Sn_9^{4-} and Pb_9^{4-}), while complexes obtained from tertiary alloys such as NaSnPb in ethylenediamine are species of much larger entropy (heteropolyanions such as $Sn_5Pb_4^{4-}$ and $Sn_4Pb_5^{4-}$) [33]. Figure 2.2 exhibits the limit proposed by Zintl with the extension when the reducing species is an alkali metal.

			\multicolumn{5}{l}{Zintl limit}
			B	C	N	O	F^-
			Al	Si_4^{4-}	P_{11}^{3-}	S	Cl^-
$In_{10}(Ni)^{10-}$	$(CuP)^{2-}$	$(ZnCd)^{2-}$	Ga_{11}^-	Ge_9^{4-}	As_7^{3-}	Se_6^{2-}	Br^-
$Tl_{10}(Ni)^{10-}$	$(CuAs)^{2-}$	$(ZnGe)^{2-}$					
$(PdSn)^{3-}$	$(AgSb)^{2-}$	$(CdPb)^-$	In_{11}^{7-}	Sn_9^{4-}	Sb_7^{3-}	Te_3^-	I^-
$In_{10}(Pd)^{10-}$	$(AgCl_2)^-$						
$(PtSn)^{3-}$	$(AuCl_2)^-$	$(HgO_2)^{2-}$	Tl^{7-}	Pb_9^{4-}	Bi_4^-	Po	At^-
$In_{10}(Pt)^{10-}$	$(AuGe)^{2-}$						

Anions complexes of transition metals and meta-metals — Zintl polyanions of AX compounds (with A = Li, Na, K, Cs) — Ionic crystals

FIGURE 2.2 Zintl classification.

2.1.3   What Is the Cathodic Reactivity of Transition Metals Used as Cathode Materials?

Platinum, palladium, copper, and tin are generally considered as relatively inert materials for cathodic electrochemical processes when experiments are performed under regular experimental conditions such as in an aqueous media or some organic solutions with a large (or fairly large) amount of moisture. Gold was also found to react electrochemically with K^+ in liquid ammonia to form the auride ion Au^-. The reduction occurs at very negative potential, and the formation of soluble Au^-, which could be deposited as Au at the anode, was observed [34]. In general, the use of Pt, Pd, Cu, and Sn as cathodic material is limited (especially platinum and palladium) by their weak hydrogen overvoltage. Thus, with technical or commercial weakly acidic solvents such as dimethylformamide (with tetrabutylammonium tetrafluoroborate as supporting electrolyte), the cathodic limit is always larger than –2 V versus SCE. Under such conditions, the cathodic boundary simply corresponds to hydrogen evolution.

2.2   EXPERIMENTAL CONDITIONS REQUIRED TO DEMONSTRATE CATHODIC CHARGING OF PLATINUM

2.2.1   Preparation of the Electrolyte, Electrodes Used, and Voltammetry and Coulometry

For most of the experiments related to this chapter devoted to the reduction of platinum, the experimental conditions are based on obtaining the so-called super-dry conditions described by Hammerich and Parker [35] and Heinze [36]. These conditions were additionally redefined in previous reports [37–39]. Even though most of the experiments described in this chapter have been performed in dimethylformamide (DMF), it should be emphasized that acetonitrile (AN) or propylene carbonate (PC) could be used as well. An important and essential condition is the use of extremely dry solvents. A simple method for obtaining dry solvents (with less than 50 ppm of water traces) consists in the direct in situ addition of activated neutral alumina in the electrochemical cell. For this, the activation of alumina is achieved by heating at 300°C for 4 h under vacuum. In particular, DMF was almost constantly checked (by the Karl Fischer method) to ascertain that it contained less than 50 ppm of water. A permanent storage of solvent over alumina gave best results.

The same precautions versus the presence of water traces have to be taken for the used electrolyte. In most of the experiments, the electrolyte concentration was 0.1 mol.L^{-1}. Potassium, lithium, sodium, and cesium iodides were used (because of their large solubilities). Tetraalkylammonium salts (puriss grade) were also used with purity >99.7%. It is required that all salts be dried by conventional methods before use [40].

Cyclic voltammetry investigations were carried out in a standard three-electrode cell. For analytical studies, a typical working electrode was a smooth disk of metal (Pt, Pd, and others) with a surface area of 8×10^{-3} cm^2. The counterelectrode was a glassy carbon rod. It is important to underline the required use of an organic reference (Ag–AgI 0.1 mol.L^{-1} NBu$_4$I system in DMF) to avoid any possible water diffusion in the cell. All potentials are in principle referred to the aqueous SCE. The potential shift in DMF with this referenced electrode is −0.52 V versus SCE.

Prior to experiments, platinum electrodes were carefully polished with silicon carbide paper of successively smaller size (18 to 5 μm), then by diamond powder (6 and 3 μm). Finally, electrodes were rinsed with ethanol and acetone, and then dried with a hot airflow. Between each scan, the electrode surface was thoroughly polished according to the procedure given earlier.

Lastly, for macro electrolysis investigations, Pt sheets (99.99% purity with a surface area of about 1 cm^2 and a thickness 0.05 mm) were used. They were employed only once for SEM investigations without further treatment.

Coulometric and electrochemical quartz microbalance (EQCM) experiments were carried out on thin metallic deposits prepared by deposition of metals from solutions of 10 g L^{-1} H$_2$PtCl$_6$ in 0.1 mol.L^{-1} HCl onto polished gold disks (2×10^{-3} cm^2). The plating was achieved in a galvanostatic mode (current 10^{-2} A cm^{-2}). All the experiments were performed with gold substrates that were experimentally found to be very weakly reactive toward salts within the potential ranges used for platinum. The procedure allows the easy preparation of different amounts of electrodeposited metal. Depending on the type of experiment, the gold substrate was a gold microelectrode polished before each deposition. When EQCM experiments were carried out, a much larger electrode of gold-coated quartz crystal was used (see the following text).

2.2.2 EQCM INSTRUMENTATION

EQCM was found to be a valuable method to quantify, under given experimental conditions, the mass cathodic change of platinum in the presence of a large variety of salts [41]. Mass balance experiments were carried out with an oscillator module quartz crystal analyzer connected to a potentiostat. Mass changes could be achieved potentiostatically at fixed potentials or in the conditions of voltammetry (linear variation of potential upon time). The EQCM device was computer controlled. In the experiments described here, 9 MHz AT-cut gold-coated quartz crystals were electrochemically plated with a thin film of metal (principally platinum, but other metals such as palladium or nickel could also be deposited in thin layers). Plating was achieved by using the galvanostatic procedure. The deposited mass was additionally checked by EQCM. EQCM measurements were performed in a Teflon cell equipped with a glassy carbon counterelectrode, a reference electrode, and the metallized quartz crystal working electrode. The apparent area of the quartz crystal was about 0.2 cm^2.

Microgravimetric data, reported in terms of mass change, were calculated using the Sauerbrey equation, which links resonant frequency and mass change. In most cases, it was found that the mass deposit of platinum (or that of other metal tested) was not totally reversible and, consequently, a small excess of mass remained at the start of each experiment. However, the amount of extra charge gained during the deposition process remained the same.

2.2.3 ELECTROCHEMICAL ATOMIC FORCE MICROSCOPE (EC-AFM) AND SCANNING ELECTROCHEMICAL MICROSCOPE (SECM) IN DIMETHYLFORMAMIDE

Because of the reactivity of the electrogenerated Pt phase, the electrochemical "reduction" of platinum was investigated by in situ EC-AFM in dry and deoxygenated DMF. AFM requires the use of a very flat sample for allowing the scanner to follow surface morphology, and this condition renders difficult experiments with native samples of platinum. Our samples were prepared by d.c. sputtering of Pt onto (100)MgO, leading to a distribution of (100)Pt platelets (around 100–200 nm size and an average thickness around 50 nm; see References 42–44 for details). The quality of the sample was always checked by x-ray diffraction. An θ–2θ x-ray diffraction showed the (100) orientation of the Pt films, and the narrow rocking curve recorded on the 200 reflection (full width at half maximum of 0.33°) showed the high crystalline quality of the film. X-ray diffraction φ-scan of the (220)Pt reflections evidenced the epitaxial growth of the Pt films (Figure 2.7).

By adjusting the experimental conditions, different general patterns can be obtained, whether the platinum areas are connected or not. The preparative conditions were selected in order to get connected or unconnected Pt plates. Under such conditions, the sample could be either macroscopically conducting (behaving like a single, flat native electrode) or, on the contrary, macroscopically insulating and composed of independent Pt plates. Samples prepared by this technique have several advantages: (1) the average roughness of the platelets is low (around 2 nm), and their crystallographic orientations correspond to a fully defined (100)Pt surface; (2) the sample behaves like a native electrode; and (3) patterns are easily recognizable, which make the observations easier by comparison with an unmodified sample. The morphology of the Pt modifications was characterized by contact-mode atomic force. For EC-AFM experiments, the cell was a three-electrode type setup. The working electrode that is also the AFM sample has a 5×5 mm² size and is fixed onto the bottom of the cell by two wires to avoid any movement during AFM scanning. The reference electrode was an Ag quasi-reference obtained with an Ag wire covered with $AgNO_3$. The counterelectrode was made by twisting a 50 μm platinum wire and placing it all around the cell. This geometry produces an almost homogeneous modification of the sample. All experiments were performed under inert gas (argon) to limit the introduction of water and oxygen because the electromodifications of platinum are highly sensitive to the presence of water and oxygen.

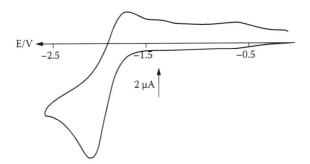

FIGURE 2.3 Voltammetric response of a smooth platinum disk electrode (area: 0.8 mm²) in contact with a superdry solution of 0.1 mol.L^{-1} NaI in DMF. Scan rate: 0.2 V s^{-1}. Potentials are referred to the SCE reference electrode (Cougnon, C., Ph.D. thesis, Université de Rennes 1, 2002).

2.3 THE PLATINUM CHARGE UNDER SUPERDRY CONDITIONS

2.3.1 ELECTROANALYTICAL EVIDENCE OF PLATINUM CHARGE IN THE PRESENCE OF ALKALI SALTS

Initially, the first experiments [45] concerning the charge of platinum were performed in the presence of monovalent cations (Li$^+$, Na$^+$, K$^+$, and Cs$^+$) associated with iodide anion, mainly for the purpose of solubility of these salts [40]. As will be explained later, in the study relative to tetraethylammonium salts, bromide, chloride, tetrafluoroborate, hexafluorophosphate, and several other ions could be used as well. With these experimental conditions, where no electroactive species are introduced in the solution, except the alkali metal iodides that was originally chosen as inert supporting electrolyte, a quasi-reversible peak is observed at the Pt microcathode (Figures 2.3 and 2.4). This unexpected cathodic current, I$_{pc}$,

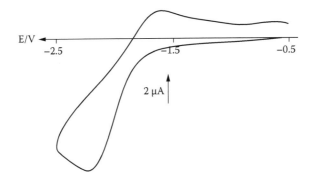

FIGURE 2.4 Voltammetry for a smooth platinum disk (surface area 0.8 mm²) in a superdry solution of 0.1 mol.L^{-1} CsI in DMF. Scan rate: 0.2 V s^{-1}. Potentials are referred to the SCE electrode (Cougnon, C., Ph.D. thesis, Université de Rennes 1, 2002).

TABLE 2.1

Voltammetric Data Related to a Stationary Platinum Cathode of Area 8×10^{-3} cm^2 [a]

Entry	Electrolyte (0.1 M)	E_{pc} (V)	I_{pc} (µA)	Epa (V)	$E_{pc} - E_{pa}$ (V)
1	LiI	−2.84	−5	−2.42	0.42
2	NaI	−2.07	−10	−1.72	0.35
3	KI	−2.17	−6.4	−1.62	0.55
4	CsI	−2.18	−5.4	−1.77	0.41
5	Bu$_4$NI	−2.89	−11	−1.27	1.62
6	Bu$_4$NBF$_4$	−2.91	−12.5	−1.09	1.82
7	Bu$_4$NClO$_4$	−2.84	−25	−0.99	1.85

[a] Potentials are referred to the SCE. The sweep rate, v, is 0.2 Vs^{-1}. Potentials (E_{pc} and E_{pa}) and peak currents (I_{pc}) obtained from cyclic voltammetry at a platinum cathode relative to 0.1 M salt alkali halide salts and to tetra-n-butylammonium salt solutions in dry DMF.

was always associated with an anodic step peak of current, I_{pa}, and was detected for all the considered alkali cations but at different potentials. It could be shown that the anodic step corresponds to the reverse reaction of the cathodic step by simply maintaining the applied potential at its level. Charges observed with alkali metals ions (Table 2.1) require rather negative potentials (<−2 V versus SCE). All cathodic steps existing under those experimental conditions vary linearly with the square root of the scan rate, which indicates a diffusion control for the charge process.

This electrochemical behavior is very similar to what has been observed with highly oriented pyrolytic graphite (HOPG) electrodes when they are cathodically charged within comparable experimental conditions (in cases of cation insertion into the lamellar compound). By analogy with these processes, the very small currents noticed for the cathodic steps on platinum suggest that the limiting diffusion corresponds to the slow insertion into the metallic bulk. It is also important to stress that the cathodic currents do not diminish when activated alumina is progressively added to the electrolytic solution, and thus, it cannot be due to residual water reduction. Additionally, the reduction of alkali metal cations to produce the corresponding metal at the surface of platinum is an unlikely process in this case. Indeed, considering the concentrations of alkali cations (0.1 mol.L^{-1}), the observed currents would have to be much higher to bring a cathodic limit to the process. However, the intensity of the peak current has been found to depend both on the concentration of the salt and the nature of the cation. Specifically, in the presence of alkali iodides, the intensity of the currents follows the order Cs$^+$ < K$^+$ < Na$^+$ ≪ Li$^+$.

To check the specific role played by moisture (traces of water), known amounts of water were added to the solution (amounts slightly larger than 200–500 ppm).

Upon water addition, the specific redox signal corresponding to the platinum charging disappears, and the classical cathodic limit assigned to water reduction is then observed.

In order to get more information about the existence of a charge–discharge process, employing the coulometric technique appeared very helpful. The idea is that the charge amount stored in the material (or at its surface) during the cathodic process could be anodically restored. However, it must be underlined that fully reversible charge systems are difficult to obtain owing to the presence of residual water or acidic impurities, even in very small amounts. Thus, the value of the charge measured during cathodic reduction, Q_f, in the course of rather long fixed-potential electrolyses, is expected to be too large because it contains a contribution due to unavoidable hydrogen evolution. On average, the charge proportion due to hydrogen evolution is around 65% to 75%, depending on the salt used and the quality of preparation of the DMF solution. Generally, it is observed that the electric charge stored in material that is specific to the platinum redox process (named Q_a; see Figure 2.5), could not be easily correlated to the total charge amount measured during the charge process. Thus, Q_a values that become invariant with long-enough charge processes could be assigned to the reduction of platinum and the experiments described here, taking into account those maximum charge values.

In order to determine the exact proportion of metal involved in the insertion process, experiments were performed with samples containing a known amount of platinum that was electrochemically deposited onto inert substrates such as gold or glassy carbon. When gold was used as the substrate, the accuracy of the mass of Pt was also checked by the EQCM technique. Experimental conditions

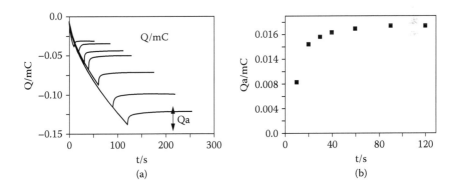

FIGURE 2.5 Successive charges and discharges of a platinized smooth gold disk (area: 0.8 mm²) at potentials of −2.15 V and −0.5 V (versus SCE), respectively. Electrolyte: 0.1 mol.L⁻¹ NaI in superdry DMF. Amount of deposited platinum: 60 µg. (a) Chronocoulometric curves for charges and discharges. (b) Variation of quantity of charge Q_a (recovered charge during the reoxidation process) upon charge times. The limit of Q_a corresponds to the saturation of the platinized layer (Cougnon, C., Ph.D. thesis, Université de Rennes 1, 2002).

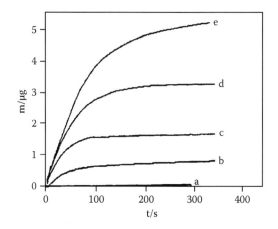

FIGURE 2.6 EQCM experiments related to the charge of platinum upon time in the presence of 0.1 mol.L⁻¹ CsI in DMF. Applied potential: –2.3 V versus SCE. (a) Gold-filmed quartz. (b, c, d, and e) the same substrate covered with platinum. Deposited masses of Pt are 0.9 µg, 1.5 µg, 3.5 µg, and 6 µg, respectively (Cougnon, C., Ph.D. thesis, Université de Rennes 1, 2002).

were always those given in Figure 2.5: the length of the electrolysis time (cathodic) was checked to be sure it was long enough to ensure the total insertion of the salt inside the platinum. Thus, as displayed in the example given in Figure 2.6, the reduction of Pt in CsI–DMF was achieved at –2.3 V versus SCE, while the reoxidation of the modified platinum was obtained at 0 V. It is remarkable that the limit, Q_a, was reached almost immediately. Additionally, the value of Q_a was found to be proportional to the electrodeposited platinum amounts. The slope of the lines $Q_a = f$ (Pt thickness) was found to be 2, with an accuracy of 2% with the tested alkali metals. The assumption relative to a reaction in mass was therefore verified, and confirms the hypothesis of an insertion of the alkali metal cations M^+ inside the platinum. Thus, an *ionometallic* structure of the general form [Pt_2^-, M^+, X] could be proposed at this stage [46].

For an exhaustive determination of the mass, it is also necessary to characterize the possible insertion of other species (X-neutral) like the solvent or the salt that was not concomitantly involved in the process. In this purpose, the variation of the mass of the platinum sample was followed during the whole charge process. Such experiments led to the data exemplified in Figures 2.6 and 2.7.

At the level of the redox step, a mass increase was found for the platinum sample in the presence of different salts. This increase was found to be proportional to the initial mass of deposited metal, and does not depend on the salt concentration. Mass increase was larger with cesium iodide than with the other alkali cations. The linear increase of mass of the layer after saturation (for experimental time >200–300 s under the conditions of Figure 2.6) on the mass of Pt allowed the proposition of a general global formula for the platinum phase. These formulas

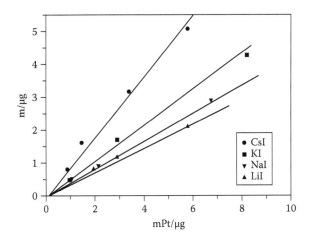

FIGURE 2.7 EQCM experiments giving the limit mass increase of platinum-covered substrates in function of Pt-deposited amounts onto gold-film quartz. Case of alkaline iodides. Applied potentials are those given for the cathodic peak potentials E_{pc} in Table 2.1 (Cougnon, C., Ph.D. thesis, Université de Rennes 1, 2002).

are displayed in Table 2.2. It is remarkable that the uptake of one electron by two atoms of platinum is always concomitant with the insertion of one alkali metal cation and one equivalent salt. Finally, the stoichiometry of the saturated phase in the presence of an alkali salt MI was established as [Pt_2^-, M^+, MI]. The good accuracy (see Table 2.2) visible from the EQCM experiments permits rejection of a significant insertion of the solvent molecule and the possible formation of another parent phase in high ratio.

One may expect a very large mass increase of the Pt interface and that the insertion of the salts will lead to very large swellings. As will be discussed, the electronic conductivity of the external layer is low and, thus, the insertion leads to a kind of passivation of the bulk material. Owing to the rapid decrease in

TABLE 2.2

Determination of Stoichiometries by EQCM Technique (Values, y) for Metal Phases [$Metal_2^-$, M^+, $(MX)_y$] Obtained with Platinum and Palladium in Super-Dry DMDF in the Presence of Alkaline Iodides

Electrolyte (0.1 mol. L^{-1})	Platinum	Palladium
LiI	[$Pt_2Li(LiI)_{1.06\pm0.06}$]	[$Pd_2Li(LiI)_{1.01\pm0.02}$]
NaI	[$Pt_2Na(NaI)_{0.98\pm0.08}$]	[$Pd_2Na(NaI)_{0.91\pm0.07}$]
KI	[$Pt_2K(KI)_{0.97\pm0.08}$]	[$Pd_2K(KI)_{1.02\pm0.03}$]
CsI	[$Pt_2Cs(CsI)_{0.88\pm0.05}$]	[$Pd_2Cs(CsI)_{0.97\pm0.05}$]

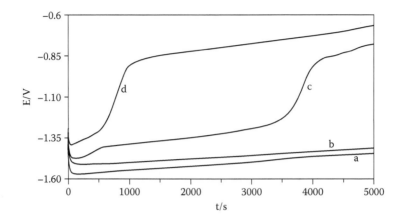

FIGURE 2.8 Chronopotentiometries at nil current for Pt deposits onto gold. Amounts of deposited Pt: 10 mC. These Pt layers were previously charged with different amounts of charge. Electrolytic solution: 0.1 mol.L^{-1} NaI in DMF. Saturation ratios are derived from theoretical values: (a): 100%, (b): 85%, (c): 75%, and (d): 50%. Potentials are referred to SCE electrode (Cougnon, C., Ph.D. thesis, Université de Rennes 1, 2002).

conductivity, the insertion remains limited within a relatively thin layer, which does not exceed a few microns. The reversibility of the charge process suggests that the reduced phase is an efficient reducing species. It is worth noting that the stability of such phases in the presence of air is weak (due to a fast reduction of dioxygen). Under an inert atmosphere (nitrogen or argon), the reduced forms of platinum are reasonably stable, and it is expected that traces of moisture in the solvent contribute to its oxidation.

To complement the coulometric measurements, chronopotentiometric experiments with these phases were performed (see Figure 2.8). It could be stressed that a saturated phase (evolution of the equilibrium potential with time for at least 1 h, curve a) appears stable, while lower levels of insertion (curves c and d, Figure 2.8) exhibit a progressive increase in the potential. This phenomenon seems provoked by a modification of the distribution of charges inside the layer. All these chronopotentiometric curves were found to depend strongly on the saturation level of charges injected in the platinum whatever the salt used.

2.3.2 Charge Processes of Platinum in the Presence of Tetraalkylammonium Salts (TAAX)

On the whole, the same general patterns were found when the charging of platinum was performed in a solution containing a quaternary ammonium salt (Table 2.3). The main difference between ammonium salts and alkaline iodides is essentially the total lack of reversibility of the charging process, especially with bulky ammonium cations. It was also observed that fairly negative potentials

TABLE 2.3

Voltammetric Data (Cathodic and Anodic Peaks) for Smooth Platinum Electrodes (Surface Area: 0.8 mm^2) Under Superdry Conditions[a]

Electrolyte (0.1 mol.L^{-1})	E_{pc}/V	$I_{pc}/\_A$	E_{pa}/V	$I_{pa}/\mu A$	$E_{pa} - E_{pc}/V$
Me$_4$NI	−2.50	−15.0	−1.82	4.9	0.66
Me$_4$NCl	−2.57	−14.5	−1.87	5.3	0.70
Et$_4$NI	−2.67	−15.7	−1.79	4.5	0.88
Et$_4$NBr	−2.64	−17.0	−1.77	4.2	0.87
Bu$_4$NI	−2.86	−12.0	−1.50	3.1	1.36
Bu$_4$NBr	−2.90	−11.8	−1.55	2.8	1.35
Bu$_4$NCl	−2.89	−13.1	−1.56	3.0	1.36
Hex$_4$NBr	<−2.92	−6.9	−1.41	1.2	>1.51
Hex$_4$NCl	<−2.92	−7.2	−1.39	1.7	>1.53
Oct$_4$NBr	<−3.02	−5.4	−1.26	0.9	>1.76
Oct$_4$NCl	<−3.02	−6.1	−1.25	0.8	>1.77

[a] Electrolyte: DMF + 0.1 mol.L^{-1} TAAX. Scan rate: 0.2 V s^{-1} versus SCE reference electrode.

are necessary to achieve most of these slow charge processes. Obviously, discharges could be obtained, but the potential differences, ΔE, between the charge–discharge processes are, in most cases, considerable, and could actually reach 2 V. As previously described for the experiments performed with alkaline iodides, coulometry and EQCM techniques were efficiently used to determine the stoichiometries of platinum phases with a large series of TAAX salts. In all cases, the results related to saturation charges agree with the molar mass of the salt. A general formula was derived from these experiments that was valid for all the considered tetraalkylammonium cations (Table 2.4) displaying different ammonium sizes and several different types of associated anions, and it has the general form [Pt$_2^-$, R$_4$N$^+$, R$_4$NX].

In general, voltammetries of the charging processes exhibit relatively slow charge processes (values of the transfer coefficient $\alpha < 0.2$). On smooth bulk platinum, the electrochemical rate seems strongly dependent on the size of the anion and that of the cation as tested with a large variety of TAAX salts. The ease of the discharge process, which could be estimated from the difference of potential between the cathodic and the anodic processes and the intensity of the reoxidation, appears to be correlated with the rate of the charging process. As an illustration, Figure 2.9 shows the comparative use of two ammonium salts of different large sizes (for clarity, the anion remains the same). Differences between the experimental data give a good idea of the dramatic difference in the charging potential, the apparent slope of the step, and the lack of reversibility obtained with a bulky cation.

TABLE 2.4

Values x and y Corresponding to Stoichiometries Relative to Phases
[Pt$_x^-$, R$_4$N$^+$, (R$_4$NX)$_y$] Obtained with Some TAAX Salts[a]

Electrolyte (0.1 mol.L^{-1})	Applied Potential/V	x-Value	y-Value
Me$_4$NI	−2.42	2.04 ± 0.12	0.97 ± 0.04
Me$_4$NCl	−2.42	2.05 ± 0.20	1.06 ± 0.15
Et$_4$NI	−2.72	2.08 ± 0.20	0.99 ± 0.06
Et$_4$NBr	−2.72	1.91 ± 0.11	1.05 ± 0.17
Bu$_4$NI	−2.82	2.06 ± 0.09	0.95 ± 0.06
Bu$_4$NBr	−2.82	1.97 ± 0.08	1.02 ± 0.12
Bu$_4$NCl	−2.87	1.96 ± 0.08	0.98 ± 0.09
Hex$_4$NBr	−2.87	2.05 ± 0.15	0.92 ± 0.06
Hex$_4$NCl	−2.87	1.98 ± 0.07	1.06 ± 0.09
Oct$_4$NBr	−2.87	2.07 ± 0.07	1.02 ± 0.05
Oct$_4$NCl	−2.87	1.99 ± 0.11	1.05 ± 0.07

[a] Potentials are referred to SCE reference electrode.

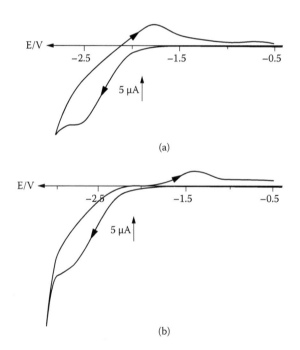

(a)

(b)

FIGURE 2.9 Compared voltammetric response of a smooth platinum electrode (area: 0.8 mm^2) in contact with a superdry solution of 0.1 M TAAX in DMF. Potentials are referred to the SCE electrode. Scan rate: 0.2 V s^{-1}. (a) Salt: Et$_4$NBr. (b) salt: n-Hex$_4$NBr (Cougnon, C., Ph.D. thesis, Université de Rennes 1, 2002).

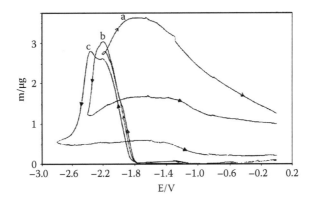

FIGURE 2.10 Variation of the mass of a film of platinum deposited onto a gold-film quartz upon potential. Scan rate: 10 mV s^{-1}. Electrolyte: 0.1 mol.L^{-1} TBAI in superdry DMF. (a) Scanning from 0 V to −2.2 V. (b) Scanning until −2.4 V. (c) scanning until −2.8 V. Potentials are referred to the SCE electrode (Cougnon, C., Ph.D. thesis, Université de Rennes 1, 2002).

As underlined by Table 2.3, the difference between the charge and discharge in terms of energy may reach very large values. It is noticeable that, with bulky TAAX salts, the intensity of the cathodic step is always significantly greater than the current observed during the corresponding discharge. Probably, mass-potential curves as displayed in Figure 2.10 could shed some additional light on the nature of the charge processes in the presence of TAA$^+$. It is evident that the charge amounts are significantly dependent on the limit boundary. Thus, if the boundary of the forward scan does not exceed −2.2 V, the mass increase is maximum, and the backward scan exhibits a progressive but effective discharge all along the scan. On the contrary, when more negative potentials are reached, the gained mass suddenly decreases, suggesting a sudden collapse of the insertion process. It was proposed that the platinum phase reaches a potential at which its electrochemical stability disappears or, in other words, the reduction of the platinum phase occurs and the insertion cannot continue. Therefore, the cathodic current observed at very negative potential would be partially due to the electrochemical degradation of the electrolyte. The following electrochemical reduction of the salt as depicted in Scheme 2.1 was proposed. The global mechanism probably could be complicated by Hofmann degradation (tetraalkylammonium cation, which plays the role of proton donor) owing to the formation of the strong base R$^-$:

$$[Pt_2^-, NR_4^+, NR_4X] \longrightarrow 2\,Pt + NR_4X + NR_3 + R^\bullet$$

$$R^\bullet \xrightarrow{\;e^-\;} R^- \xrightarrow[\text{source}]{\text{Proton}} RH$$

SCHEME 2.1

2.3.3 EVIDENCE OF SMOOTH PLATINUM SURFACE
CHANGES AFTER SALT INSERTION BY SEM

It was previously suggested that swelling could result from the insertion of cations, especially when bulky cations like Cs^+ or Bu_4N^+ are concerned. From formulas given in Tables 2.2 and 2.4, one may realize that the insertion of cations, and concomitantly of salts, leads to huge volume increases. However, samples examined by means of the SEM technique after an insertion procedure were exposed to air and were unavoidably oxidized. When the amount of electrical charge was large, huge surface changes happened that could be evidenced without any magnification of the platinum sample. Figure 2.11 demonstrates the final results of such impressive effects. The reduction of a smooth platinum plate in the presence of CsI in DMF (with amounts of charge >10 C cm^{-2}) forms a reacting layer. Then, this layer is irreversibly oxidized by contact with air to give back pure platinum. The electrochemical reaction and the kind of "corrosion" brought by the insertion are clearly shown by such experiments and are visible in Figure 2.11. In the presence of CsI, when lower amounts of charge were injected, the modified platinum surface becomes visible with unexpected patterns in which the surface appears as pierced by numerous holes and channels (Figure 2.12). It has also been confirmed that this transformation of the platinum sample brought an appreciable enlargement of the active Pt surface (checked by covering of the interface by a PtO monolayer and its coulometric determination by reduction) [47].

FIGURE 2.11 Evidence by SEM image of a huge change of smooth platinum (formation of a layer at the surface) and reoxidation by air. Superdry conditions. DMF + CsI 0.1 mol.L^{-1}. Amount of charge: 25 C cm^{-2}. At the end, there is nothing other than platinum (Cougnon, C., Ph.D. thesis, Université de Rennes 1, 2002).

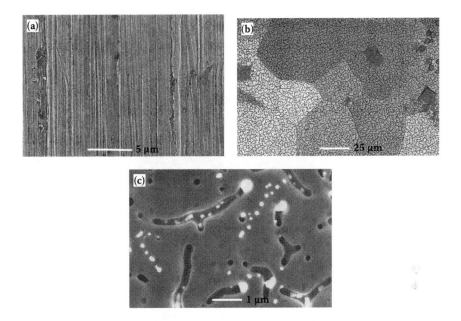

FIGURE 2.12 Behavior of platinum under reduction in a superdry DMF solution added with 0.1 mol.L^{-1} CsI. (a) Smooth platinum prior to reduction. (b) and (c): After reduction at -2 V versus SCE. Amount of charge: 10 C cm^{-2} (Cougnon, C., Ph.D. thesis, Université de Rennes 1, 2002).

Quite similarly, the same experiments performed with TAAX lead to large structural changes. As a matter of fact, structural changes are smaller than those observed with alkaline iodides, probably because of the bulkiness of the salts and their moderate level of dissociation in DMF, but also, as discussed earlier, because these phases are electroactive. Therefore, the applied potential value is of great importance. As shown on Figure 2.13, in the case of the reaction of platinum with tetra-n-butylammonium iodide (TBAI), the insertion seems focused on grain boundaries, which is certainly favored. One can also see at the grain borders the formation of cavities and channels. This method was used to separate, in a novel manner, the grains of polycrystalline platinum. However, experiments involving Pt monocrystals in order to appreciate structural changes under insertion have not been achieved so far.

The structural changes of platinized layers by TAAX (insertion shown to be more efficient owing their larger active substrate surface) are discussed in Section 2.6.

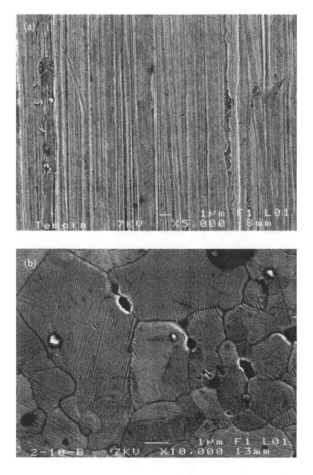

FIGURE 2.13 Surface modifications of smooth platinum under electrochemical reduction under superdry conditions. Solution: DMF + 0.1 mol.L^{-1} TBAI. (a): Surface prior to reduction. (b) At the same magnification, after reduction at -2.5 V versus SCE. Amount of charge: 10 C cm^{-2}. Note the profound changes at the level of grain boundaries with the appearance of channels (Cougnon, C., Ph.D. thesis, Université de Rennes 1, 2002).

2.4   IN SITU EC-AFM FOLLOWING OF THE PLATINUM CHARGE

2.4.1   CHARGE AND DISCHARGE OBSERVED IN REAL TIME

As discussed earlier, several effects due to cation changes have been evidenced by cyclic voltammetry and through fixed-potential coulometric investigations. The potential values required to produce at least moderately saturated forms [Pt$_n^-$, M$^+$, MX], where n < 2, are quite negative (see Tables 2.1 and 2.2). Consequently, the produced Pt phase should exhibit strong reductive properties. It was previously

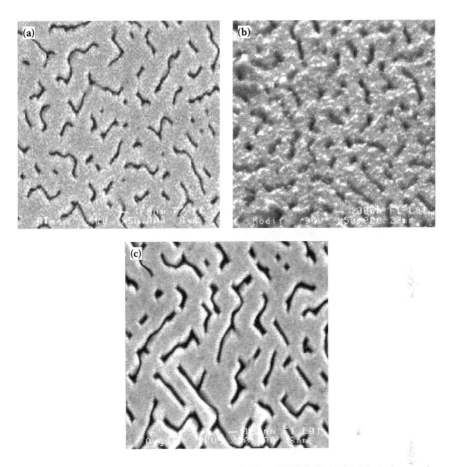

FIGURE 2.14 SEM images of platinum deposited on (100)MgO. (a) Initial platinum substrate. (b) After electrochemical modification in 0.1 mol.L^{-1} NaI in DMF, reduction with a charge density of 24 mC.cm^{-2} and then contact with air. (c) After reduction followed by an electrochemical reoxidation. (Ghilane, J., M. Guilloux-Viry, C. Lagrost, P. Hapiot, and J. Simonet, 2005, *J. Phys. Chem.* B 109: 14925. Used with permission.)

described that, after platinum charging, the initial pure platinum interface is recovered after long-term exposure under air (a case displayed in Figure 2.11), suggesting that the phase [Pt$_n^-$, M$^+$, MX] could also be electrochemically reoxidized.

To test this idea, two platinum substrate electrodes were prepared under identical conditions, and both were cathodically modified in the same DMF solution containing 0.1 mol.L^{-1} NaI. Then, one sample was immediately electrochemically oxidized without removing the sample from the solution; that is, a positive potential was applied during 10 s (more positive than the peak potential corresponding to the anodic process), typically larger than −0.5 V/ SCE. The second sample was left only in contact with air for 1 h. Figure 2.14 compares the SEM images of the two produced platinum surfaces obtained after thorough rinsing with water and acetone. (Images b and c) with that of the initial surface exempt from any cathodic treatment (Figure 2.14a). Originally, the surface free of treatment displays large flat areas of platinum with deep holes representing the surface porosity specific to the substrate Pt–MgO. On the contrary, Figure 2.14b, relative to the sample oxidized by air, shows the appearance of small grains in the platinum platelets after the cathodic charge. Lastly, the surface cathodically modified and electrochemically discharged (displayed in Figure 2.14c) exhibits a structure quasi-identical to the initial one. These experiments permit checking of the reversibility of the charge–discharge process with the recovery of the initial morphology of the sample (compare images a and c).

In order to monitor the "real" reversibility of the cathodic modification of platinum at the micrometer scale and follow the structural changes of the Pt surface in real time, the charge–discharge process was followed in situ using the EC-AFM imaging technique [42,44]. In these experiments, the electrochemical transformations of platinum surfaces and AFM imaging were simultaneously completed. The great novelty of such experiments is to be able to follow the progressive change of the interface morphology as the amount of the injected charge changes in real time. Such experiments were repeated for different supporting electrolytes. Decisive observations allow the authors to confirm the data already obtained by SEM with the in-situ EC-AFM. In particular, the latter technique permits one to follow the progressive swelling of platinum during the charge and other chemical treatments, as well as the structural reversibility of the overall process, at least for moderate injections of charge. Initial experiments have been achieved with alkaline iodides (solution: DMF + 0.1 mol.L^{-1} of the salt). The stability of the Pt–MgO sample in the contact of the liquid electrolyte was verified by AFM in the course of an immersion of 1 h. Experiments proceeded by successive charge injections performed by repetitive scans. After each scan, the sample was disconnected but maintained in the DMF solution under argon. The stability of surface changes during different observation times was checked, and no significant modifications were noticed for contact times smaller than 10 min. This confirms that the sample is stable under these conditions.

Typical AFM topography images are gathered in Figure 2.15. They stress the structural changes of Pt samples before and during scans in potential. Thus, prior to charge, bright and dark areas (displayed in Figure 2.15a) correspond to the platinum plates obtained during the epitaxial growth of the Pt deposits under well-controlled deposition conditions. In the course of charge processes, if the cathodic scan is limited at a boundary potential E_r more positive than -0.8 V, the AFM image did not exhibit any significant change. In Figure 2.3b, the boundary potential E_r corresponds to the threshold of the voltammetric step ($E_r = -1.4$ V versus SCE). Under those conditions, small grains start growing on the Pt plates, but the global structure of the platinum roughly remains unchanged. Then, when the potential boundary is shifted to a more negative potential ($E_r = -1.7$ V/SCE), a much faster insertion may occur (e.g., in the given example, injection of a charge density of around 70×10^{-3} C.cm^{-2}), and consequently,

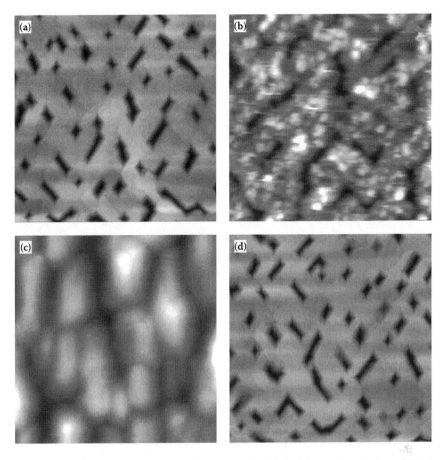

FIGURE 2.15 In-situ EC-AFM images of the platinum modification in NaI (0.1 mol.L^{-1}) in DMF. (a) In solution before any charge injection. (b) After reduction with a charge density 3.2 mC.cm^{-2}. (c) After reduction with a charge density of 70 mC.cm^{-2}. (d) After electrochemical reoxidation. Image scans size 3 × 3 μm^2. (Ghilane, J., M. Guilloux-Viry, C. Lagrost, P. Hapiot, and J. Simonet, 2005, *J. Phys. Chem.* B 109: 14925. Used with permission.)

huge changes appear in the metal surface (see Figure 2.15c). Large spheroids with diameters around 150 nm replace the initial structure. Progressively, the surface roughness increases until values of 70–100 nm are obtained. Such very large morphology changes obtained by surface swelling, clearly depending on the injected charge, underline the insertion of species into platinum through the formation of ionometallic phases [Pt_n^-, Na$^+$, NaI] as previously discussed in Section 2.3.1.

Even if crystallographic data are not available for the totally reduced [Pt_2^-, Na$^+$, NaI], its crystallographic parameters should be different from those of (100) Pt, with the result that the electrogenerated phase is not compatible with the MgO substrate. This causes irreversible layer desegregation and dispersion in solution when large amounts of charge are used. If the injected reductive charge is made sufficiently low to avoid layer desegregation, a (0.0 V) potential applied to the sample permits it to oxidize the weakly (n << 2) charged product [Pt_n^-, Na^+, NaI]. Figure 2.15d exhibits the surface morphology after this anodic treatment. The most remarkable feature is that the morphology of the sample is almost identical to the initial pattern of the virgin sample and indicates a real reversibility of the chemical process.

2.4.2 OTHER ELECTROLYTES

Similar EC-AFM experiments were performed with CsI and KI. With these electrolytes, a remarkable reversibility of the process was also observed. For all alkaline iodide salts, it was possible to recover the initial structure of the Pt sample modification after electrochemical reoxidation, or after exposure of the modified sample for a few hours under air. In order to check the magnitude of the reversibility, an additional experiment was finally performed with NaI, in which the maximum charge (120 mC.cm^{-2}) was injected, and it has been verified that no sample desegregation occurs. This injected charge value in fact globally corresponds to the limit of insertion (saturation obtained with n = 2 as defined in Section 2.3). Experiments to record AFM images in the course of the charge–discharge processes, which focused on the same microarea of the surface, were totally successful, and therefore verified the good material cohesion, even under a relatively large charge (see Figure 2.16).

As another way to confirm the reversibility of the platinum modification, x-ray diffraction analyses were additionally performed on Pt films deposited on (100) MgO (Figure 2.17). After total charge–discharge cycles, all samples exhibited similar diffraction patterns, that is, no variation was observed either for the 2θ diffraction angles on the θ-2θ diffraction patterns, or for the width of the ω-scan and φ-scan. These results confirmed that the Pt lattice parameter was not changed,

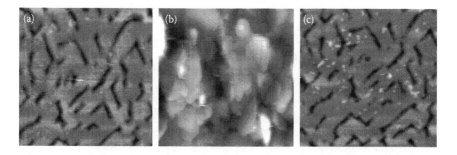

FIGURE 2.16 In-situ EC-AFM images of the platinum structural changes in NaI (0.1 mol.L^{-1}) in DMF. (a) In solution before any cathodic treatment. (b) After reduction with charge density of 120 mC.cm^{-2}. (c) After sample electrochemical reoxidation in situ. Image size 3 × 3 μm^2. (Ghilane, J., M. Guilloux-Viry, C. Lagrost, P. Hapiot, and J. Simonet, 2005, *J. Phys. Chem.* B 109: 14925. Used with permission.)

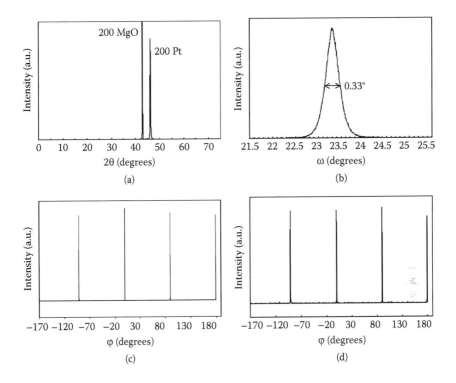

FIGURE 2.17 X-ray analyses of platinum sample after a full electrochemical cycle of reduction and reoxidation in DMF containing NaI. (a) θ-2θ x-ray diffraction pattern evidencing the (100) orientation of the Pt film, after the cycle of reduction–oxidation. (b) X-ray diffraction ω-scan performed on the 200 reflection of Pt after one cycle of reduction–oxidation (as a standard, the FWHM of the ω-scan performed on the single crystal substrate was $0.027°$). (c,d) X-ray diffraction φ scans recorded for the 220 reflections of MgO (c) and Pt after cycle of reduction–oxidation (d). (Ghilane, J., M. Guilloux-Viry, C. Lagrost, P. Hapiot, and J. Simonet, 2005, *J. Phys. Chem.* B 109: 14925. Used with permission.)

and that the crystalline quality of the epitaxial film was not globally affected by the insertion of ions and the electrochemical oxidation of the phase. In particular, there was no misorientation along the growth direction detected, and the width of the ω-scan peak was found to be the same.

The charging processes in the presence of tetraalkylammonium salts were similarly examined in the EC-AFM experiments by dissolving quaternary ammonium salts in superdry DMF. Surface modifications were carried out with the same substrate used earlier. It is known that reduction potentials and insertion conditions (as discussed previously) depend on the nature and size of the tetraalkylammonium cation. Quite specifically, it is interesting to see how the cation size can influence both the structural changes upon its insertion and the structural reversibility of the whole process. For that to be so, reduction potential values

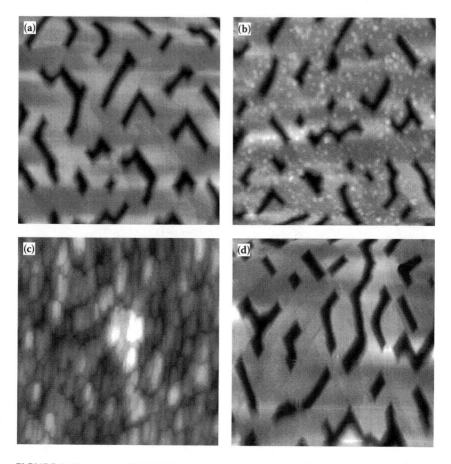

FIGURE 2.18 In-situ EC-AFM images of the platinum modification in Me_4NI (0.1 mol. L^{-1}) in superdry DMF. (a) In solution before any modification. (b) After reduction with a charge density 0.43 C.cm^{-2}. (c) After reduction with a charge density of 3.86 C.cm^{-2}. (d) Restored structure after electrochemical reoxidation. Image scans size 3×3 μm^2. (Ghilane, J., M. Guilloux-Viry, C. Lagrost, P. Hapiot, and J. Simonet, 2005, *J. Phys. Chem. B* 109: 14925. Used with permission.)

must account for the bulkiness while the cation size increases ($E_{pMe4NI} > E_{pBu4NI} > E_{pHex4NI}$). Figure 2.18 shows EC-AFM images of a platinum electrode under electrochemical treatment in solutions containing tetramethylammonium salts. First, the modification of the platinum electrode at the threshold of cathodic polarization yields small grains on the platinum terraces (Figures 2.18b). When larger charge injections are completed, the holes and flat platelets specific to the initial pattern are totally replaced by spheroids. The average roughness then increases to reach values around 120–150 nm (Figure 2.18c). After charging, the material is

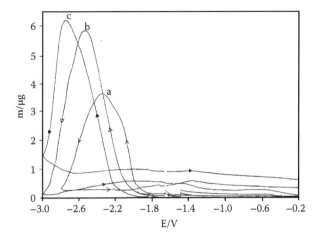

FIGURE 2.19 EQCM measurements: mass changes of a gold filmed quartz covered with a thin platinum layer as a function of the applied potential for several tetraalkylammonium salts. (a) tetra butylammonium, (b) tetrahexylammonium, (c) tetraoctylammonium. Salts concentration, 0.1 mol.L^{-1}, solvent DMF. Scan rate 10 mV.s^{-1}. Potential are referred to the SCE electrode. (Ghilane, J., M. Guilloux-Viry, C. Lagrost, P. Hapiot, and J. Simonet, 2005, *J. Phys. Chem.* B 109: 14925. Used with permission.)

oxidized at a potential around 0 V for a few seconds, and its structure is reexamined by AFM. Thus, as displayed in Figure 2.18d, the cathodic charging process, at least with Me$_4$N (which is known as the most stable TAA$^+$ under electron transfer and Hofmann degradation), is also reversible.

Similar additional experiments were done with a more bulky TAA$^+$ such as tetra-*n*-hexylammonium iodide. It must be stressed that the nature of the cathodic reduction when using TAAX salts certainly depends on the applied potential. Thus, as displayed in Figure 2.19, the insertion process rapidly vanishes beyond –2.3 V to –2.7 V. As shown by EQCM experiments, platinum phases beyond very negative potentials collapse, and no modification of the platinum occurs at a very negative potential. This phenomenon was observed with the tetraalkylammonium salts, and did not appear with the alkaline salts. The cathodic dissolution of the platinum phases at more reducing potential seems to be specifically bound to the cleavage of NR$_4$$^+$. Actually, tetraalkylammonium cations were reported to yield scissions at very negative potentials, according to the following mechanism:

$$NR_4^+ + e^- \rightarrow R_3N + R^{\cdot}$$

$$R^{\cdot} + e^- \rightarrow R^-$$

$$R^- + H^+ \rightarrow RH$$

The loss of mass is apparently due to the cathodic reduction of the ionometallic phase following a similar reaction scheme inside the modified platinum:

$$[Pt_n^-, NR_4^+, NR_4X] + e^- \rightarrow n\, Pt + R^- + R_3N + R_4NX$$

As a consequence, it can be deduced that the chemical structure of the tetraalkylammonium cation is not modified in the phase $[Pt_n^-, NR_4^+, NR_4X]$ and that the negative charge is principally localized on the platinum atoms (valence bond). An opposite situation would be the localization charge excess on the cation (which happens at very negative potentials under an additional electron transfer) and yields the degradation of the ammonium with the formation of the corresponding amine.

Recently, electrochemical reduction of a redox ionic liquid (1-ferrocenylethyl-3-methylimidazolium bis(trifluoromethylsulfonyl)imide [FcEMIM][TFSI] on platinum has shown the formation of similar Pt phases where the cation contains an electroactive ferrocene $[Pt_n^-, FcEMIM^+]$ group [48]. Moreover, the reduced electrode exhibits the presence of the ferrocene even after contact with air or different cleaning treatments as ultrasonics or physical polishing. This work shows that a similar reduced Pt phase could also be prepared in media such as ionic liquids, where the cation of the ionic liquids is inserted in the phase as in the case of an electrolyte in a classical organic solvent.

2.5   PLATINUM PHASES AS REDUCING REAGENTS

2.5.1   EVIDENCE FOR REDUCTION REACTIONS INDUCED BY THE PLATINUM PHASES

The influence of dioxygen (presence of air) on the stability of the platinum phases (formed with alkaline and TAAX salts) has already been discussed. It is expected that the platinum phase reacts with dioxygen as shown in Scheme 2.2:

$$[Pt_2M, MX] + O_2 \longrightarrow O_2^{\bullet-} + M^+ + 2\,Pt + MX$$

$$[Pt_2M, MX] + O_2^{\bullet-} \longrightarrow O_2^{2-} + M^+ + 2\,Pt + MX$$

SCHEME 2.2

There are few indications about the redox potential of the phases either with alkaline salts or with the TAAX series, even if, in the precise case of dioxygen, there is proof that superoxide is specifically produced. Additionally, preliminary experiments were conducted with a large palette of organic π-acceptors. The most impressive experiment was completed with classical π-acceptors (see Figure 2.20 for the experiment performed with a nitroaromatic compound) that are known to

FIGURE 2.20 Reduction of 2,4-dinitro-toluene by platinum prior charged in NaI/DMF at a potential of −1.8 V versus SCE. Maximum charge 52 C. Dinitrotoluene (concentration 10 mmol.L^{-1}) was dissolved in dry DMF maintained under insert atmosphere (Simonet, J., Unpublished results).

produce stable, deeply colored anion radicals. In the example given, a deep blue cloud comes out from the electrode surface (Scheme 2.3).

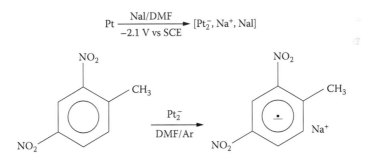

SCHEME 2.3

Additionally, some experiments permit one to reach conclusions about the nature of the color produced at the modified platinum interface (Scheme 2.4). Paramagnetic signals were obtained [40] by dipping a charged platinum wire into a vessel located inside the electron spin resonance (ESR) spectrometer. The ESR spectra produced were identified as those of the anion radicals (Figure 2.21) and were checked to ensure that they were strictly identical to responses obtained from π-acceptors under electrolyses conducted with platinum wires within (*intra muros*) the electron paramagnetic resonance (EPR) apparatus.

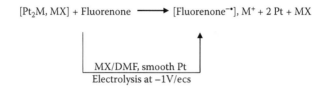

SCHEME 2.4

In this connection, organic redox couples with a standard potential in the range of 0 V to about −1.5 V versus SCE are good candidates to chemically react (via electron transfer) with platinum phases. It is quite obvious (but still not demonstrated) that the capability of [Pt₂M, MX] to react under those conditions depends on its standard potential, $E°_{MX}$, which is strongly influenced by the nature of the salts. This reducing capacity could be appreciated by means of chronocoulometry as depicted in Figure 2.22, which shows the charge decay, Qa, of the material with time after addition of an organic π-acceptor. More precisely, the rate of reduction can be subtracted from the data given in Table 2.5. After a short time, all π-acceptors for those with an E° larger than −0.5 V are readily and quantitatively reduced. Indeed, chloroaniline reacts very fast, but benzophenone appears to give

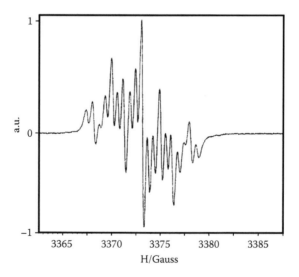

FIGURE 2.21   ESR response of a solution of 5×10^{-3} mol.L⁻¹ 9-fluorenone in DMF (under inert atmosphere) after dipping of a charged Pt wire. The charge of Pt was first achieved at −1.8 V in a solution of 0.1 mol.L⁻¹ in DMF ($Q_c = 135$ C cm⁻²). (Cougnon, C., and J. Simonet, 2002, *J. Electroanal. Chem.* 531: 179. Used with permission.)

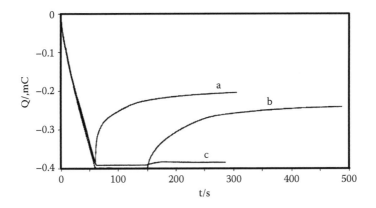

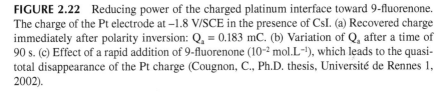

FIGURE 2.22 Reducing power of the charged platinum interface toward 9-fluorenone. The charge of the Pt electrode at −1.8 V/SCE in the presence of CsI. (a) Recovered charge immediately after polarity inversion: Q_a = 0.183 mC. (b) Variation of Q_a after a time of 90 s. (c) Effect of a rapid addition of 9-fluorenone (10^{-2} mol.L^{-1}), which leads to the quasi-total disappearance of the Pt charge (Cougnon, C., Ph.D. thesis, Université de Rennes 1, 2002).

much slower electron transfer. However, the reducing power of platinum phases is difficult to estimate since the electron exchange will become reversible for the acceptors with the most negative potentials. The question of the behavior of anion radicals toward platinum will be discussed in Section 2.7.

TABLE 2.5
Reduction of Some Organic Acceptors (Concentrations: 5 mmol.L^{-1}) by the Pt Phase[a]

π-Acceptor	Standard Potential (V versus SCE)	% of the Phase [Pt$_2$Cs(CsI)] Reacted
Chloranil	−0.27	100
Phenanthrenequinone	−0.58	95
p-Dinitrobenzene	−0.61	100
3,5-Dinitrobenzyle chloride	−0.80	100
Acenaphquinone	−0.83	98
2,4-Dinitrotoluene	−0.91	96
Nitrobenzene	−1.10	92
9-Fluorenone	−1.28	67
Benzophenone	−1.75	6

[a] A platinum sheet (area: 3.5 cm^2) is used after a charge of 1 C, which corresponds to an excess of reactant to the acceptor. The right column displays the percentage of the organic substrate reduced through electron transfer after 90 s. Ratios were calculated by means of voltammetry at a micro Pt electrode.

As other test systems, molecules containing carbon–heteroatom bonds are good candidates because such bonds are irreversibly cleaved after an electron transfer and thus could irreversibly react with the platinum phases. In fact, it was found that alkyl and aryl iodides are reduced by $[Pt_2M, MX]$. Aryl-sulfones react similarly. In Table 2.5, it is clear that 2,5-dinitrobenzyl chloride is readily and quantitatively cleaved under these conditions. Many compounds possessing leaving groups (LG) are capable of being cleaved according to Scheme 2.5 (the case of a two-electron reduction owing to the reducibility of the transient radical $R^{•}$).

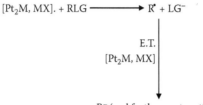

SCHEME 2.5

2.5.2   The Case of the Reductive Cleavage of Aryldiazonium Cations: Platinum Surface Modification

In this part, the reducing properties of $[Pt_nM, MX]$ phases for achieving grafting by diazonium salts were investigated [49] in the absence of an externally applied potential. Two sets of experiments were performed, the first with a bulk platinum electrode and a second with the microstructured Pt layer deposited on (100)MgO. Bulk platinum electrodes allow the injection of large and different quantities of charges. The second type of substrate was the same as described in previous in situ investigations. In the present reported experiments, NaI was chosen as the supporting electrolyte because the potential required for phase formation is not too negative and is easily handled. Previous electrochemical investigations and quartz microbalance have shown that the formula of the saturated form of the phase corresponds to $[Pt_2^-, Na^+, NaI]$. Even if the peak potentials are not thermodynamic data, the value of Ep_{NaI} indicates that $[Pt_2^-, Na^+, NaI]$ should react with molecules such as aryldiazonium salts that display reduction peak around 0.0 V/SCE.

The procedure used consists first in the electrogeneration of $[Pt_2^-, Na^+, NaI]$ at -1.7 V/SCE during 100 s, and immediately after, plunging the charged platinum sample into a deoxygenated DMF solution containing 0.1 mol.L^{-1} 4-nitrophenyl diazonium tetrafluoroborate (O_2N-Ph-N_2^+, BF_4^-). The system was allowed to react for 10 min under argon to allow sufficient time for the reaction to be totally completed. Complementary experiments have shown, however, that such reduction is almost instantaneous.

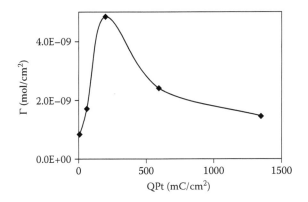

FIGURE 2.23 Surface concentration as a function of the charge injection used to modify the 1-mm-diameter Pt disk electrode. (Ghilane, J., M. Delamar, M. Guilloux-Viry, C. Lagrost, C. Mangeney, and P. Hapiot, 2005, *Langmuir* 21: 6422. Used with permission.)

The modified electrode, after treatment with the diazonium salt, could be examined by cyclic voltammetry as the NO_2^- group present on the diazonium salt serve as an internal redox probe. A linear dependence of the current versus the scan rate is observed in the 0.05–1 $V.s^{-1}$ range. It indicates an electrode process in which the electroactive species do not diffuse to and from the electrode. This reveals that nitroaryl moieties resulting from the cleavage of 4-nitrophenyl diazonium were immobilized onto the platinum electrode surface. The electrochemical signal of the nitrophenyl groups was shown to remain after an ultrasonic cleaning, confirming the attachment stability of the nitrophenyl groups at the platinum surface. Assuming that the reduction of the attached NO_2-aryl group stays monoelectronic under those experimental conditions, the integration of the cathodic current provides an estimation of the aryl group surface concentration on the platinum surface (Γ in $mol.cm^{-2}$). This data treatment assumes that the considered active surface does not appreciably vary in the course of different grafting experiments. The surface coverage (derived as the integration of the peak current of the grafted NO_2-phenyl) was plotted as a function of the quantity of the injected charge (Figure 2.23) from [Pt_2^-, Na^+, NaI] used as an interfacial reducing species (Scheme 2.6).

Even if absolute values obtained by this method are relatively inaccurate (estimation of the real active surface, assumption that all the nitroaryl groups are grafted *directly* onto the Pt surface), the dependency remains valid to follow the evolution of the coverage. Thus, the NO_2-aryl surface concentration first increases upon the amount of injected charge up to about 0.2 $C.cm^{-2}$ and then decreases slowly beyond this maximum value.

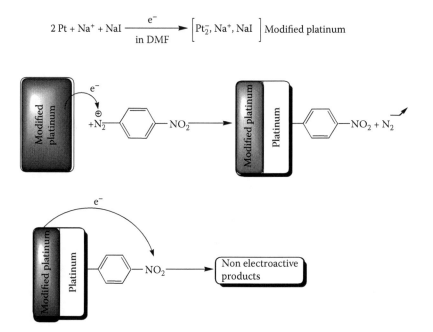

SCHEME 2.6 Schematic mechanism involved in the activation of the grafting of phenyl diazonium salts from cathodically modified platinum surface showing the competition between the grafting and the NO_2 reduction of the already immobilized layers.

The maximum coverage of active NO_2-aryl groups is 4–5×10^{-9} mol.cm^{-2}. These Γ values appear too large when compared with the expected value for a compact monolayer (Γ should be in the order of some 10^{-10} mol.cm^{-2}), indicating that phenyl multilayers are probably formed onto the platinum surface. To explain the maximum, one must take into account the possibility of the concomitant reduction of the NO_2-aryl groups by [Pt_2^-, Na^+, NaI]. The reduction of the NO_2 will lead first to a radical anion that will evolve on a longer time scale to the irreversible formation of nonelectroactive products (such as hydroxylamine and amine). The observation of a maximum in Figure 2.24 suggests that the competition between the two processes varies during the grafting. At the beginning of the process, there is no nitrophenyl group on the Pt surface and by comparison with nitroaryl groups; the reduction of the diazonium moieties on a clean surface is favored by more than 0.8 eV (based on the respective reduction potentials). Thus, while the radical addition to the surface is progressively completed, the reduction of the diazonium salt becomes more difficult as the electrons have to overpass the organic layer to reach the electroactive species in solution. This phenomenon that corresponds to a blocking effect of the electrode surface is well known with aryldiazonium

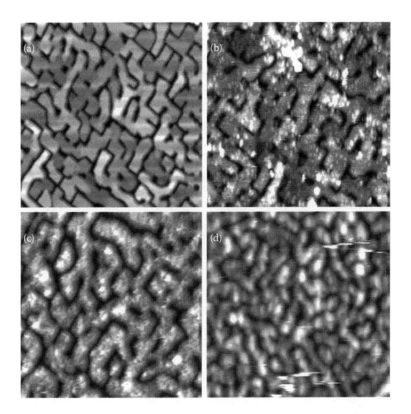

FIGURE 2.24 Reaction of $[Pt_2^-, Na^+, NaI]$ with 4-nitrophenyl diazonium salt. EC-AFM images of a (100)Pt sample. Pt deposited onto (100)MgO in DMF. (a) Initial sample. (b) After cathodic modification (passage of 15 $\mu C/cm^2$). (c) After addition of 0.01 mol.L^{-1} 4-nitrophenyldiazonium tetrafluoroborate. (d) Reexamination of the sample in a blank solution and after several reduction–oxidation cycles. Image scans size 3×3 μm^2. (a, b, c) in a deoxygenated DMF solution containing 0.1 mol.L^{-1} NaI. (Ghilane, J., M. Guilloux-Viry, C. Lagrost, P. Hapiot, and J. Simonet, 2005, *J. Phys. Chem.* B 109: 14925. Used with permission.)

salts. It is commonly observed that the formation of the layer by direct electrochemical reduction of phenyl diazonium is accompanied by a rapid decrease of the electrochemical signal.

A second series of experiments was performed on the microstructured platinum deposited on (100)MgO substrate in order to follow the cathodic charge of platinum and its specific reaction with the aryldiazonium by EC-AFM. For this, the platinum electrode is first immersed in a solution of DMF containing only the supporting electrolyte NaI and then reduced under the conditions already reported in Section 2.4. At the start of the cathodic polarization, small grains appeared on the platinum terraces. A small amount of charge was injected to create only small modifications that remain compatible with a good AFM imaging. As discussed already, such modification indicates the formation of an interfacial ionometallic phase $[Pt_n^-, Na^+, NaI]$ where n would depend on the level of the charge.

After formation of $[Pt_n^-, Na^+, NaI]$, considered now as a purely reducing species (the electrode is disconnected), 4-nitrophenyl diazonium bromide is thoroughly mixed with DMF (the solution remains under strictly inert atmosphere), and the surface changes are examined by AFM. As seen in Figure 2.24, spheres corresponding to the swelling of platinum during the cathodic charging become smaller, and the roughness of the platinum terraces decreases from 10 to 5 nm. It is noticeable that this topography and the one already described after the direct reduction of 4-nitrophenyl diazonium salt on the platinum electrode (discussed earlier) are very similar. Additionally, there is no modification of this latter surface when a positive potential is applied. Thus, there are obviously two kinds of layer modifications before and after exposure of $[Pt_n^-, Na^+, NaI]$ to the 4-nitrophenyl diazonium salt solution. Afterward, the sample was rinsed and analyzed electrochemically in a solution containing only the supporting electrolyte.

Cyclic voltammetry of the sample clearly reveals characteristic reduction peaks of immobilized nitroaryl groups. Consequently, all these experiments indicate that nitrophenyl groups have been attached on the Pt surface after chemical reduction of the 4-nitrophenyl diazonium by charged platinum. Additionally, the use of $[Pt_2^-, Na^+, NaI]$ as reducing species toward diazonium salts was checked thanks to XPS analyses. Under similar conditions, the reduction of 4-bromophenyldiazonium bromide reveals, in XPS spectroscopy, signals corresponding to Br $3p_{3/2}$ and $3p_{1/2}$ at 183.6 and 190.8 eV, respectively. Furthermore, the Br 3p signal intensity displays a maximum when plotted against the charge injected. This evolution suggests a competition between the reduction of 4-bromophenyl diazonium and the reduction of the bromophenyl substituent, leading partly to the scission of C-Br bonds.

2.6 CHARGE–DISCHARGE PROCESSES OF PLATINIZED PLATINUM LAYERS

It has been shown that the charge is significantly increased when thin platinum galvanostatic deposits are achieved onto platinum plates [50]. Two main results are presented in this chapter: the reversible charge of the layers in the presence of tetramethylammonium salts and the evidence of large swellings with a wide palette of TAAX salts [51].

2.6.1 CHARGE–DISCHARGE IN THE PRESENCE OF TETRAMETHYLAMMONIUM TETRAFLUOROBORATE

If the platinum electrode is galvanostatically platinized so as to possess a much larger active surface than smooth platinum substrate, the morphology of the voltammetric processes is dramatically changed. Thus, for a relatively thick film of electrodeposited platinum (with, for example, an average thickness equal to

0.1 μm), the use of a superdry solvent–electrolyte allows one to obtain a quite pure quasi-reversible step. For example, platinized platinum in 0.1 mol.L^{-1} TMAClO$_4$ exhibits a pair of broad peaks whose half-peak potentials are $E_{0.5pc} = -1.90$ V and $E_{0.5pa} = -1.74$ V, respectively, versus SCE. At scan rates greater than 50 mV s^{-1}, the areas of these two peaks are roughly the same. Coulometry, for the charge and discharge, shows that the charge amounts are equivalent. At potentials more negative than -2.2 V versus SCE, the cathodic rise in current does not appear; reaching these very negative potentials revealed a process that strongly resembles a self-inhibition phenomenon. It was proposed that this pure charging–discharging process is specific to the platinized layer. Possibly, the compactness associated with a structural change in the platinized layer creates a further control on the movement of species taking part in this charging phenomenon. For times longer than the duration of voltammetric experiments, the charging process depends on the thickness of plating. With thicknesses of the order of 0.1 to 2 μm, the system becomes quasi-reversible, with current efficiencies smaller than 60%. Figure 2.25 displays the two branches of chronocoulometric responses of platinized platinum in 0.1 mol.L^{-1} TMABF$_4$ in DMF; the very sharp slope at the beginning of the discharge process is worth noting.

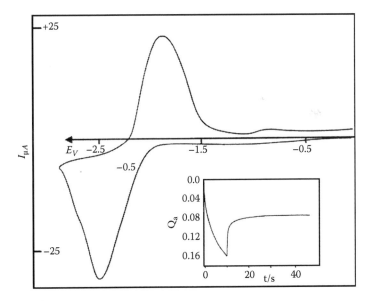

FIGURE 2.25 Voltammetric responses of a platinum electrode in superdry DMF containing 0.1 mol.L^{-1} TMAClO$_4$. Scan rate: 0.1 Vs^{-1}. Apparent electrode area: 0.8 10^{-2} cm^2. Potentials are referred to the SCE electrode. Platinized electrode (area of platinum substrate: 0.8 10^{-2} cm^2) with a plating of average thickness of $\delta = 0.1$ μm. In the insert, chronoamperometric curve with a reduction at -2.6 V (10 s) followed by an oxidation at -0.5 V is shown. (Simonet, J., 2006, *J. Electroanal. Chem.* 93: 3. Used with permission.)

Coulometric and ECQM measurements have revealed that tetramethylammonium salts behave quite differently from other TAAX salts. In fact, the obtained phases appeared to be the following when the associated anions are ClO_4^-, or $BF4^-$: [Pt_4^-, TMA^+, TMAX]. Under these conditions, the amount of charge inside Pt would be significantly smaller than with the other TAAX.

2.6.2 SWELLING OF PLATINIZED LAYERS UPON ELECTRIC CHARGE

It was previously stressed that the insertion of ions in bulk platinum leads to impressive changes of the surface morphology. Platinized layers deposited onto a conducting material are of interest owing to a better permeability and a larger specific surface than those of typical smooth platinum interfaces. Consequently, it is relevant here to show the swelling of such platinum layers and briefly discuss the morphology changes according the nature of the salts and the conditions of the charge–discharge reaction. As a typical example, Figure 2.26 features the changes of a platinum layer structure upon charging. The cathodic charge of the layer, in particular by bulky ammonium cations, leads after oxidation (here by air) to the formation of spherical nanostructures. In the case where the layer is thin enough, the produced spheres are not always adjacent. Therefore, such changes show that there is a profound modification via the reduced phase (sites of swelling) and also discharge of these spheres. This is schematically shown in Figure 2.27.

Why a quasi-uniform size of spheres? For that, it may be proposed that these structures grow under thermodynamic conditions. This assumption is supported by the quasi-reversibility of the electrochemical process as depicted in Figure 2.25 with TMAX salts. The total conformational energy per sphere, E_{sp}, may be split into two terms corresponding to the bulk (ρ_v) and surface (ρ_s) energies, respectively, as given in the following equation:

$$E_{sp} = (4/3) \pi R^3 \rho_v - 4\pi R^2 \rho_s$$

where R is the radius of each sphere. At equilibrium, the equation can be written as

$$(dE/dR)_{eq} = 4\pi R^2 \rho_v - 8\pi R \rho_s$$

Therefore, the equilibrium radius, R_{eq}, of spheres is given by

$$R_{eq} = 2\rho_s/\rho_v$$

Thus, the radii of the spheres depend only on energy factors that are specific to the experimental conditions (such as the nature of the salt used, its concentration, its level of dissociation, and also the amount of platinum available in the whole deposited layer). The intriguing formation of contiguous spheres of very similar volume suggests that the ionometallic layer (at the quasi-maximum of the charge)

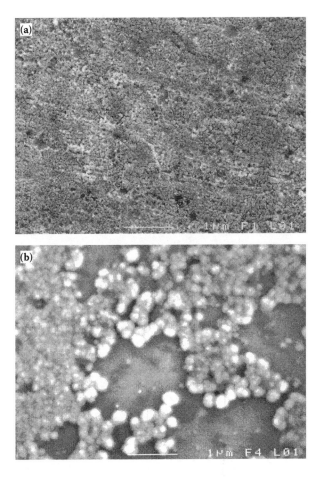

FIGURE 2.26 Morphology change in the course of a charge–discharge process. (a) Galvanostatically deposited platinum layer onto platinum substrate. (b) The same layer after a charge process in 0.1 M tetra-*n*-hexyl ammonium bromide in DMF until saturation and oxidation by air (Simonet, J., Unpublished results).

turns out to be very mobile on the conducting substrate. Thus, a model taking into account the growth of swollen structures from randomly distributed activated centers onto the substrate surface is certainly wrong. Experiments allow visualizing the occurrence of the platinum swelling. The degree of swelling in the case of TMAX could be estimated from the phase formula [Pt_4^-, TMA^+, TMAX]. The example of TMAI could be considered as representative. From the values of ionic radii for the phase constituents (Pt = 0.80 Å, I^- = 2.2 Å, and TMA^+ = 3.01 Å), it is possible to assess the degree of expansion of the platinum layer. A rapid

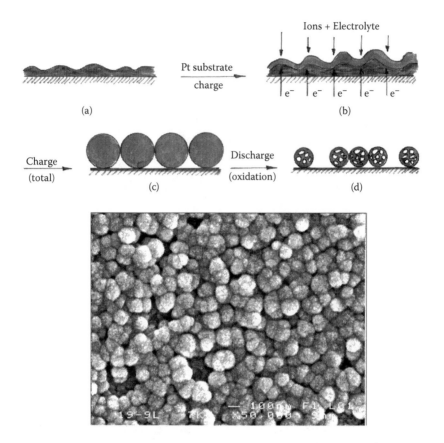

FIGURE 2.27 Schematic representation of the morphology change of a Pt–Pt layer (shown in a) during the charging process by TMAX (processes b and c), followed by air oxidation (see d). The ionometallic layer (adjacent spheres) leads to a totally reduced form (shown in d). The oxidation by dioxygen affords smaller spongy spheres, of which the shrinking factor here is totally arbitrary. (Cougnon, C., and J. Simonet, 2002, *J. Electroanal. Chem.* 531: 179. Used with permission.)

calculation (assumption done for closely adjacent ions in the resulting structure) leads to a volume increase of about 30 times, which is considerable. Similar calculations confirm the swelling factors for tetrafluoroborate and perchlorate ions. By contrast, the swelling limited to the insertion of cation alone (without that of the salt) would appear smaller but remains theoretically important (14 times with the iodide).

Similarly, the discharge process through chemical (in the presence of dioxygen with the concomitant diffusion of the superoxide) or electrochemical (withdrawing of inserted electrons in the structure and the simultaneous ejection of

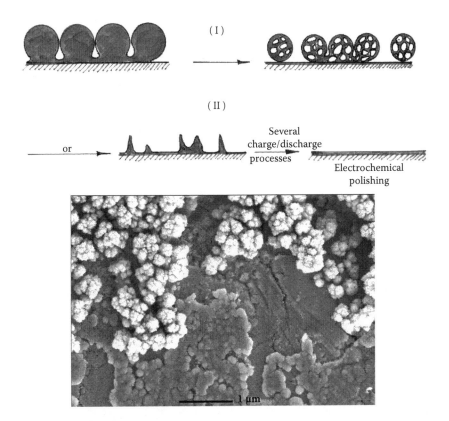

FIGURE 2.28 Progressive formation of smooth platinum surface by oxidation by air (I) or anodic oxidation (II) of a totally reduced layer. In general, several reduction–oxidation cycles are necessary to obtain an almost perfect annealing of the original surface. The image (bottom) clearly shows the crushing of spheres or their corrosion (top) in the course of the first anodic oxidation. Conditions: TMAClO$_4$ in DMF; reduction at −2.5 V versus SCE; oxidation at 0 V. (Cougnon, C., and J. Simonet, 2002, *J. Electroanal. Chem.* 531: 179. Used with permission.)

salt ions) processes is also intriguing. Indeed, the structure in layers of spheres is maintained. It is expected that produced spheres will be affected by the discharge due to the regeneration of the platinum metal as becoming spongy or of smaller size. As seen in Figure 2.28, they totally collapse upon repetitive charge–discharge processes until a quasi-smooth Pt surface is obtained. Among the remarkable morphologic changes observed with TMAX salts, one could observe the formation of holes or caldera-like profiles. Such structures are provoked by a sudden collapse of the spheres due to a concomitant ejection of the salt and cations (TMA$_+$) during the reoxidation of the layer through

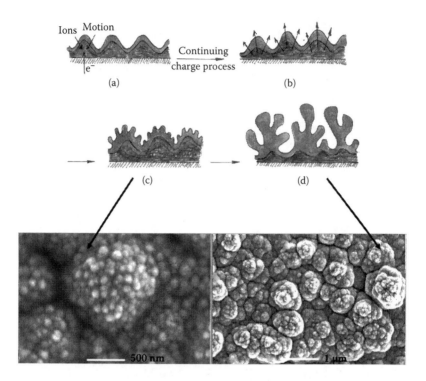

FIGURE 2.29   Schematic presentation of the swelling of a platinum layer (E = −2.5 V versus SCE) of platinized platinum in 0.1 mol.L⁻¹ TMAI. (a) Initial surface. (b) During the charge process. Thickness of platinum deposit: 0.2 μm. In (c) the quantity of charge corresponds to about half of the maximum charge, and in (d) the full charge process was achieved, which permits then to get the "cauliflower"-like structure. Oxidation by air (Simonet, J., Unpublished results).

injection of electrons from the substrate as in the case of a chemical oxidation (see Figures 2.29 and 2.30).

For practical applications of such structures, it is likely that such layers composed of empty spheres could present a strong surface activation. For this purpose, the surface increase provoked by the layer swelling was estimated by coulometry of the anodic oxidation of the platinum external layer (coverage of a monolayer of PtO). It was established that the surface increase is large, until three to five times, when a few repetitive charge–discharge scans are achieved. It is also worth underlining that swellings described here appear to be very similar to the observations described in the EC-AFM experiments (see Section 2.4) performed on microsized platinum particles. In conclusion, the swelling process is produced only during the charge process (see, for example, Figure 2.18) and not in the course of the layer reoxidation.

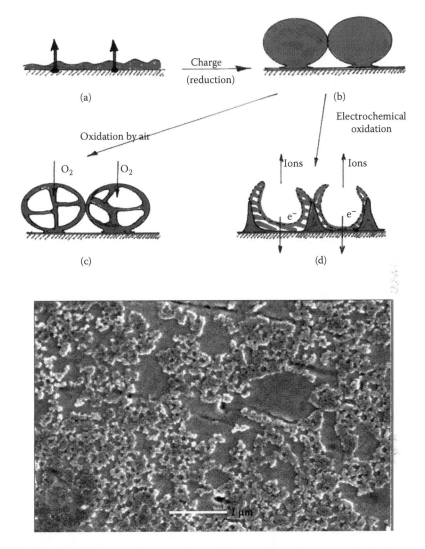

FIGURE 2.30 Schematic changes of morphology provoked by electrochemical reduction of the Pt–Pt layer followed by two kinds of oxidation processes. (a) Initial surface. (b) A reduction. It is proposed here that the swelling into adjacent spheres occurs at activated centers at the Pt surface. The oxidation may be achieved by air (leading to spongy spheres) (c) or by anodic oxidation which implies a specific motion of electrons (d). This difference would afford the formation of a caldera-like morphology, which progressively vanishes in the course of recurrent reduction–oxidation cycles (case of TMA tetrafluoroborate). (Simonet, J., 2006, *J. Electroanal. Chem.* 93: 3. Used with permission.)

2.7   IS PLATINUM REALLY INERT TOWARD ELECTROGENERATED ORGANIC ANIONS AND Π-ACCEPTOR-REDUCED FORMS?

When a platinum phase of general formula $[Pt_n^-, M^+, MX]$ is formed in the presence of a salt MX under the experimental conditions described earlier, several questions arise: What are the species that could react with the platinum phase? Any kinds of anions, cations—organic or inorganic? These questions are fundamentally important. Such reduced phases of platinum prepared in the presence of complex salts could be used to elegantly immobilize some special bulky oligomeric anions or cations at the surface that will be definitively blocked inside the Pt matrix at this end of the reoxidation process. Therefore, it should be possible to use the platinum charge for the immobilization of large ions in a way similar to the numerous procedures where the conducting organic polymers (such as, for example, polythiophene or polypyrrole) play the role of a conductive matrix. So far, this preliminary study was done essentially on platinum and does not totally answer all these questions. In the present chapter, we will focus on the role of some π-acceptors (and their electrogenerated forms) toward unreduced and reduced platinum.

2.7.1   REACTIVITY OF ACCEPTORS TOWARD PLATINUM-REDUCED PHASES

An intriguing experiment was first done with reduced platinum (prepared in the presence of CsI), which was placed in an acenaphthenoquinone (AcQ) ($E^\circ = -0.31$ V versus SCE) solution. No diffusion of the anion radical was observed around the reduced Pt. On the contrary, an intense blue color stays at the platinum surface. One can expect the formation of an insoluble radical-anion–cesium salt that covers the surface. However, after rinsing the sample with acetone and water in an ultrasonic bath, it appears that the Pt surface is deeply attacked [52]. This preliminary experiment suggests that AcQ is reduced by the cesium–platinum phase and that the reduced form of the acceptor reacts with the still unreacted platinum phase. In order to simply mimic this kind of chemical route, the concomitant reduction of the acceptor at a platinum sheet was realized under superdry conditions with a salt such as CsI. One may see a noticeable increase of the mass of the electrode after a sufficient charge was passed, accompanied by a huge modification of the surface as depicted by SEM images, exemplified in Figure 2.31.

A close examination of the surface by the SEM technique shows the presence of large crystals of AcQ, or of one of its reduced forms, clearly embedded in the platinum plate. Thorough rinsing by ether and acetone leads to elimination of most of the organic acceptor. Additionally, very thin platinum needle-shaped crystals are also formed. Such an experiment is a striking example of the attack of the electrode material; the polarized platinum is reactive toward an organic acceptor! It is quite reasonable to believe that the concomitant reduction of platinum *and* AcQ produces another platinum phase at the interface *including* the $AcQ^{-\bullet}$ anion-radical salt. The organic anion radical could concomitantly react

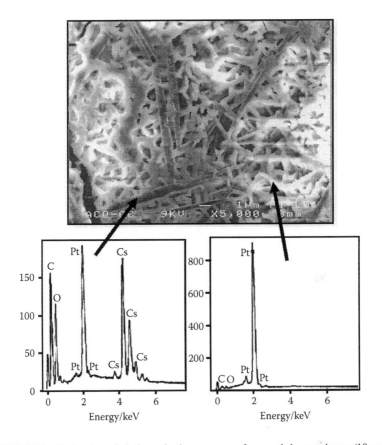

FIGURE 2.31 Reduction of platinum in the presence of acenaphthenoquinone (10 mmol. L^{-1}) superdry DMF + 0.1 $molL^{-1}$ CsI. Reduction potential: -2.1 V versus SCE electrode. Amount of electricity: 25 C cm^{-2}. Surface structure after oxidation by air and rinsing with water and alcohol. EDS spectra show differences between the dark zones (essentially carbon and oxygen) and light zones (only platinum needles). One notes the embedding of organic crystals inside the platinum bulk. (Cougnon, C., Ph.D. thesis, Université de Rennes 1, 2002).

with the polarized platinum to yield a new phase or simply give an exchange with the iodide anion inside the Pt/CsI phase:

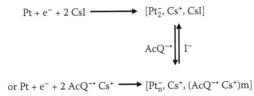

$$Pt + e^- + 2\,CsI \longrightarrow [Pt_2^-, Cs^+, CsI]$$

$$AcQ^{-\bullet} \Updownarrow I^-$$

$$\text{or } Pt + e^- + 2\,AcQ^{-\bullet}\,Cs^+ \longrightarrow [Pt_n^-, Cs^+, (AcQ^{-\bullet}\,Cs^+)m]$$

SCHEME 2.7

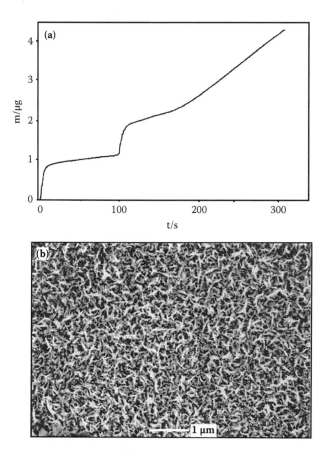

FIGURE 2.32 Electrochemical reduction of platinum in the presence of an organic π-acceptor. Superdry conditions. (a) Electrolysis of 2.5 μg of Pt deposited onto a gold-filmed quartz. EQCM analysis upon time in 0.1 molL^{-1} in DMF at −2.1 V versus SCE. After 100 s (obtaining of [Pt_2^-, Na^+ NaI]), 1,4-diacetyl benzene (to lead in total to a solution of 10 mmol.L^{-1}) is suddenly added to the solution. (b) SEM image of the resulting surface after an electrolysis at −2.1 V/SCE until 5 C cm^{-2}, and the oxidation by air followed by a thorough rinsing with alcohol and acetone. (Cougnon, C., Ph.D. thesis, Université de Rennes 1, 2002).

Thus, the oxidation process could restore in volume the platinum metal and AcQ. In order to verify the validity of the mechanism given in Scheme 2.7, which supposes the concomitant insertion of AcQ via its reduced form, EQCM experiments have been performed. For example, as displayed in Figure 2.32, the addition of a π-acceptor to the saturated phase of platinum provides a new mass increase. Moreover, it was established that the mass increase specifically due to the π-acceptor, is proportional to the initial amount of platinum.

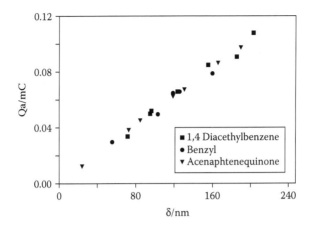

FIGURE 2.33 Charge–discharge process of platinum under superdry conditions in the presence of π-acceptors. Solvent–electrolyte: DMF/0.1 mmol.L^{-1} NaI. Concentration of organic compounds: 10 mmol.L^{-1}. Values of Q_a (see definition in Section 2.3) as a function of the amount of deposited Pt onto a gold surface. Potentials of charge and discharge: -2.1 V and -0.5 V versus SCE, respectively. (Cougnon, C., and J. Simonet, 2002, *J. Electroanal. Chem.* 531: 179. Used with permission.)

In Figure 2.32, after a careful rinsing, the appearance of a very porous platinum surface is noticeable. The structure is apparently produced by the dissolution of the π-acceptor by means of an appropriate solvent.

$$\text{Phase } [Pt_n^-, M^+, (\pi\text{-acceptor}^-, M^+)] \xrightarrow{\text{Ox}} [Pt, \pi\text{-acceptor}] \xrightarrow{\text{Solvent}} \text{divided Pt}$$

Attempts to estimate the stoichiometry of the new inserted phases involving π-acceptors were achieved. To this purpose, coulometry and ECQM experiments were performed. With several π-acceptors, it was additionally shown (Figure 2.33) that the stored charge Q_a is proportional to the thickness of the platinum layer; the ratio of stored charge toward the number of available platinum atoms is now equal to one-fourth. Thus, the soft-donor effect of the anion implied in the phase structure could significantly decrease the global acceptor capacity of platinum. If we assume that the organic substrate is involved in the form of its anion radical salt, the global formula of these new organometallic phases (determined thanks to a series of EQCM data) fits well with the following formula: $[Pt_4^-, M^+, \pi\text{-acceptor}^{\bullet-}, M^+]$.

2.7.2 Reactivity of Reduced Organic Acceptors toward Platinum

In the first part of Section 2.7, it was shown that the reduced platinum phase could specifically react with acceptors. In this connection, is pure platinum able to provide an electron exchange with the reduced forms of acceptors, essentially organic? An intriguing observation concerns the reduction of some π-acceptors

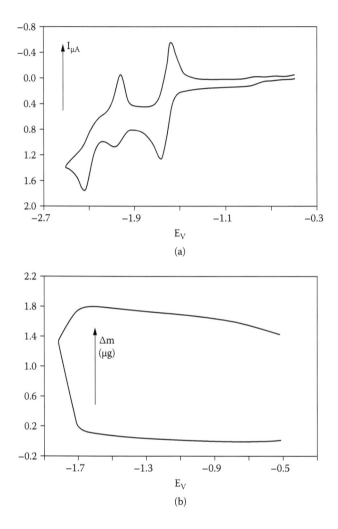

FIGURE 2.34 Reduction of acenaphthylene (concentration: 5 mmol.L⁻¹) in superdry DMF + 0.1 molL⁻¹ M TBABF₄ added with 7.5 mmol.L⁻¹ NaI. Potentials are referred to the SCE electrode. Scan rate: 0.1 V s⁻¹. (a) Voltammetry at a platinum microelectrode. (b) EQCM experiment at gold film quartz covered by platinum. Experimental conditions are strictly the same for (a) and (b) (Simonet, J., Unpublished results).

under superdry conditions at a platinum sheet electrode: if the anion radical is not readily protonated, the reaction does not end. Why? The (conventional) answer is that diffusion from the anodic compartment interferes with the normal path of the experiment. However, if the voltammetric current of a π-acceptor of $E° < -1.5$ V versus ECS is compared to the change of mass measured by EQCM, it was found [53] that the threshold of mass rather coincides with the first electron transfer to the acceptor (Figure 2.34).

Furthermore, experiments performed in different media, in particular in [DMF + TAAX + NaI] have shown that Na⁺ is consumed (by means of an indirect insertion), while the acceptor concentration stayed almost constant in the course of electrolyses. All these observations suggest that the acceptor plays the role of a mediator or an electron carrier, and permits, by indirect means, to reduce the platinum plate to provide a platinum phase that is identical to those described in Section 2.4. Moreover, a platinum sheet not connected to the electrolysis circuit could also react with some acceptor-reduced forms. The validity of these assumptions was clarified and fully checked by the use of the SECM technique (see the following text).

2.7.3 Indirect Reduction of Platinum Followed by SECM Experiments

SECM offers a different and unique way of investigation, as the reduction process is induced from the solution side through a chemical reaction between the Pt metal and an electrogenerated radical anion [54]. The key step is the reaction between the electrogenerated radical-anion with the metallic surface. To study such effect, we used the basic principle of the feedback mode of the SECM in the steady state as presented in Figure 2.35.

In this setup, the Pt layer is not electrically connected (unbiased substrate). Effectively, to achieve the SECM experiments, it is not possible to polarize the Pt substrate: a polarization of the substrate at a negative potential would result in direct electrochemical reduction, and, on the contrary, a polarization at a positive potential would lead to the reoxidation of the electrogenerated phase. The probe tip (10-μm-radius ultramicroelectrode, UME) is placed in a deoxygenated solution containing a redox mediator (M/M⁻), and different quaternary ammonium salts NR_4BF_4 (R = ethyl, n-butyl, n-octyl) under superdry conditions were examined. Six different mediators were chosen (as displayed in Figure 2.35). The standard potentials of the first three mediators were more positive than E_T, while the other

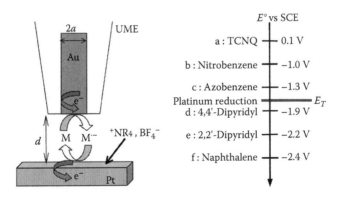

FIGURE 2.35 Schematic of SECM configuration for showing localized modification of platinum substrate. E_T: threshold potential of platinum reduction of the order −1.6 V/SCE. (Ghilane, J., M. Guilloux-Viry, C. Lagrost, J. Simonet, P. Hapiot, 2007, *J. Am. Chem. Soc.* 129: 6654. Used with permission.)

were more negative than E_T. Later on, E_T is considered as the threshold of the platinum reduction process. This series of experiments provide a unique way of investigating the influence of the driving force through the choice of the redox potential of the mediator and the nature of the tetraalkylammonium salts. The approach curves represent the variations of the normalized current I_T (I_T = approach current tip i_T/steady-state current i_{inf}) versus the normalized distance $L = d/a$, where d is the distance between the tip and the substrate, and a is the radius of the tip (feedback mode). The steady-state current i_{inf} is given by the following equation: $i_{inf} = 4\,nFDCa$ (n represents the number of charges transferred per species, F is the Faraday constant, and D and C are the diffusion coefficient and concentration of the electrochemically organic species in solution, respectively). When the distance between the tip and the Pt substrate is close to the order of the electrode radius (a few micrometers), the reduced mediator (M^-) diffuses to the liquid–solid interface, where it can react with the Pt surface, inducing a modification of the faradic tip current. For the weaker reducing M^- (mediators **a, b,** and **c** in Figure 2.35, $E_T < E^\circ_{M/M}{}^-$), one can observe that the dimensionless current $I_T = i_T/i_{T,-}$ increases corresponding to the classical positive feedback. These responses (depicted in Figure 2.36) are in

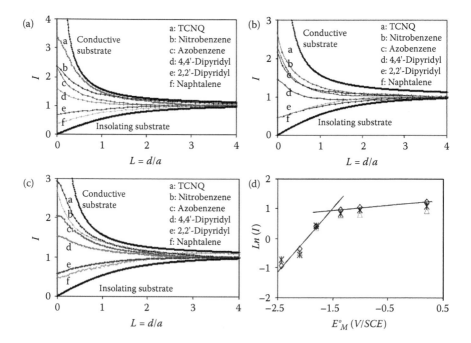

FIGURE 2.36 (a) Approach curves using Et_4NBF_4. (b) Approach curves using n-Bu_4NBF_4. (c) Approach curves using n-Oct_4NBF_4. (d) Variation of the substrate reduction current I, obtained at a tip–substrate distance of 0.1a, with the standard reduction potential of the redox mediator, ($\Diamond$) tetraethylammonium, ($\triangle$) tetrabutulammonium, (.) tetraoctylammonium tetrafluorobate salts. (Ghilane, J., M. Guilloux-Viry, C. Lagrost, J. Simonet, P. Hapiot, 2007, *J. Am. Chem. Soc.* 129: 6654. Used with permission.)

agreement with the "normal" behavior predicted for conductive metallic platinum, and simply correspond to the rapid oxidation of the radical anion on a metallic substrate (regeneration of the mediator). On the contrary, for the strong reducing $M^{\cdot-}$ (mediators **d**, **e**, and **f** with $E^{\circ}_{M/M^{-}} < E_T$), an unexpected decrease of I_T is observed. The current attenuation is proportional to the separation distance d, and the overall approach curves evidence a negative feedback.

The drop off of the current indicates the absence of regeneration of the mediator at the platinum surface and tends to the behavior expected for an insulator surface. This tendency is more pronounced when the standard potential of the mediator becomes more negative (i.e., when the radical anion exhibits a stronger reducing character). This phenomenon leads to a considerable modification of the chemical nature of the Pt substrate that becomes insulating. These observations can be rationalized with the following set of reactions (Scheme 2.8):

$$\text{At the tip UME: } M + e^- \longrightarrow M^{\cdot-}\ E^{\circ}_{M/M}{}^{\cdot-} \qquad (1)$$

$$\text{At the platinum substrate: } Pt + M^{\cdot-} \longrightarrow Pt + M + e^- \qquad (2)$$

$$E^{\circ}_{M/M}{}^{\cdot-} > E_T \text{ with Mediators } \mathbf{a}, \mathbf{b}, \text{ and } \mathbf{c}$$

$$nPt + M^{\cdot-} + 2R_4N^+ + BF_4^- \longrightarrow [Pt_n^-, R_4N^+, R_4NBF_4] + M \qquad (3)$$

$$E^{\circ}_{M/M}{}^{\cdot-} < E_T \text{ with Mediators } \mathbf{d}, \mathbf{e}, \text{ and } \mathbf{f}$$

SCHEME 2.8

In this chemical scheme, the radical anion of the mediator is electrogenerated at the UME (reaction 1). When the distance between the UME and the substrate of platinum is small enough (few μm $d < 3a$), the radical anion interferes with the platinum substrate. For the mediators (**a**, **b**, and **c**), no changes of the chemical nature of the platinum surface are observed as indicated by the positive feedback; the radical anion is simply exchanging an electron with a metallic surface and is not reactive enough for inducing the modification (reaction 2). If the reducing power of the radical anion is sufficient to induce the reduction of Pt (mediators **d**, **e**, and **f**), the cathodic modifications of the platinum surface become possible, which leads to the formation of the reduced organometallic phases (reaction 3). As attested by our experiments, this phase is electrically insulating and, under steady-state conditions, rapidly leads to a blocking effect of the oxidation of $M^{\cdot-}$ to M (scheme shown in Figure 2.37). From this point, one may conclude that an organic radical anion with a reducing power stronger than E_T is able to "reduce" the platinum sample. This reduction process, as a consequence, changes the chemical nature of the platinum from a conducting material to an insulating one.

To complete the experiments, the topography of the platinum substrate was imaged by AFM after the SECM approach experiments. During these experiments, the substrate was kept in the original solution under an argon atmosphere

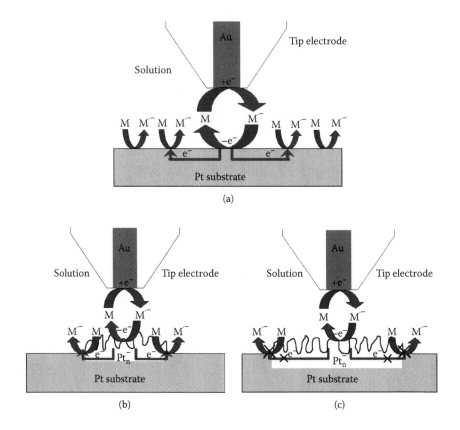

FIGURE 2.37 Scheme showing the competition between **Pt** reduction and mediator **M** regeneration. (a) Initial metallic with larger reacting zones, the surface turns to be nonconductive (b) Formation of the platinum phase, and the reduction of the mediator ceases (c). (Ghilane, J., M. Guilloux-Viry, C. Lagrost, J. Simonet, P. Hapiot, 2007, *J. Am. Chem. Soc.* 129: 6654. Used with permission.)

to avoid the chemical oxidation of the "reduced Pt" by air. Figure 2.38 displays the topography of the freshly prepared platinum substrate on a large area of 100×100 µm. At this scale, the sample appears as a flat surface and, as expected, demonstrates the stability of platinum after contact with the solution containing the mediator and the supporting electrolyte. After the SECM experiments, when the mediator was the naphthalene–naphthalene radical anion couple, and the supporting electrolyte the tetrabutylammonium salt (Figure 2.38b) or tetraoctylammonium (Figure 2.38c), considerable changes in the surface topography are visible and obviously localized on roughly circular areas. The estimated diameter of these circular zones was about 80 µm, which corresponds to four times the UME diameter. Limits of such modified zones depend essentially on the formation of insulating areas around the tip and allow confirmation that the platinum phase is not conducting.

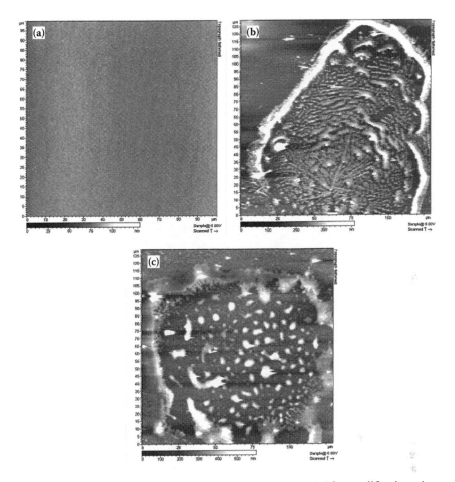

FIGURE 2.38   AFM images (a) Initial platinum substrate. (b, c) After modification using SECM in mode feedback, in the presence of naphthalene as mediator and tetrabutylammonium and tetraoctylammonium tetrafluoroborate, respectively. Scan size 100×100 μm. (Ghilane, J., M. Guilloux-Viry, C. Lagrost, J. Simonet, P. Hapiot, 2007, *J. Am. Chem. Soc.* 129: 6654.)

2.8   CATHODIC CHARGE OF PALLADIUM IN SUPERDRY ELECTROLYTES

Similar experiments were done with palladium under the conditions described in Section 2.3. First of all, a charged sheet (at a potential of the order of that used for platinum reduction, that is, −2 V versus SCE) of palladium in superdry DMF containing CsI appears to play the role of a reducing material. When this piece of palladium after chemical treatment is put in contact with a solution of AcQ or 1,4-dinitrobenzene, the corresponding color assigned to that of the anion radical

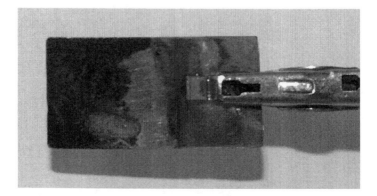

FIGURE 2.39 Charge of a palladium sheet (surface area: 4 cm²) superdry solution of CsI 0.1 molL⁻¹ in DMF. Charge 10 C. Contact of a solution I0 mmol.L⁻¹ AcQ during 5 min (Simonet, J., Unpublished results).

rapidly appears (Figure 2.39) from the palladium interface. Scheme 2.9 checks that an electrochemical charge of the material is possible and that the radical anion is formed (verified by ESR spectroscopy).

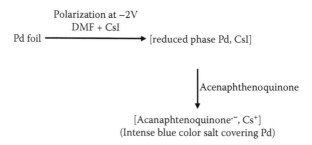

SCHEME 2.9

Charge and discharge processes were achieved with palladium and palladized interfaces (mainly onto glassy carbon, gold, and platinum). In voltammetry, smooth palladium exhibits a reduction during the charge and an anodic peak. The whole feature looks totally irreversible (Figure 2.40).

The cathodic peak current is diffusion controlled, and the reduction potentials in the case of alkaline iodides are close to those already measured with the platinum interfaces. Also with TAAX salts, palladium shows a behavior similar to platinum. TMAX (X = halide) displays a quasi-reversible system. Contrarily, bulky TAA⁺ cations considerably enlarge the irreversibility of the charge–discharge process. Thus, while the potential difference ΔE between the cathodic and anodic peaks is about 0.7 V when the cation is tetramethylammonium, it reaches almost 1.8 V when the cation is a tetra-*n*-octyl salt (Table 2.6).

Global formulas for the reduced forms of palladium were established from the limiting charges with a large family of salts in DMF. In the case of alkaline

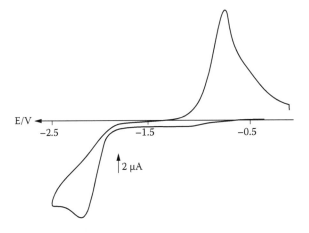

FIGURE 2.40 Voltammetric response of a smooth palladium electrode in contact with a superdry solution (0.1 mol.L^{-1} CsI in DMF). Scan rate: 0.2 V s^{-1}. Potential are referred to the SCE electrode (Cougnon, C., Ph.D. thesis, Université de Rennes 1, 2002).

metals, charges-discharge processes were evidenced in coulometry (Figure 2.41 with the use of CsI). The values of charge amounts, Q_a, corresponding to the number of electrons effectively inserted in the metallic substrate, were found to be proportional to the number of palladium atoms (see Figure 2.42). With Li$^+$, Na$^+$, K$^+$, and Cs$^+$, the stoichiometry is always one charge element per two

TABLE 2.6

Voltammetric Behavior of a Smooth Palladium Electrode (Surface Area: 0.8 mm^2) in the Presence of Different TAAX Salts Dissolved in Superdry DMF[a]

Electrolyte (0.1 molL^{-1})	E_{pc} (V)	I_{pc} (µA)	E_{pa} (V)	I_{pa} (µA)	$E_{pa} - E_{pc}$ (V)
Me$_4$NI	−2.45	−16.9	−1.77	5.7	0.68
Me$_4$NCI	−2.45	−20.0	−1.79	6.0	0.66
Et$_4$NI	−2.66	−15.0	−1.73	4.3	0.93
Et$_4$NBr	−2.69	−16.0	−1.74	4.5	0.94
Bu$_4$NI	−2.84	−13.0	−1.33	2.3	1.510
Bu$_4$NBr	−2.81	−11.2	−1.46	2.1	1.34
Bu$_4$NCI	−2.83	−11.7	−1.44	2.3	1.39
Hex$_4$NBr	−2.87	−8.0	−1.16	0.8	1.71
Hex$_4$NCl	−2.85	−9.1	−1.15	0.9	1.70
Oct$_4$NBr	−2.90	−7.5	−1.04	0.4	1.86
Oct$_4$NCl	−2.89	−8.0	−1.08	0.5	1.81

[a] Scan rate: 0.2 Vs^{-1}. Potentials are referred to SCE electrode.

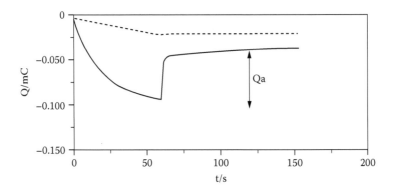

FIGURE 2.41   Palladium electrode (metal deposited onto gold). Coulometric determinations in a solution 1 molL^{-1} CsI in DMF for a surface polarized at -2.1 V versus SCE electrode during 60 s (full line) and then discharge at -0.5 V. The discharge amount Q_a is quasi-instantaneous. By comparison, the broken line shows the same chronocoulometry at the bare gold electrode (Cougnon, C., Ph.D. thesis, Université de Rennes 1, 2002).

palladium atoms. With these four different iodide salts, EQCM experiments, as with platinum, were completed (Figure 2.43). The concomitant insertion of the salt MI with M^+ is then obvious, and stoichiometries obtained from the limiting charges injected to produce the phases (Table 2.7) correspond to the global formula: $[Pd_2^-, M^+, MI]$

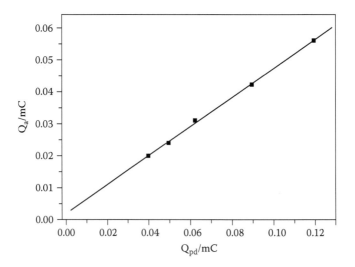

FIGURE 2.42   Coulometric experiments on deposited amounts of Pd on a gold substrate. Variation of Q_a (stocked amounts of charge at -2.3 V) upon deposited amounts of Pd (by means of galvanostatic deposits from PdCl$_2$). Solution: 0.1 mol.L^{-1} CsI in DMF. (Cougnon, C., Ph.D. thesis, Université de Rennes 1, 2002.)

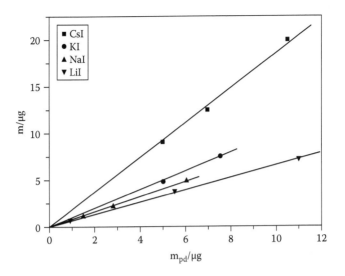

FIGURE 2.43 EQCM. Increase of mass deposited amount of Pd on a gold-film quartz for limit charges achieved at –2.3 V versus SCE reference system with several alkaline iodides at the concentration 0.1 molL^{-1} (Cougnon, C., Ph.D. thesis, Université de Rennes 1, 2002).

TABLE 2.7

Values of Parameters x and y Corresponding to Stoichiometries Obtained by Coulometry (x-Values) and Gravimetry (y-Values) by Means of Potentiostatic Electrolyses in Superdry DMF

Electrolyte (0.1 molL^{-1})	E$_{applied}$ (V versus. SCE)	x-Values for (R$_4$N) Pd$_x$	y-Values for [Pd$_x$NR$_4$(R$_4$NX)$_y$]
Me$_4$NI	–2.42	2.13 ± 0.09	0.92 ± 0.02
Me$_4$NCl	–2.42	4.02 ± 0.10	1.01 ± 0.03
Et$_4$NI	–2.72	2.01 ± 0.08	0.95 ± 0.05
Et$_4$NBr	–2.72	4.01 ± 0.16	0.98 ± 0.03
Bu$_4$NI	–2.82	2.10 ± 0.15	0.92 ± 0.04
Bu$_4$NBr	–2.82	4.07 ± 0.16	0.98 ± 0.04
Bu$_4$NCl	–2.82	4.05 ± 0.13	0.95 ± 0.03
Hex$_4$NBr	–2.87	4.09 ± 0.18	0.99 ± 0.03
Hex$_4$NCl	–2.87	4.04 ± 0.04	1.02 ± 0.08
Oct$_4$NBr	–2.87	4.04 ± 0.10	1.00 ± 0.08
Oct$_4$NCl	–2.87	4.18 ± 0.09	0.98 ± 0.09

TABLE 2.8

Consumption of the Electrogenerated Palladium Phase in Firm Contact with 0.1 mol.L^{-1} CsI after Contact Duration of the Phase with the π-Acceptor of 90 s^a

Organic π-Acceptor (10 mmol.L^{-1})	Standard Potential (V versus SCE)	Consumed [Pd$_x$Cs(CsI)] (%)
Chloranile	−0.27	100
Phenanthrenequinone	−0.58	85
p-Dinitrobenzene	−0.61	73
3,5-Dinitrobenzyle chloride	−0.80	68
Acenaphthenoquinone	−0.84	64
2,4-Dinitrotoluene	−0.91	65
Nitrobenzene	−1.10	69
9-Fluorenone	−1.28	23
Benzophenone	−1.75	0

^{a} Excess of charge: about fivefold.

The accuracy of the experiments is reasonably good. The data exclude the possibility of an insertion of solvent molecule in the obtained phases, at least at the level of saturation.

All these results obtained with palladium samples confirm the stoichiometries already found with platinum with alkaline iodides salts. When charge processes are completed in the presence of TAAX salts, some important discrepancies are noted for stoichiometries derived from the measurement performed at saturation (masses relative to plateau level in EQCM). Table 2.8 exhibits the phase formulas with a large number of tetraalkylammonium salts (Scheme 2.10), and it appears that the nature of the anion of the salt can modify the charge process: the harder the anion (strong donor effect), the weaker the energy storage inside palladium.

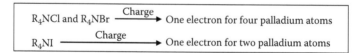

SCHEME 2.10

These data are striking since the level of insertion depends not only on the bulkiness of the TAA$^+$ cation but also on the donor efficiency of the anion. Unfortunately, we have not yet obtained data concerning complex anions such as BF$_4^-$ and ClO$_4^-$. Nevertheless, the good reproducibility of the data gathered in Table 2.8 allows us to expect that the anion hardness may compete with the electron during the charge process. Up to now, there are no available theoretical arguments to rely

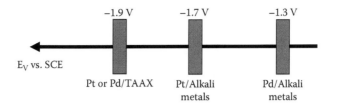

SCHEME 2.11 Limits of reducibility by electron transfer, estimated by the relaxation potentials specific to concerned phases in contact with solutions used for their formation. Errors on given potentials are about 0.1 V.

the level of empty orbitals of palladium and the influence of halide ions on the amplitude of the charge.

Palladium phases, like those formed from platinum, were found to be efficiently reactive by electron transfer toward acceptors as displayed by Table 2.8. The reducing efficiency depends on the conditions of the charge. In particular, the charge with TAAX salts at more negative potentials, permits, in return, more efficient reductions. (See Scheme 2.11)

Lastly, it is certainly of interest to underline the EQCM process concerning the charge of palladium (Figure 2.44). As a typical variation of mass versus the applied potential (here in the presence of TBAI), a sudden increase of mass was observed at potential around −1.7 V, which is close to the threshold of the reducing capacity of the phase as it was estimated from the relaxation potentials. After a large increase (zone A), a maximum is reached, and we may expect that the saturation at the interface occurs. Thereafter, and quite suddenly, a decrease in

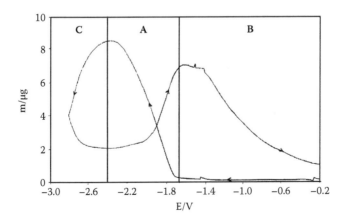

FIGURE 2.44 EQCM experiments on the mass change of a deposited amount of palladium on a gold-film quartz. Solution: superdry solution of 0.1 mol.L⁻¹ TBAI in DMF. Scan rate: 10 mV s⁻¹. Potential are referred to SCE (Cougnon, C., Ph.D. thesis, Université de Rennes 1, 2002).

mass is observed (zone C). Such decay was observed with platinum within this potential range and could be ascribed to the reduction of the phase. At very negative potentials (−2.9 V vs. SCE), the reduced phase had totally disappeared. A new increase of mass only occurs in the course of the reverse scan (zone A) with a new maximum happening at the level of the charge threshold. Then (zone B) one observes a progressive decay of mass until about 0 V. This mass decay is assigned to the diffusion of ions out of the material until the original mass is refound. Such behavior could be observed along repetitive scans. With palladium, a relatively slow loss of mass during the oxidation process could attest a weaker sensitivity to air. This is illustrated in Figure 2.45: after a deep charge and contact with air, the salt (TBAI) merges out of palladium, and its presence is underlined by mean of SEM (crystals of TBAI decorating the metal interface).

2.9 CONCLUDING REMARKS

Platinum and palladium, both belonging to column VIII of Mendeleev's classification, can absorb cathodic charges efficiently under specific experimental conditions. These charged products, named in this chapter as ionometallic complexes, resemble and behave, to a certain extent, like some heteropolyanions, implying either Pt or Pd with anions derived from metalloids as displayed in Figure 2.1. The insertion was followed by coulometry and EQCM (nanobalance) techniques. When mass of the substrate attained a maximum, it was established that insertion might concern up to two metal atoms per stored electron, which is a high value. Additionally, the salt itself (the cation for stabilizing the coulombic neutrality inside the bulk matrix) and the cation–anion pair were found to be concomitantly involved, at least at the final state of the insertion. The main consequence of this complex insertion is the swelling of the reaction layer at the interface. The large insertion of ions compared to the weak amount of platinum or palladium was expected to lead progressively, in the course of the charge, to the formation of a nonconducting layer, which contributed to cessation of the insertion process. Consequently, the thickness of the transformed metal at the electrolyte interface is small—no more than a few microns. The formation of this nonconducting layer was confirmed without ambiguity by the use of SECM. A better reactivity may exist at platinized and palladized interfaces, where porous substrates of large specific surface areas contribute to enlarging the extent of this reduction reaction.

A very recent XPS spectroscopy study confirmed that platinum could incorporate cations from the salt, in particular Cs^+, when reduced in the presence of CsI under the conditions given in Section 2.3 [43]. However, the data did not confirm the concomitant insertion of the salt and the anion (here iodide) that was not detected in the bulk metal. It must be stressed that conditions of the experiments require movement of the sample from the electrolysis solution to the XPS instrument, that is, in contact with air, even if this operation is short. All the formulas of platinum and palladium phases were established in situ at the strict saturation state (obtaining of a plateau in mass analyses by EQCM). Currently, we do not possess any information about the kinetics of the collapse of saturated phases by

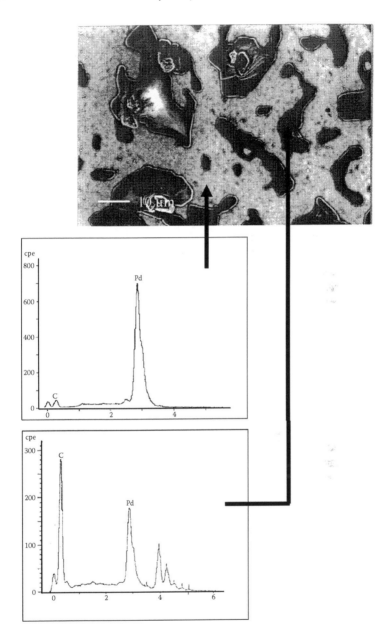

FIGURE 2.45 Charge of smooth palladium in the presence of 0.1 mmol.L^{-1} TBAI in superdry DMF. Charge: 10 C cm^{-2}. SEM image after rinsing and contact with air during more than 10 min. Appearance of TBAI crystals at the Pd surface and EDS analysis of light zones (pure Pd) and dark zones assigned to TBAI (presence of carbon and iodine) (Cougnon, C., Ph.D. thesis, Université de Rennes 1, 2002).

contact with dioxygen, but it is certainly possible that transient states of the form $[Pt_n^-, M^+]$ with $n \ll 2$ might exist as intermediate forms, both at the charge and at the discharge (chemical and electrochemical) stages.

In this approach concerning Pt and Pd, electrochemistry appears to be a suitable tool in the formation and detailed study of these phases. It is certain that phases that imply TAAX salts cannot be produced without polarization of the metal. Reactions of these phases with bulky ions and possible exchanges (or incorporations) of a large palette of π-acceptors or free radicals at the metal surface or inside a reaction layer adjacent to the interface certainly provide rich possibilities for decorating functionalized metals. Preliminary tests on nickel showed that charges in the presence of alkaline iodides were successful and led to electron storage.

ACKNOWLEDGMENT

The authors wish to thank Charles Cougnon and Jalal Ghilane for the pioneering work they accomplished during their Ph.D. work devoted to the study of platinum and palladium as new electrode materials, especially under superdry conditions. Most of results described in this chapter were published, and those data are fully available in the recent literature (see reference list).

REFERENCES

1. Vocke, J. Orientation of substances in solution in neighbourhood of mercury electrodes. Chem. *Listy* 62 (1968): 497.
2. Vocke, J. and A. M. Kardos, Change of mechanism with electrode potential in electrocemical reduction of cyanopyridines. *Coll. Czch. Comm. Commun.* 33 (1968): 2560.
3. Bard, A. J., and L. R. Faulkner. *Electrochemical methods: Fundamentals and applications. 2nd ed.* John Wiley & Sons, New York (2001).
4. Brown, O. R., in A. K. Covington. *Electrode Processes 2: Electrode reactions. Physical chemistry of organic solvent systems.* T. Dickinson (ed.).Plenum Press, New York (1973): 747.
5. Ross, S. D., M. Finkelstein, and R. C. Petersen. Mechanism of the electroreduction of benzyltriethylammonium nitrate in dimethyl-formamide at aluminum and platinum cathodes. *J. Am. Chem. Soc.* 92 (1970): 6003.
6. Coulon, E., J. Pinson, J.-D. Bourzat, A. Commerçon, and J.-P. Pulicani. Electrochemical attachment of organic groups to carbon felt surfaces. *Langmuir* 17 (2001): 7102.
7. Pinson, J., and F. Podvorica. Attachment of organic layers to conductive or semiconductive surfaces by reduction of diazonium salts. *Chem. Soc. Rev.* 34 (2005): 439.
8. Adenier, A., C. Combellas, F. Kanoufi, J. Pinson, and F. I. Podvorica. Formation of polyphenylene films on metal electrodes by electrochemical teduction of benzenediazonium salt. *Chem. Mater* 18 (2006): 2021.
9. Schâfer, H. J. Organic electrochemistry, *Encyclopedia of Electrochemistry Vol. 8.* A. J. Bard and M. Stratmann (eds). Wiley-VCH (2004).
10. Baizer, M. M. Electrolytic reductive coupling 1: Acrylonitrile. *J. Electrochem. Soc.* 111 (1964): 215.
11. Feoktistov, L. G., A. P. Tomilov, and I. G. Sevast'Yanova. Relation between the acrylonitrile electroreduction products and the proton donating properties of the solution. *Sov. Electrochem* 1 (1965): 1300.

12. Braauer, G., and G. Düsing, Z. Zur kenntnis der ammoniumamalgame. *Aorg. Allg. Chem*, 328 (1964): 154.

13. Littlehailes, J. D., and B. Woodhall. Quaternary ammonium amalgams. *J. Chem. Soc. Chem. Commun.* 14 (1967): 665.

14. B. Southworth, K. D. Fleischer, F.C. Nachod and R. Osteryoung. Polarography of quaternary ammonium compounds. *Anal. Chem.* 33 (1961): 208.

15. Horner, L., and H. Neumann. Studien zum vorgang der wasserstoffubertragung. 12. Hydrierende soaltung von sulfonen mit tetramethylmethylammonium als elektronenubertrager. *Chem. Ber.* 98 (1965): 1715.

16. Garcia, E., A.H. Cowley and A.J. Bard. Quaternary ammonium amalgams as Zintl ion salts and their use in the synthesis of novel quaternary ammonium salts. *J. Am. Chem. Soc.* 108 (1986): 6082.

17. E. Kariv-Miller, P. D. Christian, and V. Svetlicic. The first cathodically generated tetraalkylammonium-tin compounds. *Langmuir* 11 (1995): 1817.

18. Kariv-Miller, E., V. Stetlicic and P.B. Lawin. Electrogenerated $R_4N(Hg)_5$ films. *J. Chem. Soc., Faraday Trans.*, 1 83 (1987): 1169.

19. Buqa, H., A. Würsig, J. Vetter, M. E. Spahr, F. Krumeich, and P. Novák. SEI film formation on highly crystalline graphitic materials in lithium-ion batteries. *J. Power Sources* 153 (2006): 385.

20. Buqa, H., D. Goers, M. E. Spahr, and P. Novak. The influence of graphite surface modification on the exfoliation during electrochemical lithium insertion. *J. Solid State Electrochem.* 8 (2003): 79.

21. Simonet, J. and H. Lund. Electrochemical behaviour of graphite cathodes in the presence of tetraalkylammonium cations. *J. Electroanal. Chem.* 75 (1977): 719.

22. Besenhard, J. O., and H. P. Fritz. Cathodic reduction of graphite in organic solutions of alkali and NR_4^+ salts. *J. Electroanal. Chem.* 53 (1974): 329.

23. Besenhard, J. O., E. Theodoridou, H. Mohwald, and J. J. Nickl. Electrochemical applications of graphite intercalation compounds. *Synth. Met.* 4 (1982): 211.

24. Berthelot, J., M. Jubault, and J. Simonet. Tetra-alkylmmonium-graphite lamellar compounds—A new class of reducing agent in organic chemistry. *J. Chem. Soc., Chem. Commun.* (1982): 759.

25. Zintl, E., J. Goubeau, and W. Z. Dullenkopf. Metals and alloys I. Salt-like compounds and intermetallic phases of sodium in liquid ammonia. *Z. Phys. Chem.*, Abst. A 154 (1931): 1.

26. Adolphson, D. G., J. D. Corbett, and D. J. Merryman. Stable homopolyatomic anions of the post-transition metals. "Zintl ions". The synthesis and structure of a salt containing the heptantimonide(3-) anion. *J. Am. Chem. Soc.* 98 (1976): 7234.

27. Edwards, P. A., and J. D. Corbett. Stable homopolyatomic anions – Synthesis and crystal structures of salts containing pentaplumbide (2-) and pentastannide(2-) anions. *Inorg. Chem.* 16 (1977): 903.

28. Cisar, A., J. D. Corbett, and D. J. Merryman. Polybismuth anions – Synthesis and crystal structure of a salt of tetrabismuthide(2-) ion, bi-4(2-) basis for interpretation of structure of some complex intermetallic phases. *Inorg. Chem.* 16 (1977): 2482.

29. Corbett, J. D., D. G. Adolphson, P. A. Edwards, and F. J. Armatis. Synthesis of stable homopolyatomic anions of antimony, bismuth, tin, and lead. Crystal structure of a salt containing the heptaantimonide(3-) anion. *J. Am. Chem. Soc.* 97 (1975): 6267.

30. Dietrich, B., J. M. Lehn, and J. P. Sauvage. Diaza-polyoxa-macrocycles et macrobicycles. *Tetrahedron Lett.* 34 (1969): 2885.

31. Pons, B. S., D. J.Santure, R. C. Taylor and R. W. Rudolph. Electrochemical generation of the naked metal anionic clusters, $Sn_{9-x} Pb_x^{4-}$ (x = 0 to 9). *Electrochim. Acta.* 26 (1981): 365.

32. Haushalter, R. C., C. J., D. Warren, A. B. Bocarsly and M. Ho. Electrochemical synthesis of new Sb[bond]Te Zintl anions by catholic dissolution of Sb_2Te_3 electrodes: Structures of $[Sb_4Te_4]^{4-}$ and $[Sb_9Te_6]^{3-}$. *Angew. Chem. Int. Ed. Engl.* 32 (1993): 1646.

33. Rudolph, R., W. Wilson, F. Parker, R. C. Taylor, and D. Young. Nature of naked-metal-cluster polyanions in solution. Evidence for (Sn9-xPbx)4-(x = 0-9) and tin-antimony clusters. *J. Am. Chem. Soc.* 100 (1978): 4629.

34. Teherani, T. H., W. J. Peer, J. J. Lagowski, and A. J. Bard. Electrochemical behavior and standard potential of gold(1-) ion in liquid ammonia. *J. Am. Chem. Soc.* 100 (1978): 7768.

35. Hammerich, O., and V. D. Parker. The reversible oxidation of aromatic cation radicals to dications. Solvents of low nucleophilicity. *Electrochim. Acta* 18 (1973): 537.

36. Heinze, J. Cyclic voltammetry – Electrochemical spectroscopy. *Angew. Chem.* 96 (1984): 823.

37. Simonet, J., E. Labaume, and J. Rault-Berthelot. On the cathodic corrosion of platinum in the presence of iodides in dry aprotic solvents. *Electrochem. Commun.* 1 (1999): 252.

38. Dano, C. Insertion cathodique des ions tétraalkylammonium dans le graphite et le platine: réactivité, fonctionnalisation et applications., *Ph.D. thesis, Université de Rennes 1*, 1998.

39. Cougnon, C. Réactivité cathodique du platine et du palladium en présence d'électrolytes en mileiur super anhydre. Formation de nouvelles phases réductrices. *Ph.D. thesis, Université de Rennes 1*, 2002.

40. Cougnon, C. and J. Simonet. Are tetraalkylammonium cations inserted into palladium cathodes? Formation of new palladium phases involving tetraalkylammonium halides. *J. Electroanal. Chem.* 507 (2001): 226.

41. Buttry, D. A. Applications of the quartz crystal microbalance to electrochemistry. *Electroanalytical chemistry*, A. J. Bard (ed.), Vol.17. Marcel Dekker, New York (1991): p. 1.

42. Bergamini, J.-F., J. Ghilane, M. Guilloux-Viry, and P. Hapiot. In situ EC-AFM imaging of cathodic modifications of platinum surfaces performed in dimethylformamide. *Electrochem. Commun.* 6 (2004): 188.

43. Ghilane, J., C. Lagrost, M. Guilloux-Viry, J. Simonet, M. Delamar, C. Mangeney, and P. Hapiot. Spectroscopic evidence of platinum negative oxidation states at electrochemically reduced surfaces. *J. Phys. Chem. C* 111 (2007): 5701.

44. Ghilane, J., M. Guilloux-Viry, C. Lagrost, P. Hapiot, and J. Simonet. Cathodic modifications of platinum surfaces in organic solvent: reversibility and cation type effects. *J. Phys. Chem. B* 109 (2005): 14925.

45. Cougnon, C., and J. Simonet. Cathodic reactivity of alkaline metal iodides toward platinum bulk. The formation of new reducing phases. *Electrochem. Commun.* 4 (2002): 266.

46. Cougnon, C., and J. Simonet. Cathodic reactivity of platinum and palladium in electrolytes in superdry conditions. *Platinum Metals Rev.* 46 (2002): 94.

47. Simonet, J. The mild anodic oxidation of platinum in organic solvents of low acidity. *J. Electroanal. Chem.* 578 (2005): 79.

48. Ghilane, J., O. Fontaine, P. Martin, J.-C. Lacroix, and H. Randriamahazaka. Formation of negative oxidation states of platinum and gold in redox ionic liquid: Electrochemical evidence. *Electrochem. Commun.* 10 (2008): 1205.

49. Ghilane, J., M. Delamar, M. Guilloux-Viry, C. Lagrost, C. Mangeney, and P. Hapiot. Indirect reduction of aryldiazonium salts onto cathodically activated platinum surfaces: formation of metal–organic structures. *Langmuir* 21 (2005): 6422.

50. Simonet, J. *Platinum Metals Rev.* 50 (2006): 180.
51. Simonet, J. The platinized platinum interface in super-dry solvents: Cathodic reversible reactivity and morphology modifications in the presence of tetramethylammonium salts. *J. Electroanal. Chem.* 93 (2006): 3.
52. Cougnon, C., and J. Simonet. Cathodic immobilization of ð-acceptors such as aromatic ketones onto platinum interfaces under superdry conditions. *J. Electroanal. Chem.* 531 (2002): 179.
53. Simonet, J., *Unpublished results.*
54. Ghilane, J., M. Guilloux-Viry, C. Lagrost, J. Simonet, P. Hapiot. Reactivity of platinum metal with organic radical anions from metal to negative oxidation states. *J. Am. Chem. Soc.* 129 (2007): 6654.

3 Impact of Metal–Ligand Bonding Interactions on the Electron-Transfer Chemistry of Transition-Metal Nanoparticles

Shaowei Chen

CONTENTS

3.1 INTRODUCTION

The research interest in nanosized metal particles has been primarily motivated by the unique material properties that exhibit vast deviations from those of their constituent atoms or bulk forms. Consequently, these materials may be exploited as new functional building blocks in diverse applications, such as nanoelectronics, (electro)catalysis, chemical and biological sensing, etc.[1–4] Of these, monolayer-protected nanoclusters (MPCs) represent an intriguing class of nanomaterials. Because of their inorganic/organic nanocomposite nature, the materials' properties can be manipulated readily and independently by the structures of the metal cores as well as the organic capping shells.[1,2,5,6] In fact, a great deal of research effort has been devoted to the examination of the effects of these two structural parameters on the nanoparticle physical and chemical properties, in particular, electron-transfer characteristics, which have been studied both at the single particle level as well as with particle ensembles.

The studies of individual nanoparticles are mainly carried out with scanning probe microscopy. For instance, scanning tunneling microscopy (STM) and spectroscopy (STS) have been used to investigate the charge-transfer dynamics of

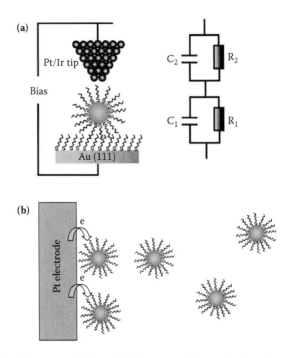

SCHEME 3.1 Schematic of (a) an STM setup and the equivalent circuit of a double-barrier tunneling junction, and (b) electron transfer of nanoparticles at the electrode–electrolyte interface. (Xu, L. P., and S. W. Chen, 2009, *Chem Phys Lett* 468: 222–226. Used with permission.)

nanoparticles entrapped between an STM tip and a conductive substrate, where the tip–nanoparticle–electrode junction may be modeled with an equivalent circuit of a double barrier tunneling junction (DBTJ), as depicted in Scheme 3.1 (A). One junction is formed between the metal nanoparticle and the conductive substrate (R_1 and C_1), and the other between the nanoparticle and the STM tip (R_2 and C_2). The variation of the tunneling current as a function of the applied bias (I–V) typically exhibits a coulomb blockade region centered around zero bias beyond which coulomb staircases may emerge. Such I–V characteristics may be tuned by varying the structure of the nanoparticles (i.e., R_1 and C_1) as well as the STM tip–nanoparticle distance (i.e., R_2 and C_2). In order to observe both the coulomb blockade and staircase features, two necessary conditions have to be satisfied.[7] First, the charging energy, $E_c = e^2/2C$, determined by the addition of one electron (e) to the metal particle with capacitance $C (= C_1 + C_2)$, must exceed the thermal kinetic energy $k_B T$ (k_B the Boltzmann constant and T the temperature). Consequently, the tunneling of electrons through the nanoparticle becomes discrete, as the charging of a second electron to the nanoparticle has to overcome this energetic barrier, leading to the appearance of a staircase feature in the I–V measurement. If this condition is not fulfilled, thermally activated electrons will overcome the coulombic barrier, leading to a nonzero tunneling current even at zero bias and, consequently, only featureless I–V profiles. Second, the electrical contacts to the particles should have tunneling resistances larger than the resistance quantum, $h/4e^2$ (~6.5 kΩ, h the Planck's constant), in order to suppress quantum fluctuations of the electron charge and hence to confine electrons within the particle unless the applied bias is sufficiently large to initiate tunneling across the junction. Typically, nanometer-sized particles with a capacitance of the order of attofarad (10^{-18} F) satisfy both of these conditions and exhibit single electron transfer (SET) phenomenon even at room temperature.[7]

Furthermore, using the DBTJ model, one can quantify the junction electronic properties by using the equivalent circuit for curve fitting. For a given nanoparticle, the capacitance C_1 is fixed, and the variation of tip–particle distance with different set-point currents can change C_2. By treating the particle as a sphere, the overall geometric capacitance[8] of the nanoparticles can be estimated ($C = C_1 + C_2$) as $C = 4\pi r \varepsilon_0 \varepsilon$, where r is the nanoparticle core radius, ε is the (effective) dielectric constant of the nanoparticle-protecting layer, and ε_0 is the vacuum permittivity. Note that C may be determined experimentally by the width of the coulomb blockade ($C = e^2/2E_c$). In addition, since the STM tip is at a fixed distance with respect to the center of the nanoparticles, $R_2 C_2 \gg R_1 C_1$. From the analytical expression derived by Hanna and Tinkham,[9] the slope on the steps (constant number of electrons tunneling, n_0) in the I–V curve is given by $C_1/R_2 C$, and the current step (where $n_0 \to n_0 + 1$) is equal to $e/R_2 C$. These two data can then be used to extract the values of R_2 and C_1, as C has been determined (see the following text), from which C_2 can also be calculated, whereas resistance R_1 is typically evaluated by curve fitting of the I–V profiles, based on the orthodox theory. For instance, curve fitting of the I–V profile of a 1-octanethiolate-passivated gold nanoparticle (core diameter 1.5 nm) immobilized onto an Au(111) surface by 1,8-octanedithiol bifunctional linkers yields $R_1 = 0.1$ GΩ, $C_1 = 1.2 \times 10^{-19}$ F, $R_2 = 4.6$ GΩ, and $C_2 = 0.6 \times 10^{-19}$ F, with a fractional residual charge of $Q_0 = 0.05$e.[10] Note that the resistance R_1 of this junction is in good agreement with the resistance of alkanethiol molecules in a standing-up configuration.[11–13]

On a complementary front, nanoparticle charge-transfer properties can also be evaluated in solution where the nanoparticles undergo reversible charging and discharging at the electrode–electrolyte interface (Scheme 3.1, panel b). Depending on the specific particle structure and hence molecular capacitance, voltammetric features of nanoparticle quantized charging may be observed. This phenomenon has been extensively exemplified by alkanethiolate-passivated transition-metal nanoparticles, and the details can be readily varied by the particle structures, namely, core size, ligand chemical details, as well as solvent media.[14–16] Similar to the coulomb staircase observed with individual nanoparticles,[7] the electrochemical quantized-charging phenomenon only occurs when the energetic barrier for a single-electron transfer is substantially greater than the thermal kinetic energy.[17] This means that, at ambient temperature ($k_B T \approx 26$ meV), the particle molecular capacitance (C) must be of the order of (sub)attofarad, as $C = 4\pi \varepsilon \varepsilon_0 (r/d)(r + d)$, with d being the chain length of the particle-protecting layer. Therefore, for alkanethiolate ligands, the particle core dimensions must fall within a very small range, typically less than 3 nm in diameter, and the particle core size dispersity must be sufficiently small so that the discrete charging features may be revealed and resolved by conventional voltammetric measurements.[18] This is primarily because, in electrochemical quantized charging, an ensemble of nanoparticles are charged or discharged concurrently at the electrode–electrolyte interface, with the "formal" potential ($E_{z,z-1}^{o'}$) determined by the nanoparticle capacitance (and hence molecular structure), $E_{z,z-1}^{o'} = E_{PZC} + (z - \frac{1}{2})e/C$, where z represents the nanoparticle charge state, and E_{PZC} is the particle potential of zero charge (PZC).[18]

More intriguing, for ultrasmall (e.g., subnanometer-sized) nanoparticles, semiconducting characteristics emerge, reflected by the unique optical and luminescent properties that are generally associated with semiconductor quantum dots, because of the emergence of a HOMO-LUMO bandgap.[19–21] Experimentally, a featureless region starts to emerge around the potential of zero charge, analogous to the coulomb-blockade phenomenon that is observed in STS measurements. In addition, for nanoparticle assemblies, the charge-transfer behaviors also vary drastically with the solvent media and the specific electrolyte ions, as a result of particle solvation and ion-paring for charge neutralization. A series of review articles on nanoparticle charge-transfer chemistry in solution can be found in the latest literature and will be not be repeated here.[14,15]

In addition, a great deal of research has also been directed toward the investigation of solid-state conductivity of nanoparticle ensembles. Here, in addition to the structures of the nanoparticle core and surface-protecting ligands, interparticle interactions also play a critical role in regulating interparticle charge-transfer dynamics, thanks to the close proximity of nanoparticles in solid state. In fact, three important factors have been identified that are responsible for the control of nanoparticle electronic conductivity:[22,23] (1) dipolar coupling interactions between electrons of adjacent particles that arise from the overlap of electronic wavefunctions; (2) particle structural dispersity that results in disordered domains within the nanoparticle ensemble and hence enhanced impedance to interparticle charge transfer; and (3) coulombic repulsion that hinders the charging of more than one electron onto an individual nanoparticle.

Experimentally, interparticle charge transfer is found to be governed by the thermal activation mechanism where the particle film conductivity exhibits clear Arrhenius dependence on temperature. Additionally, an exponential decay is generally observed with increasing interparticle separation as a result of the hopping process where the organic matrix between adjacent particles serves as the barrier for interparticle charge transfer.[24] Because of the organic–inorganic composite nature, an increase of the particle core size is found to enhance the dipolar interactions between neighboring particles, leading to enhancement of the particle ensemble conductivity, which can also be effected by the incorporation of more conductive aromatic moieties into the particle-protecting ligand shells. More intriguingly, the interactions between the π electrons of the aromatic moieties may also be exploited as a molecular switch to regulate interparticle charge transfer.[25] This has been demonstrated recently with gold nanoparticles stabilized by a monolayer of phenylethylthiolates (Au-PET). By deliberate variation of the interparticle separation using the Langmuir–Blodgett (LB) technique, the electronic conductivity of the corresponding particle monolayer (Figure 3.1) exhibits a volcano-shaped dependence instead of the exponential decay profile that has been observed typically with nanoparticle dropcast thick films. Interestingly, the maximum conductance of the nanoparticle monolayer coincides with an interparticle distance where the π–π stacking of phenyl electrons from adjacent particles is maximized. This observation is analogous to the periodic oscillation of charge transport when the inner and outer tubes of a carbon nanotube telescope are sliding along the tube long axis.[26]

It can be seen that all these studies share a common theme where the particle charge-transfer chemistry is closely correlated to their molecular structure. In fact, a host of nanoparticle structural parameters have been identified, such as core size, ligand chemical structure, and interparticle separation. These have been the main focus of most earlier work,[14,22–24] and several rather comprehensive reviews have recently been published.[15,27] Interested readers may consult these references for details.

Importantly, these fundamental insights may be exploited for the development of functional nanodevices based on nanoparticle ensembles. For instance, interparticle separation, and hence, charge transfer dynamics, may be readily manipulated when a nanoparticle assembly is exposed to volatile organic vapors, as the penetration of organic solvents into the particle protecting layer varies with the nanoparticle core size as well as the specific molecular structure of the protecting ligands. Such a correlation has indeed been used as the fundamental basis for nanoparticle-based chemical sensors of organic vapors.[28–33]

Yet, despite the large body of work, little attention has been paid to the impact of core–ligand interfacial bonding interactions on the particle charge-transfer properties. This is largely because, in these early studies, mercapto-derivatives have been used extensively as the ligands of choice in nanoparticle passivation by virtue of the strong affinity of thiol groups to transition-metal surfaces. However, the metal–sulfur bonds typically lack interesting chemistry. With the recent emergence of new chemistry in the stabilization of nanoparticle materials, in particular by metal–carbon covalent bonds, it now becomes possible to examine and compare nanoparticle charge-transfer properties within the context of metal–ligand bonding interactions.

For instance, recent high-resolution crystallographic studies have shown that, with a thiolate-protecting shell, gold nanoparticle cores actually consist of at least two types of gold atoms. While the interior gold atoms forms crystalline structures that are consistent with the bulk form, the top layer of gold atoms form orthogonal semirings (i.e., staple moieties) with the thiolate ligands.[20,21,34] Because of this poly(Au(I)-S) complex, it is presumed that the strong dipoles at the core–ligand interface lead to confinement of the conducting electrons within the metal cores, and the interparticle charge transfer dynamics may be controlled

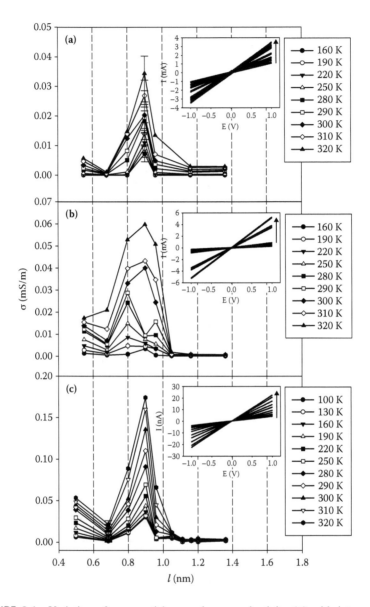

FIGURE 3.1 Variation of nanoparticle monolayer conductivity (σ) with interparticle edge-to-edge separation (l) at different temperatures for three Au-PET particles of different core diameters: (A) 1.39 ± 0.73 nm, (B) 1.64 ± 0.79 nm, and (C) 2.97 ± 0.62 nm. Error bars reflect statistical average of at least three measurements. Insets show the representative I–V profiles of the corresponding nanoparticles at $l = 0.90$ nm with the temperature increased as indicated by the arrow. Potential scan rate 20 mV/s. (Pradhan, S., D. Ghosh, L. P. Xu, and S. W. Chen, 2007, *J Am Chem Soc* 129: 10622–10623. Used with permission.)

separately by the conducting metal cores as well as the insulating organic protecting layers. In contrast, the enhanced metal–ligand bonding interactions as a consequence of the metal–carbon covalent bonds may lead to extended spilling of the nanoparticle core electrons into the organic protecting matrix. Such intraparticle charge delocalization is anticipated to affect not only the charge-transfer properties of functional moieties bound onto the particle surface but also the interparticle charge-transfer dynamics. We will examine these two aspects in this chapter in detail.

It should be noted that the impact of metal–organic interfacial contacts on the charge-transfer properties of functional nanostructures have been recognized, for instance, in carbon nanotube-based nanoelectronics,[35–39] where the nanotubes are generally bonded with metal leads, and the interfacial bonding interactions may dictate the overall conductivity behaviors. For instance, for ohmic contacts, the charge transport is mainly controlled by the intrinsic property of carbon nanotubes, whereas, for those contacts that behave as a Schottky barrier, the interface plays a dominant role in controlling the overall conductivity. Fundamentally, such a discrepancy may be accounted for by the different degree of chemical bonding at the interface. Thus, an examination of the metal–organic interfacial contacts, in particular the nature of the bonding interactions, is of critical importance in advancing our understanding of nanoscale charge-transfer chemistry. This will be the focal point of this chapter.

The chapter will be structured as follows. We will begin with some highlights of the synthetic routes for nanoparticle preparation, where key experimental variables will be identified in the manipulation of the structures of resulting nanoparticles. We will then discuss the electrochemistry of two types of nanoparticles. The first refers to solid-state conductivity of a series of metal nanoparticles that are passivated by metal-carbon single bonds. The effects of the chemical details of the capping fragments on interparticle charge-transfer dynamics will be examined. The other involves ruthenium nanoparticles stabilized by ruthenium–carbene π bonds, where intraparticle charge delocalization leads to nanoparticle-mediated intervalence transfer, as illustrated by ferrocene functional moieties.

3.2 NANOPARTICLE PREPARATION

There have been a number of synthetic protocols for the preparation of transition-metal nanoparticles, for example, vapor condensation,[40] sonochemical reduction,[41,42] chemical liquid deposition,[43] reflux alcohol reduction,[44–47] decomposition of organometallic precursors,[48] hydrogen reduction,[49] etc. Of these, the colloidal reduction route provides a powerful platform for the ready manipulation of particle structure and functionalization.[50–56] One excellent example is the biphasic Brust method,[57] in which nanoparticles are formed by chemical reduction of a metal salt precursor in the presence of stabilizing ligands. In a typical reaction, a calculated amount of a metal salt precursor is dissolved in water, and the metal ions are then transferred into the toluene phase by ion-pairing with a phase-transfer catalyst such as tetraoctylammonium. A desired amount of passivating ligands is then added into the toluene phase, followed by the addition of a strong reducing reagent (e.g., $NaBH_4$), leading to the formation of monolayer-protected nanoparticles. This protocol has been adopted and employed extensively in the synthesis of a wide variety of transition-metal nanoparticles such as Au, Pd, Cu, Ag, Ru, etc.[58–62]

In this colloidal approach, the growth dynamics of the particles has been found to be controlled by at least two competing processes:[63] nucleation of zero-valence metal atoms to form the cores, and passivation of surfactant ligands to limit the growth of the cores. Thus, it can be envisaged that the eventual size of the resulting nanoparticles may be readily varied by the initial metal:ligand feed ratio as well as temperature. Typically, the larger the excess of the protecting ligands, the smaller the particle core size. This has been very nicely demonstrated by Murray and coworkers. They employed this simple strategy to vary the core dimensions of gold nanoparticles from 1 to 5 nm in diameter by systematically changing the initial Au:thiol feed ratio and reaction temperature, as determined by transmission electron microscopy (TEM), nuclear magnetic resonance (NMR), and x-ray measurements.[58] Similar behaviors are also observed with other transition metals.

By adopting this biphasic route, transition-metal nanoparticles protected by other capping groups have also been recently reported. For instance, using diazonium derivatives both as metal ion phase-transfer reagents and precursors of protecting ligands, a variety of transition-metal nanoparticles have been prepared by virtue of the formation of metal–carbon covalent bonds, e.g., Au, Pt, Ru, Pd, Ti, etc.[64–67] In this approach, the diazonium compounds are first prepared by oxidation of aniline precursors with sodium nitride in cold fluoroboric acid, and then form ion-pairs with metal salt ions, which helps transfer the metal ions into an organic phase. Upon the introduction of a reducing agent, the metal salts are reduced to zero-valence metal atoms that aggregate into nanoclusters; concurrently, nitrogen is released from the diazonium compound, and the resulting phenyl radicals bind to the metal nanocluster surface forming metal–carbon single bonds.

Nanoparticles passivated by metal–carbon double bonds have also been achieved and exemplified by ruthenium nanoparticles.[68,69] The synthetic procedure is somewhat different from the biphasic route detailed. Here, ruthenium colloids are first prepared by thermolytic reduction of ruthenium chloride in 1,2-propanediol in the presence of sodium acetate. A toluene solution of diazo derivatives is then added, where the strong affinity of the diazo moiety to a fresh ruthenium surface leads to the formation of ruthenium–carbene π bonds and the concurrent release of nitrogen. The resulting particles become soluble in toluene and can be purified in a typical manner.[68]

In these studies, it has been found that the choice of the organic capping ligands and hence the metal–ligand bonding interactions play a critical role in regulating the nanoparticle core dimensions. For instance, gold nanoparticles passivated by Au-C covalent bonds are more than 8 nm in diameter,[64] markedly larger than that of those stabilized by alkanethiolates prepared in a similar fashion. This may be accounted for by the fact that Au-C bonds (<200 kJ/mol) are significantly weaker than Au-S (418 kJ/mol). In contrast, for Pd nanoparticles, the stronger Pd-C bonds (436 kJ/mol) lead to the formation of somewhat smaller particles than the Pd-S linkage (380 kJ/mol).[70] Overall, the experimental observation is consistent with the nanoparticle growth mechanism.[63]

3.3 SOLID-STATE ELECTRONIC CONDUCTIVITY OF NANOPARTICLE ENSEMBLES

In this section, we will examine the electronic conductivity properties of transition-metal nanoparticles that are passivated by metal–carbon covalent bonds by using the nanoparticles obtained, and compare the results with those from similarly sized nanoparticles that are protected by thiolate ligands. Experimentally, a (micron) thick film of particles is dropcast onto an interdigitated array (IDA) electrode, and the electronic conductivity characteristics are evaluated by electrochemical measurements in vacuum and at controlled temperatures. We will use Pd-C nanoparticles as the initial example.[65] Seven types of Pd-C nanoparticles are prepared with varied protecting ligands: biphenyl (Pd-BP), ethylphenyl (Pd-PhC$_2$), butylphenyl (Pd-PhC$_4$), hexylphenyl (Pd-PhC$_6$), octylphenyl (Pd-PhC$_8$), decylphenyl (Pd-PhC$_{10}$), and dodecylphenyl (Pd-PhC$_{12}$). Table 3.1 lists the nanoparticle core sizes, which are all very close to 3 nm in diameter, as estimated by TEM measurements. Figure 3.2 (a) shows the current-potential (I–V) profiles of the Pd-BP nanoparticles within the temperature range of 80 to 300 K. It can be seen that the I–V responses exhibit a linear (ohmic) character within the entire temperature range under study, suggesting very efficient interparticle charge transfer. This is not surprising, considering the short and aromatic capping ligands. As compared to the control measurements with the same blank electrode (current of the order of pA; dotted line), the currents of the particle solid films are significantly greater,

---

TABLE 3.1

Summary of Conductivity Profiles of Solid Films of Palladium Nanoparticles Stabilized By Varied Ligands

Pd Particle[a]	Core Diameter (nm)	Ensemble Conductivity Profile[b]
Pd-BP	3.1	Metallic
Pd-PhC$_4$	4.8	Metallic
Pd-PhC$_6$	1.9	Conductivity virtually independent of temperature
Pd-PhC$_8$	3.1	Metal–semiconductor transition at 200–240 K
Pd-PhC$_{10}$	3.0	Metal–semiconductor transition at 160–180 K
Pd-PhC$_{12}$	3.3	σ Too low to determine temperature dependence

[a] Abbreviations: BP = biphenyl, PhC$_4$ = phenylbutyl, PhC$_6$ = phenylhexyl, PhC$_8$ = phenyloctyl, PhC$_{10}$ = phenyldecyl, and PhC$_{12}$ = phenyldodecyl.

[b] Within the temperature range of 80–320 K.

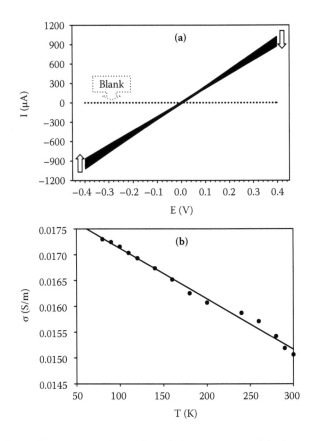

FIGURE 3.2 (a) Current-potential profiles of a Pd-BP nanoparticle-dropcast thick film at varied temperature. Potential scan rate 20 mV/s. Arrows signify the increase of temperature from 80 K to 300 K at an increment of 20 K. (b) Variation of the ensemble conductivity with temperature. (Ghosh, D., and S. W. Chen, 2008, *J Mater Chem* 18: 755–762. Used with permission.)

of the order of mA even at a potential bias of only ±0.4 V. More important, the currents are found to decrease with increasing temperature (80 K to 300 K), a behavior actually anticipated for metallic materials.[70] From the slope of the I–V curves, one can calculate the conductivity of the particle solid films, which is of the order of 10^{-2} S/m (Figure 3.2b). This is about nine orders of magnitude smaller than that of bulk palladium (ca. 0.93×10^7 S/m at room temperature[70]), which can be ascribed to the metal–organic nanocomposite nature of the particles. Additionally, the particle conductivity decreases (almost linearly) with temperature in the range of 80 to 300 K (Figure 3.2b). While the decrease is less dramatic than that of the bulk metal, the observation deviates significantly from previous studies[24,32,71–73] of metal nanoparticles protected with thiol derivatives, where the conductivity is generally found to increase with increasing temperature; that is, the particle ensembles behave as semiconductor materials, and the interparticle charge transfer (hopping) is interpreted on the basis of a thermal activation mechanism.

As the average core diameter of the Pd-BP nanoparticles is ~ 3 nm (Table 3.1), the particle cores are anticipated to behave similarly to bulk palladium. Thus, the ensemble conductivity is anticipated to be mainly determined by the Pd-ligand contact and the ligand matrix itself. Yet, as mentioned earlier, palladium has been used extensively as the metal of choice to create contacts with carbon nanotubes because of the strong Pd-C bonding interactions and low contact resistance.[74–76] Furthermore, the π–π stacking as a result of ligand intercalations between adjacent particles in the solid films may serve as an effective pathway for interparticle charge transfer.[25] Thus, it is most probable that the combination of all these factors gives rise to the metal-like conductivity properties of the Pd-BP particle solids. Similar metallic characteristics are also observed with Ti-BP,[66] as well as Pd-PhC$_2$ and Pd-PhC$_4$ nanoparticles where the particles are protected by short aliphatic fragments (Table 3.1).

By introducing a relatively long saturated spacer into the capping ligands, the resulting particles exhibit drastic differences in the I–V measurements. Figure 3.3 (panels a and b) depicts the I–V data of the Pd-PhC$_{10}$ particles. First, linear I–V profiles can also be seen within the entire temperature range of 80 to 320 K, but the magnitude of the currents (of the order of tens of nA) is substantially smaller than that observed with the Pd-BP particles (Figure 3.2). Nevertheless, the currents are still significantly larger than the background currents (dotted line in panel (a)). This can be ascribed to the long decyl spacer that impedes interparticle charge transfer.

Second, in contrast to the Pd-BP particles, which exhibit a monotonic decrease of the ensemble conductivity with increasing temperature (Figure 3.2), the behaviors of the Pd-PhC$_{10}$ particles are somewhat more complicated. From panel (a) of Figure 3.3, one can see that the current first increases with increasing temperature from 80 to 180 K, exhibiting semiconductor characters; yet, with further increase in the temperature (from 180 to 320 K), the current actually starts to decrease (panel b), akin to that observed in Figure 3.2 in which the particle films behave like metallic materials instead. Panel (c) summarizes the ensemble conductivity with temperature, and it clearly depicts a transition temperature around 180 K where the interparticle charge transfer evolves from semiconducting to metallic characters.

A similar semiconductor–metal transition is also observed with Pd-PhC$_8$ nanoparticles (Table 3.1). For Pd-PhC$_6$ nanoparticles, however, the conductivity is found to be virtually invariant with temperature within the same temperature range of 80–320 K, whereas, for Pd-PhC$_{12}$ nanoparticles, the low ensemble conductance resulting from the long aliphatic barriers renders it difficult to resolve any clear transition.

It has to be noted that, in the aforementioned studies, the particles may be recovered completely from the IDA electrode surface by dissolving them in CH$_2$Cl$_2$ at the end of the measurements, and the I–V responses of the IDA electrode remain virtually invariant to that prior to the deposition of the particle films. This suggests that the nanoparticles remain well passivated, and little ripening of the particles occurs after repetitive potential cycling. Thus, it is highly unlikely that the observed metallic characters arise from ligand desorption and, consequently, direct contact of the particle cores and shorting of the IDA fingers.

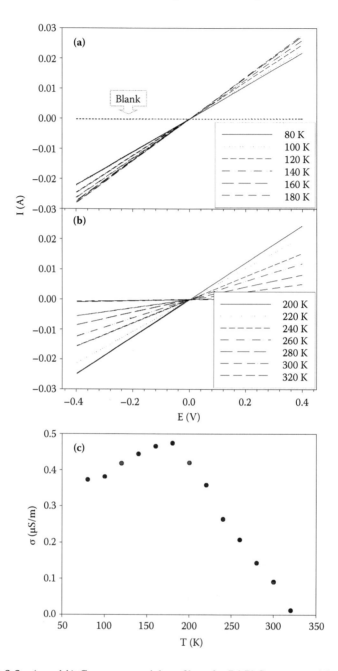

FIGURE 3.3 (a and b) Current-potential profiles of a Pd-PhC$_{10}$ nanoparticle-dropcast thick film at varied temperature. Potential scan rate 20 mV/s. (c) Variation of the ensemble conductivity with temperature. (Ghosh, D., and S. W. Chen, 2008, *J Mater Chem* 18: 755–762. Used with permission.)

Similar transitions have also been reported in resistivity measurements of monolayers of much larger Ag nanoparticles (diameter 7.8 nm) passivated by octanethiolates or decanethiolates,[77] where at temperatures above 200 K, the film resistance exhibits metal-like dependence on temperature, whereas at lower temperatures (60 to 200 K), the conductance is mainly driven by thermal activation and the films behave as a semiconductor (Figure 3.4). Heath and coworkers explained this behavior by using the Mott–Hubbard band model.[78] That is, at low temperatures, the particle super lattice is a Mott insulator with a coulomb gap that originates from the single-particle charging energy (i.e., the Hubbard energy); whereas, at higher temperatures, metallic behaviors start to emerge, where the interparticle charge transfer is facilitated as a result of the temperature-induced band widening. Using a scattering formalism, Remacle and Levine[79] carried out a computational study to examine the current-potential-temperature characteristics of these nanoparticle ensembles, and the temperature-dependent conductivity was attributed to the increase in the number of energy states that contributed to interparticle charge transfer.

In another study, Snow and Wohltjen[32] measured the conductivity of dropcast solid films of dodecanethiolate-stabilized gold (Au:C12) nanoparticles where the core sizes were varied by the initial feed ratios of Au:S: (1:3), 1.72 nm; (1:1), 2.28 nm; (3:1), 3.12 nm; (4:1), 4.58 nm; (5:1), 5.94 nm; and (8:1), 7.22 nm (Figure 3.5). For particles with core diameter <5.0 nm, only semiconductor-like temperature-dependence was observed in the ensemble conductivity within the temperature range of 190 to 300 K, whereas, for larger particles (5:1) and (8:1), the ensemble conductivity was found to evolve from semiconductor-to-metallic behaviors with increasing temperature, and the transition temperature decreased with increasing particle core size, from 260 K (5:1) to 240 K (8:1). Two mechanisms are employed to account for such a transition. One is related to the shrinking of the particle core to the domain of the de Broglie wavelength; and the other is the diminishment of the charging-energy barrier as a result of increasing core size, which facilitates interparticle charge transfer.

Furthermore, the solid-state conductivity of nanoparticle-disordered films might also exhibit a clear metal-insulator transition by the variation of the length of alkanedithiols that chemically crosslinked the nanoparticles. This has been demonstrated by Zabet-Khosousi and coworkers with tetraoctylammonium-stabilized gold nanoparticles (diameter 5.0 ± 0.8 nm).[80] In this study, they found that, when the dithiol linker was longer than a hexyl spacer, the particle film exhibited semiconductor-like temperature dependence of the ensemble conductivity, whereas, at shorter chain lengths, a metallic character was observed (Figure 3.6). Again, the observations were accounted for within the context of the Mott–Hubbard model as a consequence of the competing thermally activated processes and strong temperature-independent elastic scattering.

The manipulation of interparticle separation may also be achieved readily by the Langmuir technique, as demonstrated by Heath and coworkers.[81] In this study, they measured the linear and nonlinear ($\chi^{(2)}$) optical responses of Langmuir monolayers of alkanethiolate-passivated silver nanoparticles as a continuous function

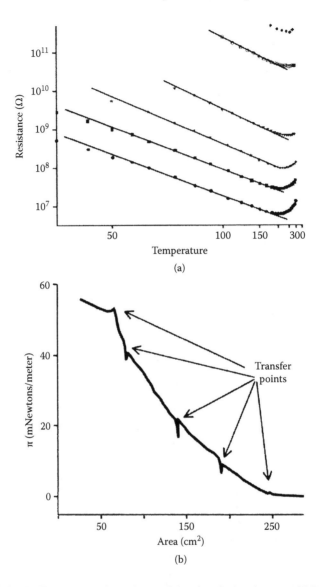

FIGURE 3.4 (a) Temperature dependence of the electrical resistance of LB monolayer films of 1-decanethiolate-protected Ag nanoparticles (7.8 nm in diameter), with the interparticle distance decreasing from top to bottom as determined by the Langmuir isotherm shown in panel (b). (Sampaio, J. F., K. C. Beverly, and J. R. Heath, 2001, *J Phys Chem B* 105: 8797–8800. Used with permission.)

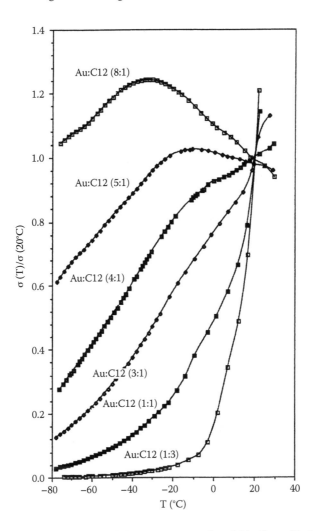

FIGURE 3.5 Linear-scale conductivity–temperature plot of thin films of 1-dodecanethi-olate-passivated gold nanoparticles prepared at different Au:thiol ratios (Au:C12, (X:Y)) where the data for each individual cluster is normalized to its value at 20°C. (Snow, A. W., and H. Wohltjen, 1998, *Chem Mater* 10: 947–950. Used with permission.)

of interparticle separation under near-ambient conditions. It was observed that an insulator-to-metal transition occurred when the nanoparticle monolayers were compressed to a sufficiently small edge-to-edge interparticle distance (e.g., 0.5 nm).

More recently, Choi et al.[82] reported that temperature-induced core motion could also lead to drastic enhancement of the ensemble conductivity as a consequence of the thermal modulation of interparticle separations. Thus, it is very plausible that the aforementioned transitions observed may also be interpreted

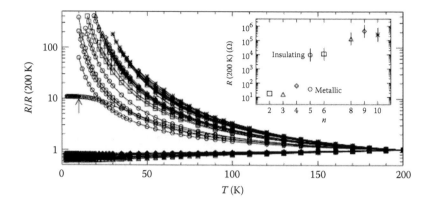

FIGURE 3.6 Variation of the normalized resistance (R) with temperature (T) of 43 nanoparticle film samples at various chain lengths (n) of the alkanedithiol linkers (HS(CH$_2$)nSH). The resistances (R) are normalized to their respective values at 200 K. Inset shows R of the samples at 200 K as a function of n. R was measured after adding indium contacts onto the films on gold electrodes. (Zabet-Khosousi, A., P. E. Trudeau, Y. Suganuma, A. A. Dhirani, and B. Statt, 2006, *Phys Rev Lett* 96: 156403–1. Used with permission.)

in a similar fashion by taking into account the contributing effects of thermally induced core motions as well as thermal activation of interparticle charge transfer. Both are clearly dependent upon specific particle structures such as core size and protecting ligands (more discussion follows).

Control experiments with palladium nanoparticles passivated by alkanethiolates (Pd-SR) further illustrate the significant impacts of metal–ligand contacts on particle conductivity properties (Figure 3.7). Two aspects warrant special attention here. First, the conductivity of Pd-SR nanoparticles is significantly smaller than that observed earlier with the Pd-C nanoparticles, despite the comparable (or even shorter) chain lengths of the organic capping ligands and somewhat larger nanoparticle cores (Table 3.1). The conductivity exhibits an exponential decay with ligand chain length, signifying a hopping process in interparticle charge transfer as observed previously.[24,73,83] In addition, the I–V characteristics were found to exhibit semiconductor behaviors within the same temperature range in which conductivity increased with increasing temperature.

In another study with gold particles of similar sizes but protected by arenethiolates, for example, benzylthiolate (SCPh), phenylethylthiolate (SC$_2$Ph), phenylbutanethiolate (SC$_4$Ph), and cresolthiolate (SPhC), the ensemble conductivity exhibits semiconductor characters within a much narrower temperature range of 223 K to 323 K, analogous to those particles that are passivated by saturated alkanethiolates, for example, butanethiolates (C$_4$), and hexanethiolates (C$_6$). That is, conductivity increases with increasing temperature (Figure 3.8), in sharp contrast to Pd-C particles.[24] Similar semiconductor behaviors have also been observed with *p*-toluenethiolate-protected Pd nanoparticles.[24]

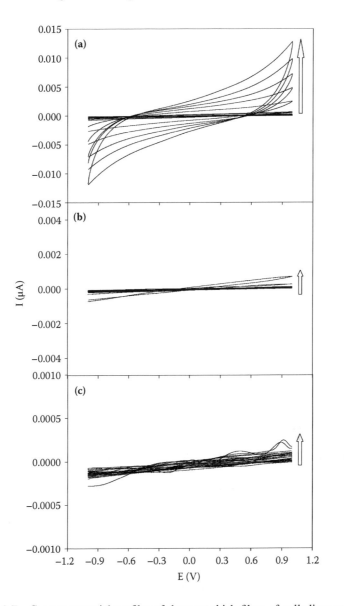

FIGURE 3.7 Current-potential profiles of dropcast thick films of palladium nanoparticles passivated by varied alkanethiolates at controlled temperature: (a) Pd-SC$_6$ (diameter 3.5 ± 0.6 nm), hexanethiolate; (b) Pd-SC$_{10}$ (diameter 3.9 ± 0.9 nm), decanethiolate; and (c) Pd-SC$_{12}$ (diameter 5.0 ± 1.4 nm), dodecanethiolate. Potential scan rate 20 mV/s. Note that, in panel (c), the curves are all smoothed because of low signal-to-noise ratio, and the significance of the wavy features is unknown. Arrows signify the increase of temperature from 80 K to 320 K at an increment of 20 K. (Ghosh, D., and S. W. Chen, 2008, *J Mater Chem* 18: 755–762. Used with permission.)

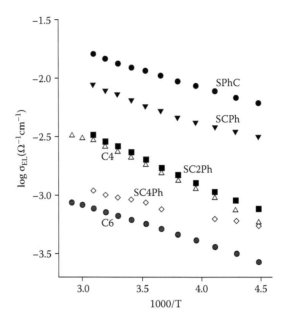

FIGURE 3.8 Arrhenius plot of the electronic conductivities of dropcast films of gold nanoparticles passivated by varied thiolate ligands: SPhC (4-cresolthiolate), SCPh (benzyl-thiolate), SC_2Ph (phenylethylthiolate), SC_4Ph (phenylbutylthiolate), C_4 (1-butanethiolate), and C_6 (1- hexanethiolate). The gold particle core size is 2 to 3 nm in diameter. (Wuelfing, W. P., and R. W. Murray, 2002, *J Phys Chem B* 106: 3139–3145. Used with permission.)

Taken together, the foregoing observations can be rationalized by Mott's model for metal–insulator transition.[78] According to this model, a compressed lattice of hydrogen atoms is anticipated to exhibit metallic behavior when the interatomic distance approaches $\sim 4.5 a_0$, where a_0 is the Bohr radius. By approximating nano-particles as artificial atoms, $a_0 \sim \frac{\hbar}{\sqrt{2m^*\phi}}$, where m^* is the effective mass of elec-trons, ϕ the barrier height, and $\hbar$ the Planck's constant. Thus, the transition of the conductance of a nanoparticle ensemble into the metallic domain may occur when the interparticle charge transfer is sufficiently enhanced. This can be achieved by (1) increasing particle core size (thus enhancing interparticle dipolar coupling), (2) decreasing interparticle separation (thus lowering the tunneling barrier for interparticle charge transfer), (3) decreasing particle–ligand contact resistance (hence enhancing electronic spilling beyond the metal cores), and/or (4) raising ensemble temperature (by virtue of thermal activation and core thermal motions). Thus, whereas the aforementioned Pd-C particles (e.g., Pd-PhC$_8$, and Pd-PhC$_{10}$ in Table 3.1) are substantially smaller than the Au-SR[32] and Ag-SR[77] particles that exhibit similar semiconductor–metal transitions with increasing temperature, the Pd-C particles benefit from the strong Pd-C bonding and hence markedly reduced interfacial contact resistance that offset the core size effects.

Mott's model may also be used to account for the impact of the chain length of the aliphatic spacer on particle ensemble conductivity (Table 3.1). For relatively short aliphatic fragments, the energetic barrier for interparticle charge transfer is anticipated to be low so that a metallic character is observed, whereas, with a longer spacer, semiconducting behaviors start to show up at low temperatures. Interestingly, Pd-PhC$_6$ appears to be the transition point. Although the mechanistic details remain elusive, the hexyl spacer appears to reflect the threshold-length scale that divides the semiconducting and metallic domains of these nanoparticle materials.[84]

More recent studies show that, with ever stronger metal-carbon covalent bonds, the metallic behavior can be observed with even longer aliphatic spacers. This is illustrated by Ru-C nanoparticles.[67] The motivation is primarily twofold. First, because of the significantly stronger Ru-C bond (616.2 kJ/mol) compared to the Pd-C (436 kJ/mol) and Ti-C (423 kJ/mol) bonds,[70] the electronic conductivity of the Ru-C nanoparticles is anticipated to be far greater than that of the Pd-C and Ti-C counterparts, and hence shows a more prominent metallic character within the same temperature range. Second, as interparticle charge transfer is facilitated by the extended spilling of core electrons into the organic protecting layer, the electronic coupling coefficient (β) is anticipated to be substantially smaller than that in the presence of a more localized metal–ligand bonding contact (e.g., metal-sulfur bonds).

The Ru-C particles are prepared and purified in a similar fashion. TEM studies show that the size of the particles is reasonably monodisperse, with the majority of particles falling within the narrow range of 2 to 3 nm in diameter. The results are summarized in Table 3.2. Figure 3.9 shows the I–V profiles of these Ru-C nanoparticles, with the temperature varied within the range of 80 to 300 K. There are at least three aspects that warrant special attention here. First, it can be seen that, similar to those of the aforementioned Pd-C particles,[65] regardless of the chemical structure of the aliphatic ligands, the I–V curves all exhibit a linear (ohmic) profile within the entire temperature range under study, suggesting very efficient interparticle charge transfer. Second, film conductivity (σ) decreases with increasing temperature (highlighted by arrows in Figure 3.9, and further illustrated in panel (A) of Figure 3.10), a behavior typically observed with metallic materials. In previous studies with Pd-C and Ti-C nanoparticles,[65,66] the metallic characters are generally observed only with particles passivated with short

---

TABLE 3.2

Summary of the Average Core Diameter of the Ruthenium Nanoparticle Estimated From TEM Measurements

Nanoparticle	Ru-BP	Ru-PhC$_4$	Ru-PhC$_6$	Ru-PhC$_8$	Ru-PhC$_{10}$	Ru-PhC$_{12}$
Diameter (nm)	3.06 ± 0.84	2.68 ± 0.66	2.95 ± 0.73	3.15 ± 0.70	3.39 ± 0.87	3.22 ± 0.78

Source: Ghosh, D., and S. W. Chen, 2008, *Chem Phys Lett* 465: 115–119. Used with permission.

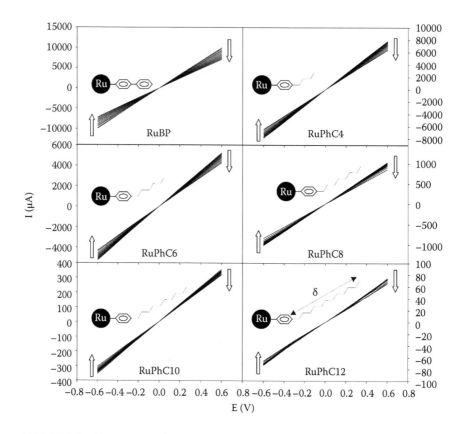

FIGURE 3.9 Current-potential (I–V) profiles of solid films of ruthenium nanoparticles passivated by Ru-C bonds at varied temperatures (shown as figure legends). Potential sweep rate 20 mV/s. Arrows signify the decrease of film conductance with temperature. The respective nanoparticle structure is also included in each panel, and the length of the saturated alkyl spacer is denoted as δ. (Ghosh, D., and S. W. Chen, 2008, *Chem Phys Lett* 465: 115–119. Used with permission.)

aliphatic ligands (e.g., biphenyl and butylphenyl). In the present study, such metallic characteristics remain well defined even with particles protected by relatively long aliphatic ligands such as the decylphenyl and dodecylphenyl fragments. This unusual behavior may be attributed to the significantly stronger Ru-C covalent bond that leads to extended spilling of the core electrons into the organic protecting shell and hence enhanced interparticle charge transfer. In other words, this observation further suggests that the ensemble conductivity may be sensitively varied by the metal–ligand interfacial contact.

Third, at any given temperature, the ensemble conductivity decreases with increasing length (δ, Figure 3.9) of the saturated alkyl spacer in the particle-protecting ligands, which is manifested in panel (b) of Figure 3.10. For instance, Ru-BP nanoparticles exhibited conductivity of the order of 10^2 mS/m, the highest

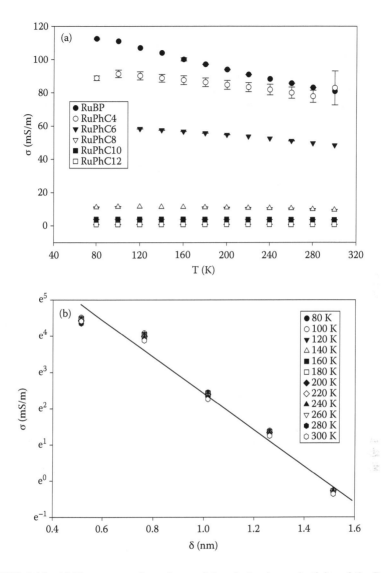

FIGURE 3.10 (a) Temperature dependence of the electronic conductivity of the Ru-C nanoparticles. (b) Variation of nanoparticle electronic conductivity with the chain length of the alkyl segment of the particle-protecting ligands (δ). Symbols are data obtained from Figure 3.9. Error bars are estimated from statistical average of at least three independent measurements. Line is linear regression. (Ghosh, D., and S. W. Chen, 2008, *Chem Phys Lett* 465: 115–119. Used with permission.)

among the series; whereas Ru-PhC$_{12}$ nanoparticles were the least conductive, with the conductivity of the order of 1 mS/m. In comparison to pure ruthenium metal (conductivity of 1.4×10^7 S/m),[70] these nanoparticle samples are 8 to 10 orders of magnitude less conductive as a result of the encapsulation of the metal cores by the insulating organic protecting layer.[65,66]

Furthermore, if one assumes that, in the solid state, the intercalation of the alkyl segments of the protecting ligands between neighboring particles occurs, the barrier of interparticle charge transfer would be effectively determined by the length of this alkyl spacer (δ), as the phenyl moiety is substantially more conductive.[24] In fact, Figure 3.10 (b) depicts an exponential decay of the particle ensemble conductivity with δ within the entire temperature range ($\sigma \propto e^{-\beta\delta}$), with an electronic coupling coefficient (β) of 0.48 Å^{-1} (for comparison, the β value of the Pd-C nanoparticles examined above in Table 3.1 was 0.90 Å^{-1}). This is surprisingly low for charge transfer through a saturated alkyl pathway because such a low value ($\beta = 0.4 \sim 0.5$ Å^{-1}) is typically observed with electron transfer through a π-conjugated linker.[85,86] This may be accounted for, again, by the strong metal–ligand covalent bonding interactions that lead to spilling of the core electrons into the organic protecting shells and hence enhanced interparticle charge transfer.

The results are drastically different when the particles are passivated by metal-sulfur bonds. For instance, for gold nanoparticles passivated by alkanethiolates where the alkyl shells serve as the insulating layer for interparticle electron transfer (hopping),[83] the β coefficient estimated from conductivity measurements of the particle solid films is about 1.2 Å^{-1}, which is very close to those measured in electron transfer through rigid alkyl bridges on an electrode surface.[87] A close value of the electronic coupling coefficient ($\beta = 0.8$ Å^{-1}) is also found with arenethiolate-protected gold nanoparticles,[24] where the alkyl spacer of the arenethiolate ligands serve as the primary barrier for interparticle charge transfer. In these nanoparticle systems, the key structural discrepancy as compared to the Ru-C particles presented earlier is the metal–ligand bonding interactions. That is, with the localized Au-S bonds, minimal impacts are observed of the core electrons on interparticle charge transfer through the organic protecting matrix.

3.4 INTRAPARTICLE INTERVALENCE TRANSFER

The unique metal-carbon covalent linkage also has a strong impact on the electron-transfer properties of redox-active moieties when they are bound onto the particle surface by conjugated linkers. This is manifested by carbene-stabilized ruthenium nanoparticles,[68] where the Ru = C π linkage may be utilized as a unique mechanism for ligand-core charge delocalization, leading to nanoparticle-mediated electronic communication between particle-bound functional moieties and, consequently, the emergence of new optoelectronic properties. Such an unprecedented degree of control of the particle materials properties is further powered by the olefin metathesis reactions of these carbene-stabilized nanoparticles with vinyl-terminated derivatives, where multiple and versatile functional moieties can be incorporated onto individual nanoparticles.[68,88]

Traditionally, electronic communication refers to intervalence charge transfer between two or more identical molecular moieties at mixed valence that are linked by a conjugated molecular bridge, and has been typically observed in organometallic complexes.[89–95] Depending on the degree of charge delocalization or the extent of interactions (α) between the functional moieties, three classes of compounds have been identified by Robin and Day.[96] Class I refers to the compounds that exhibit little or no interaction ($\alpha \approx 0$), whereas in Class III compounds, extensive charge delocalization occurs ($\alpha = 0.707$), and Class II compounds fall into the intermediate range ($0 < \alpha < 0.707$). Quantum-mechanically, delocalization from one metal center to another in a complex is primarily determined by two contributing factors: (1) direct overlap of the orbitals of the two metal centers (i.e., through-space interactions) and (2) metal–ligand–metal overlap that may involve σ or π metal–ligand bonds (i.e., through-bond interactions). When the metal centers are separated by a sufficiently long distance, the contribution from the first factor will be minimal, whereas the second contribution may become predominant, which can be readily varied by the specific ligand structure and metal–ligand interactions. For instance, for biferrocene (a Class II compound),[89] intervalence charge transfer leads to the appearance of two—instead of one—voltammetric waves corresponding to the sequential redox reactions of the two iron centers (Figure 3.11), and in spectroscopic measurements, a rather intense absorption peak is observed at mixed valence in the near-infrared (NIR) region (1800 ~ 1900 nm), which is ascribed to an intramolecular electron-transfer transition. These electrochemical and spectroscopic characteristics can be readily varied by the molecular structure of the chemical bridge that links the two ferrocene moieties. With a conjugated spacer, effective intervalence transfer can

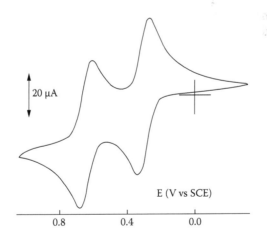

FIGURE 3.11 Cyclic voltammogram of biferrocene in 1:1 (v:v) CH_2Cl_2-CH_3CN containing 0.1 M tetrabutylammonium hexafluorophosphate ($TBAPF_6$). Potential sweep rate 200 mV/s. (Levanda, C., D. O. Cowan, and K. Bechgaard, 1975, *J Am Chem Soc* 97: 1980–1981. Used with permission.)

still occur; however, when bridged by sp³ carbons, the electronic communication diminishes drastically.[97]

It should be noted that, so far, these studies have been mostly confined to relatively simple organometallic molecular complexes, and none have involved nanoparticle molecules. Therefore, a fundamental question arises, when redox-active moieties are bound onto a transition-metal nanoparticle surface through conjugated chemical bonds: will effective electronic communication take place between these particle-bound functional moieties? This issue has been addressed in a recent study[69] in which carbene-functionalized ruthenium nanoparticles are used as the nanoscale structural scaffold, and ferrocene moieties are exploited as the molecular probe because of well-known electrochemical and spectroscopic characteristics.

Ruthenium nanoparticles are produced by thermolytic reduction of ruthenium chloride in 1,2-propanediol.[98] The resulting colloid is then mixed with octyldiazoacetate in toluene, where the strong affinity of the diazo moieties to the ruthenium surface leads to the formation of Ru=C carbene bonds and the concurrent release of nitrogen. The resulting purified nanoparticles are denoted as Ru=C8, which exhibit a core diameter of 2.12 ± 0.72 nm as determined by TEM.[68] The Ru=C8 particles thus obtained are then subject to ligand-exchange reactions with vinylferrocene (vinylFc) (Fc–CH=CH₂) by olefin metathesis reactions on the Ru surface (Scheme 3.2).[68] The resulting particles are referred to as Ru=CH–Fc. The surface coverage of the ferrocene moieties (typically 5% to 20% replacement) may be quantitatively assessed by proton nuclear magnetic resonance (¹H NMR) measurements and readily varied by the initial feed ratio between vinylFc and the Ru=C8 particles. Experimentally, the metal cores of the Ru=CH–Fc particles are dissolved by dilute potassium cyanide (KCN) before the NMR spectra of the remaining organic components are acquired. Metathesis reactions of the Ru=C8

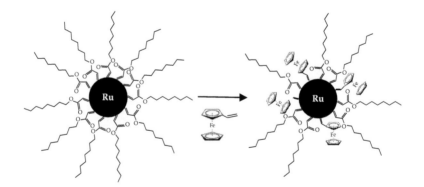

SCHEME 3.2 Preparation of ferrocene-functionalized ruthenium nanoparticles by olefin-metathesis reactions on carbene-stabilized ruthenium nanoparticle surface. (Chen, W., S. W. Chen, F. Z. Ding, H. B. Wang, L. E. Brown, and J. P. Konopelski, 2008, *J Am Chem Soc* 130: 12156–12162. Used with permission.)

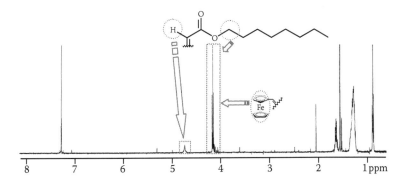

FIGURE 3.12 Representative ^{1}H NMR spectrum (in CDCl$_3$) of Ru=CH–Fc nanoparticles after the metal cores were dissolved by dilute KCN. (Chen, W., S. W. Chen, F. Z. Ding, H. B. Wang, L. E. Brown, and J. P. Konopelski, 2008, *J Am Chem Soc* 130: 12156–12162. Used with permission.)

particles with allylferrocene (CH$_2$=CH–CH$_2$–Fc) are carried out and characterized in a similar fashion.

The structural details of ruthenium nanoparticles are first characterized. Figure 3.12 depicts a representative ^{1}H NMR spectrum of the Ru=CH–Fc nanoparticles after the metal core is dissolved by dilute KCN. The singlet peak at 4.77 ppm (left dashed box) is attributed to the methine proton of the carbene ligands, and the peaks between 4.0 and 4.4 ppm (right dashed box) are ascribed to the combined contribution of the ferrocenyl protons and the α-methylene protons of the carbene ligands. The ratio between the integrated areas of these two peaks is then used to estimate the surface coverage of Fc on the particle surface, which is summarized in Table 3.3 as sample III. Nanoparticles with other ferrocene surface concentrations (5%–20%) are prepared and characterized in a similar fashion, and the results are also included in Table 3.3.

The incorporation of the ferrocene moieties into the particle-protecting layer is also confirmed by Fourier-transformed infrared (FTIR) measurements. For instance, the ferrocene C–H vibrational bands at 1091 and 1462 cm^{-1} are observed with both Ru=CH–Fc nanoparticles and vinylFc monomers. The vinyl (C=C) vibrational band observed at 1630 cm^{-1} with the vinylFc monomers disappears when the ligands undergo metathesis reaction and are bound to the Ru particle surface, forming Ru=C π bonds.

Metathesis reactions of the Ru=C8 particles with allylferrocene (Fc–CH$_2$–CH=CH$_2$) are carried out in a similar fashion as a control experiment. ^{1}H NMR measurements show that 19.8% of the protecting ligands of the resulting particles contained the ferrocene moieties (Table 3.3).

Voltammetric measurements of the ferrocene-functionalized ruthenium nanoparticles prepared as described are then carried out. Figure 3.13 depicts the representative square-wave voltammograms (SWVs) of vinylFc monomers, Ru=C8, Ru=CH–Fc (sample III), and Ru=CH–CH$_2$–Fc nanoparticles in

TABLE 3.3

Summary of Voltammetric Data of Ferrocene-Functionalized Ru Nanoparticles[a]

Particles	Fc%[b]	$E_{p,c}$ (V)	$E_{p,a}$ (V)	$E^{o'}$ (V)	E_p (V)	$\Delta E^{o'}$ (V)
Ru=CH–Fc (I)	5.9	−0.017	−0.013	−0.015	0.004	0.194
		0.177	0.181	0.179	0.004	
Ru=CH–Fc (II)	6.2	−0.014	−0.018	−0.016	0.004	0.180
		0.164	0.164	0.164	0	
Ru=CH–Fc (III)	10.5	−0.019	−0.019	−0.019	0	0.204
		0.177	0.193	0.185	0.016	
Ru=CH–Fc (IV)	12.9	−0.004	−0.007	−0.006	−0.003	0.219
		0.209	0.217	0.213	0.008	
Ru=CH–Fc (V)	13.5	−0.020	−0.014	−0.017	0.006	0.202
		0.180	0.190	0.185	0.010	
Ru=CH–Fc (VI)	21.3	0.027	0.013	0.020	0.014	0.217
		0.233	0.241	0.237	0.008	
Ru=CH–CH$_2$–Fc	17.9	0.024	0.028	0.026	0.004	
CH$_2$=CH–Fc		0.019	0.043	0.031	0.024	

[a] Data acquired from SWV measurements as exemplified in Figure 3.13. $E_{p,c}$ denotes the cathodic peak potential, $E_{p,a}$ the anodic peak potential, $E^{o'}$ the formal potential, ΔE_p the peak splitting ($=|E_{p,a} - E_{p,c}|$), and $\Delta E^{o'}$ the difference between the two formal potentials.

[b] Based on ^{1}H NMR characterizations as exemplified in Figure 3.12.

Source: Chen, W., S. W. Chen, F. Z. Ding, H. B. Wang, L. E. Brown, and J. P. Konopelski, 2008, *J Am Chem Soc* 130: 12156–12162. Used with permission.

dimethylformamide (DMF) containing 0.1 M tetrabutylammonium perchlorate (TBAP). Interestingly, two pairs of voltammetric peaks can be seen with the Ru=CH–Fc particles (short dashed curve), with the formal potentials ($E^{o'}$) at −0.019 V and +0.185 V (versus Fc$^+$/Fc), respectively (and hence a potential spacing of $\Delta E^{o'} = 0.204$ V). These voltammetric peaks are ascribed to the redox reaction of the ferrocene moieties, Fc$^+$ + e $\leftrightarrow$ Fc, and the small peak splitting ($\Delta E_p = 0$ and 16 mV) is consistent with the facile electron-transfer kinetics of the ferrocene moiety. The appearance of two pairs of voltammetric waves with the Ru=CH–Fc particles strongly suggests that intraparticle intervalence transfer occurs between the ferrocene centers through the ruthenium particle cores. Similar responses are observed for other Ru=CH–Fc particles in the series (Table 3.3). This is in sharp contrast to those of the Ru=C8 particles (solid curve) that exhibit only featureless responses, and to vinylferrocene monomers (dotted curve) that show only one pair of voltammetric peaks at +0.031 V within the same potential range.

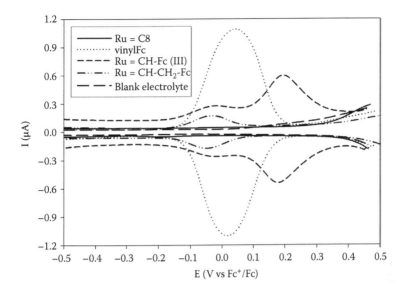

FIGURE 3.13 SWVs of vinylferrocene monomers (0.1 mM, dotted curve), Ru=C8 (4.7 mg/mL, solid curve), Ru=CH–Fc (III, 5 mg/mL, short dashed curve), and Ru=CH–CH$_2$–Fc (2 mg/mL, dashed double dotted curve) particles in DMF containing 0.1 M TBAP. Also shown is the SWV acquired after the electrode is removed from the Ru=CH–Fc particle solution, rinsed with DMF, and then immersed in a blank electrolyte (long dashed). Experimental conditions: Au disk electrode area 0.3 mm^2, increment of potential 4 mV, amplitude 25 mV, and frequency 15 Hz. (Chen, W., S. W. Chen, F. Z. Ding, H. B. Wang, L. E. Brown, and J. P. Konopelski, 2008, *J Am Chem Soc* 130: 12156–12162. Used with permission.)

Table 3.3 summarizes the voltammetric results of the six Ru=CH–Fc nanoparticles. Notably, it can be seen that the potential spacing ($\Delta E^{o\prime}$) is all close to 0.20 V in DMF with the ferrocene surface coverage varied in the range of 5% to 20%. This is very comparable to those observed with biferrocene derivatives with a conjugated spacer,[99,100] suggestive of a Class II compound as defined by Robin and Day.[96] In fact, if we treat the Ru particle core as a conducting medium with fully delocalized electrons, the ferrocene moieties that are bound onto the Ru particle surface can be considered to be approximately equivalent to Fc–CH=CH–Fc.[99] Indeed, the $\Delta E^{o\prime}$ values are comparable (0.20 V versus 0.17 V), where the small discrepancy may partly arise from the different solvent media used (Ru=CH–Fc particles are not soluble in CH$_2$Cl$_2$), as dictated by the Marcus and Hush theory.[101] Note that the $\Delta E^{o\prime}$ value for biferrocene (Fc–Fc) is typically found around 0.35 V, and a smaller value is generally observed when a chemical spacer is inserted between the two ferrocenyl moieties because of the diminishment of direct metal–metal overlap.[102–104]

The results are drastically different when the ferrocenyl groups are bound to the particle core through a saturated aliphatic spacer. This is demonstrated experimentally with the Ru=CH–CH$_2$–Fc particles (dashed double dotted curve, Figure 3.13). Here, only one pair of voltammetric waves can be seen with a formal potential of +0.026 V, very similar to that of the ferrocene monomers (Table 3.3). This suggests that the intervalence transfer is effectively impeded because of the sp^3 carbon spacer, in accord with previous studies of biferrocene derivatives linked by an ethylene spacer $(-CH_2-CH_2-)$[104,105] and metal nanoparticles functionalized with terminal ferrocene derivatives through a saturated linker and/or Au-S bond.[106–109] In these studies, the voltammetric features are found to be consistent with those of monomers of the ferrocene derivatives, indicating little electronic interactions between the ferrocene moieties.

It is unlikely that the appearance of two pairs, instead of one pair, of voltammetric peaks as observed arises from electrostatic interactions between the ferrocene moieties upon electrooxidation. Such effects have been observed in two-dimensional (2D) self-assembled monolayers of ω-ferrocenyl alkanethiols on a gold surface.[110,111] However, the potential splitting ($\Delta E^{o'}$) is typically less than 100 mV, much smaller than what we observed here with the Ru=CH–Fc particles (Figure 3.13 and Table 3.3). Furthermore, the control experiment with the Ru=CH–CH$_2$–Fc nanoparticles (Figure 3.13) strongly discounts this electrostatic hypothesis. This is because of the similar packing and proximity of the ferrocene groups on the Ru=CH–Fc and Ru=CH–CH$_2$–Fc particle surfaces. Additionally, one would anticipate a much stronger electrostatic interaction between the ferrocene groups in Murray's fully ferrocenated gold particles[109] and in structurally analogous ferrocenyl-terminated poly(propylenimine) dendrimers,[112,113] and yet, in these systems, only a single pair of voltammetric waves are observed. Thus, taken together, these results suggest that intraparticle intervalence transfer indeed occurs in the Ru=CH–Fc particles, most probably as a consequence of the π bonding linkage that binds the redoxactive ferrocene moieties onto the particle-core surface, with negligible electrostatic contributions. In other words, direct metal–metal overlap is minimal compared to the through-(ligand)-bond contributions, which is further supported in theoretical modeling and calculations (see the following text).

However, there remains a possibility that the appearance of two pairs of voltammetric peaks may arise from particle adsorption onto the electrode surface. As in solid films, the ferrocene moieties might exhibit different energetic states and accessibility to counterions because of spatial effects. This hypothesis is discounted by results from two additional experiments. First, the cathodic and anodic current density of the redox peaks was found to be linearly proportional to the square root of potential scan rates, suggesting that the charge-transfer processes were under diffusion control. Second, after the electrochemical measurements in the Ru=CH–Fc particle solution, the Au electrode was taken out and rinsed with a copious amount of DMF and then immersed into a same electrolyte solution without the nanoparticles. Only featureless voltammetric responses were observed, as shown in Figure 3.13 (long dashed curve). In short, both measurements signify minimal surface adsorption of the particles.

The notion that intervalence charge transfer occurs within the Ru=CH–Fc nanoparticles is further supported by NIR spectroscopy by using nitrosonium hexafluorophosphate ($NOPF_6$) as the oxidizing reagent. Figure 3.14 (a) shows the

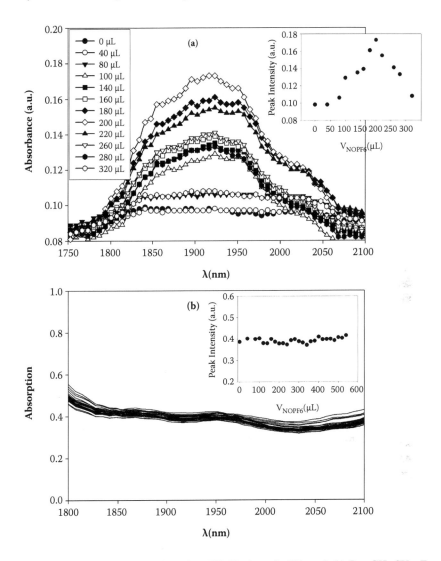

FIGURE 3.14 NIR spectra of (a) Ru=CH–Fc (sample III) and (b) Ru=CH–CH$_2$–Fc nanoparticles with the addition of varied amount of 1 mM NOPF$_6$. In (a), the amounts of NOPF$_6$ added are shown as figure legends, whereas in (b) the amount is varied from 40 to 540 μL at an increment of 20 μL. Particle concentrations ca. 0.5 mg in 3 mL of DMF. Insets show the variation of the absorbance at 1930 nm with the amount of NOPF$_6$ added. (Chen, W., S. W. Chen, F. Z. Ding, H. B. Wang, L. E. Brown, and J. P. Konopelski, 2008, *J Am Chem Soc* 130: 12156–12162. Used with permission.)

NIR absorption spectra of Ru=CH–Fc (sample III) particles with the addition of varied amount of 1 mM $NOPF_6$. A moderately intense absorption peak can be seen at ca. 1930 nm, and the peak intensity exhibits a volcano-shaped dependence on the amount of $NOPF_6$ added (inset). Again, similar responses are observed for other Ru=CH–Fc particles in the series with different ferrocene surface coverage (Table 3.3).

Such a spectroscopic signature is generally taken as a strong indicator that intervalence charge transfer occurs between the ferrocene groups at mixed valence. Importantly, it has been shown previously that intervalence transfer can exclusively take place through the ligand (bond) system,[99,100] where the spatial separation between the two iron centers is too large to have a meaningful direct overlap. Therefore, it is most probable that the intervalence transfer in the Ru=CH–Fc particles also occurs through the ligand bridges because of the Ru=carbene π bonds and the conductive particle core, which is further confirmed in quantum calculations (see the following text).

Table 3.4 lists the amount of $NOPF_6$ needed to reach maximum NIR absorption in the particle solution for the six types of Ru=CH–Fc particles. Although the maximum NIR absorption corresponds to the mixed valence of the ferrocene moieties on the nanoparticle surface (i.e., 50% oxidized to ferrocenium, and the remaining 50% at neutral state), it can be seen from Table 3.4 that the amount of $NOPF_6$ added at maximum NIR absorption is typically greater than 50% of the ferrocene in the sample (although $V_{NOPF6,max}$ does increase almost linearly with the amount of ferrocene in the solution), and the deviation increases with increasing ferrocene surface coverage on the nanoparticle surface. This may, at least in

---

TABLE 3.4

Variation of the Amount ($V_{NOPF6, max}$) of $NOPF_6$ (1 mM) Added at Maximum NIR Absorption with Ferrocene Surface Coverage on the Particle Surface

Particles[a]	Fc %	$V_{NOPF6, max}$ (µL)	Amount of Ferrocene[b] (µmol)
Ru=CH–Fc (I)	5.9	60	0.127
Ru=CH–Fc (II)	6.2	85	0.133
Ru=CH–Fc (III)	10.5	200	0.228
Ru=CH–Fc (IV)	12.9	280	0.277
Ru=CH–Fc (V)	13.5	300	0.290
Ru=CH–Fc (VI)	21.3	300	0.456

[a] The experimental conditions were all the same as those denoted in Figure 3.14.

[b] Estimated by assuming a particle composition of $Ru_{367}L_{71}$ where L denotes the surface-protecting ligands.

Source: Chen, W., S. W. Chen, F. Z. Ding, H. B. Wang, L. E. Brown, and J. P. Konopelski, 2008, *J Am Chem Soc* 130: 12156–12162. Used with permission.

part, be attributable to the capacitive nature of the nanoparticle molecules,[68] where the particle molecular capacitance is anticipated to increase with increasing surface coverage of the ferrocene moieties on the particle surface. In other words, in addition to the oxidation of ferrocene to ferrocenium, the addition of the oxidizing reagent ($NOPF_6$) will also lead to charging of the nanoparticle double-layer capacitance.[114]

In sharp contrast, when the ferrocene groups are linked by sp^3 carbons to the particle surface, no intervalence transfer is anticipated and, consequently, no transition is observed in the NIR region. Experimentally, we observed only a featureless NIR response with the $Ru=CH-CH_2-Fc$ particles (again by using $NOPF_6$ as the oxidizing reagent (Figure 3.14 (b)), and the absorbance remained virtually invariant with the addition of $NOPF_6$ (inset), in good agreement with the voltammetric results depicted in Figure 3.13.

To facilitate a deeper understanding of the mechanism of intraparticle charge delocalization, density functional theory (DFT) calculations, with hybrid B3LYP functional, are performed for a series of ruthenium cluster models capped with two ferrocene moieties via Ru=C π bonds. Figure 3.15 (a) shows the highest-occupied molecular orbitals (HOMOs) for the model system consisting of a cluster of 17 Ru atoms with two Fc–CH= groups at +1 net charge. The frontier orbitals (HOMOs and LUMOs) can be found to be delocalized Ru metal states. As evidenced in the energy diagram (panel b), the MOs energetically close to the HOMOs–LUMOs are dominated by the band states of the bridging Ru metal cluster.

This is typical for resonance electron transfer (ET) through metals (conductor) where electrons (or holes) go through delocalized metal states. In other words, the states in the Ru cluster form an effective conducting medium for the intervalence transfer between different ferrocenyl groups. A positive charge (hole) is first transferred from the Fe(III) center to an occupied Ru band state, and then reaches the Fe(II) center of the other ferrocenyl group. Electron transfer is along the opposite direction. These results suggest that the through-bond contributions play a predominant role in nanoparticle-mediated intervalence transfer, consistent with the experimental data (see the following text). In contrast, in the insulator–semiconductor

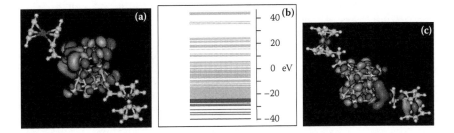

FIGURE 3.15 (a) HOMO topology, (b) energy diagram of a $Ru_{17}[=CH-Fc]_2$ cluster at +1 charge state, and (c) HOMO topology of a $Ru_{17}[=CH-CH_2-Fc]_2$ cluster at +1 charge state. (Chen, W., S. W. Chen, F. Z. Ding, H. B. Wang, L. E. Brown, and J. P. Konopelski, 2008, *J Am Chem Soc* 130: 12156–12162. Used with permission.)

bridge-assisted superexchange electron transfer, the electrons tunnel through "virtual" states.[115] This is demonstrated with $Fc(CH=CH)_3Fc$ monocation, where the HOMO is delocalized over the entire molecule. Similar behaviors are observed in monocations of other diferrocenylpolyene derivatives, $Fc(CH=CH)_nFc$.[69]

To have a more quantitative comparison to the experimental data, we have also employed the constrained DFT (CDFT) method developed by Wu and Van Voorhis.[116] In this method, charge-localized diabatic states (donor and acceptor) are explicitly constructed in the modified DFT calculations, which are performed with a patched version of NWCHEM,[117] using the same basis functions as in the aforementioned standard DFT study. First, with respect to $Fc-CH=CH-Fc$ monocation, the relative electronic coupling is estimated to be 1.36 for $Ru_{17}=[CH-Fc]_2$ monocation (Figure 3.15 (a)), suggesting that the Ru=CH–Fc particle core indeed provides an environment equivalent to the conjugated spacer in $Fc-CH=CH-Fc$. Furthermore, unlike in the diferrocenylpolyene system in which electronic coupling follows the exponential decay law, the coupling in the $Fc-CH=(Ru)_n=CH-Fc$ system changes much more slowly with respect to distance variation (e.g., by particle diameter). This finding is consistent with the fact that the Ru nanoparticle is a conductor, which is different from the insulator/semiconductor behavior of the conjugated $(CH=CH)_n$ group in diferrocenylpolyenes.[115] To further verify that the electronic communication is indeed through the Ru particles (through-bond interaction), we have performed calculations where all but the two binding Ru atoms are removed. Zero electronic coupling is observed between the two Fc moieties, indicating that through-space interaction is negligible. Again, this is consistent with the experimental observations.

We have also investigated the effects of chemical spacers on intraparticle charge delocalization by examining the HOMO topology of monocations of $Ru_{17}=[CH-X-Fc]_2$ particle models. For saturated alkyl spacers, e.g., $X = (CH_2)_n$, charge delocalization is significantly impeded at mixed valence, akin to those observed earlier with the $Ru=CH-CH_2-Fc$ particles (Figure 3.15 (c)). In contrast, with conjugated spacers such as $-CH=CH-$ and $-C\equiv C-CH=CH-$, the HOMOs for monocations of the corresponding particle models exhibit delocalization characteristics over the Ru particle core, as shown in Figure 3.14 (a).

From these analyses, it can be seen that effective electronic communication may occur between the metal centers through the band states of the Ru metal core when a conjugated spacer is used to link the ferrocene moieties onto the particle surface, whereas saturated linkers diminish drastically the intervalence transfer at mixed valence. Additionally, the nanoparticle-mediated intervalence transfer appears to be primarily dominated by through-bond contributions rather than direct overlap between the metal centers.

In a more recent study,[118] ferrocene moieties are bound onto a ruthenium thin-film surface also by olefin metathesis reactions. With a conjugated chemical linker, two pairs of voltammetric waves are observed, suggesting that interfacial intervalence transfer does take place between the ferrocene metal centers. In addition, the potential spacing between the two voltammetric waves increases with increasing ferrocene surface coverage, suggesting enhanced contributions from through-space interactions. Note that the potential spacing is somewhat larger than that observed earlier with Ru=CH–Fc nanoparticles, indicating that interfacial charge delocalization is better facilitated by the ruthenium thin-film, which behaves as a better electron sink. In contrast, upon the introduction of a sp^3 carbon into the chemical linkage, the ferrocene moieties start to behave independently of each other, a behavior similar to that of Ru=CH–CH$_2$–Fc nanoparticles.

Four decades ago, Hush made a prediction that, when redox-active moieties were bound onto an electrode surface by conjugated linkers, effective interfacial intervalence transfer might occur.[119] These studies are believed to be the first experimental verification to directly support such a hypothesis, which is anticipated to open a new chapter in electrochemical study of intervalence transfer.

3.5 CONCLUDING REMARKS AND OUTLOOK

In this chapter, the impact of metal–ligand covalent interactions on nanoparticle electron-transfer properties is examined, which provides a unique perspective in the investigation of electron transfer dynamics at nanoscale interfaces. While traditional metal-sulfur bonds lack interesting chemistry, the emergence of metal-carbon covalent bonding interactions endows unprecedented materials properties to the resulting particles. In particular, the strong metal-carbon covalent interactions lead to extensive spilling of the nanoparticle core electrons into the organic protecting matrix and hence enhanced interparticle charge transfer. In addition, metal–ligand conjugated bonds may be exploited for intraparticle charge delocalization, a behavior that is manifested by optoelectronic properties that have never been seen or accessible before.

As mentioned earlier, an insightful understanding of the metal–ligand interactions will have substantial impact on and fundamental implication in diverse areas of research such as molecular electronics and carbon nanotube-based nanoelectronics, where metal leads are an essential device component and play a critical role in determining the overall charge transport characteristics. It should be noted that numerous organometallic complexes feature covalent interactions involving a wide array of metal and other nonmetal elements. In principle, one will anticipate that the same chemistry may be applied to the chemical passivation of nanoparticles such that nanoparticle material properties can be manipulated in an unprecedented manner. The challenge mainly lies in the design and preparation of appropriate precursors that are compatible with nanoparticle growth. Thus, an interdisciplinary approach is necessary for the success of such an endeavor.

ACKNOWLEDGMENTS

I am grateful to my collaborators and students for the contributions to the work cited herein. Their names are listed in the respective references. I would also like to thank the National Science Foundation for financial support to my research program. The TEM work described in this chapter was mostly done at the Molecular Foundry and National Center for Electron Microscopy at Lawrence Berkeley National Laboratory, which is supported by the U.S. Department of Energy.

REFERENCES

1. Templeton, A. C., Wuelfing, M. P., Murray, R. W. Monolayer protected cluster molecules. *Acc Chem Res* 2000 *33*, 27–36.
2. Whetten, R. L., Shafigullin, M. N., Khoury, J. T., Schaaff, T. G., Vezmar, I., Alvarez, M. M., Wilkinson, A. Crystal structures of molecular gold nanocrystal arrays. *Acc Chem Res* 1999, *32*, 397–406.
3. Templeton, A. C., Hostetler, M. J., Warmoth, E. K., Chen, S. W., Hartshorn, C. M., Krishnamurthy, V. M., Forbes, M. D. E., Murray, R. W. Gateway reactions to diverse, polyfunctional monolayer-protected gold clusters. *J Am Chem Soc* 1998, *120*, 4845–4849.
4. Brust, M., Bethell, D., Kiely, C. J., Schiffrin, D. J. Self-assembled gold nanoparticle thin films with nonmetallic optical and electronic properties. *Langmuir* 1998, *14*, 5425–5429.
5. Collier, C. P., Vossmeyer, T., Heath, J. R. Nanocrystal superlattices. *Annu Rev Phys Chem* 1998, *49*, 371–404.
6. Shenhar, R., Rotello, V. M. Nanoparticles: Scaffolds and building blocks. *Acc Chem Res* 2003, *36*, 549–561.
7. Likharev, K. K. Correlated discrete transfer of single electrons in ultrasmall tunnel-junctions. *IBM J Res Dev* 1988, *32*, 144–158.
8. Jiang, P., Liu, Z. F., Cai, S. M. In situ CdS nanocluster formation on scanning tunneling microscopy tips for reliable single-electron tunneling at room temperature. *Appl Phys Lett* 1999, *75*, 3023–3025.
9. Hanna, A. E., Tinkham, M. Variation of the Coulomb staircase in a 2-junction system by fractional electron charge. *Phys Rev B* 1991, *44*, 5919–5922.
10. Yang, G. H., Tan, L., Yang, Y. Y., Chen, S. W., Liu, G. Y. Single electron tunneling and manipulation of nanoparticles on surfaces at room temperature. *Surf Sci* 2005, *589*, 129–138.
11. Cui, X. D., Primak, A., Zarate, X., Tomfohr, J., Sankey, O. F., Moore, A. L., Moore, T. A., Gust, D., Harris, G., Lindsay, S. M. Reproducible measurement of single-molecule conductivity. *Science* 2001, *294*, 571–574.
12. Adams, D. M., Brus, L., Chidsey, C. E. D., Creager, S., Creutz, C., Kagan, C. R., Kamat, P. V., Lieberman, M., Lindsay, S., Marcus, R. A., Metzger, R. M., Michel-Beyerle, M. E., Miller, J. R., Newton, M. D., Rolison, D. R., Sankey, O., Schanze, K. S., Yardley, J., Zhu, X. Y. Charge transfer on the nanoscale: Current status. *J Phys Chem B* 2003, *107*, 6668–6697.
13. Gomar-Nadal, E., Ramachandran, G. K., Chen, F., Burgin, T., Rovira, C., Amabilino, D. B., Lindsay, S. M. Self-assembled monolayers of tetrathiafulvalene derivatives on Au(111): Organization and electrical properties. *J Phys Chem B* 2004, *108*, 7213–7218.
14. Chen, S. W. Chemical manipulations of nanoscale electron transfers. *J Electroanal Chem* 2004, *574*, 153–165.
15. Murray, R. W. Nanoelectrochemistry: Metal nanoparticles, nanoelectrodes, and nanopores. *Chem Rev* 2008, *108*, 2688–2720.
16. Su, B., Girault, H. H. Absolute standard redox potential of monolayer-protected gold nanoclusters. *J Phys Chem B* 2005, *109*, 11427–11431.

17. Chen, S. W., Ingram, R. S., Hostetler, M. J., Pietron, J. J., Murray, R. W., Schaaff, T. G., Khoury, J. T., Alvarez, M. M., Whetten, R. L. Gold nanoelectrodes of varied size: Transition to molecule-like charging. *Science* 1998, *280*, 2098–2101.

18. Chen, S. W., Murray, R. W., Feldberg, S. W. Quantized capacitance charging of monolayer-protected Au clusters. *J Phys Chem B* 1998, *102*, 9898–9907.

19. Yang, Y. Y., Chen, S. W. Surface manipulation of the electronic energy of subnanometer-sized gold clusters: An electrochemical and spectroscopic investigation. *Nano Lett* 2003, *3*, 75–79.

20. Zhu, M., Aikens, C. M., Hollander, F. J., Schatz, G. C., Jin, R. Correlating the crystal structure of a thiol-protected Au-25 cluster and optical properties. *J Am Chem Soc* 2008, *130*, 5883–5885.

21. Heaven, M. W., Dass, A., White, P. S., Holt, K. M., Murray, R. W. Crystal structure of the gold nanoparticle [N(C8H17)(4)][Au-25(SCH2CH2Ph)(18)]. *J Am Chem Soc* 2008, *130*, 3754–3755.

22. Remacle, F. On electronic properties of assemblies of quantum nanodots. *J Phys Chem A* 2000, *104*, 4739–4747.

23. Beverly, K. C., Sampaio, J. F., Heath, J. R. Effects of size dispersion disorder on the charge transport in self-assembled 2-D Ag nanoparticle arrays. *J Phys Chem B* 2002, *106*, 2131–2135.

24. Wuelfing, W. P., Murray, R. W. Electron hopping through films of arenethiolate monolayer-protected gold clusters. *J Phys Chem B* 2002, *106*, 3139–3145.

25. Pradhan, S., Ghosh, D., Xu, L. P., Chen, S. W. Interparticle charge transfer mediated by pi-pi stacking of aromatic moieties. *J Am Chem Soc* 2007, *129*, 10622–10623.

26. Grace, I. M., Bailey, S. W., Lambert, C. J. Electron transport in carbon nanotube shuttles and telescopes. *Phys Rev B* 2004, *70*, 153405–1.

27. Liu, Y. H., Nalwa, H. S. *Handbook of electrochemical nanotechnology*. American Scientific Publishers: Los Angeles, California, CA, 2009, Vol. 1.

28. Ahn, H., Chandekar, A., Kang, B., Sung, C., Whitten, J. E. Electrical conductivity and vapor-sensing properties of ω-(3-thienyl)alkanethiol-protected gold nanoparticle films. *Chem Mater* 2004, *16*, 3274–3278.

29. Grate, J. W., Nelson, D. A., Skaggs, R. Sorptive behavior of monolayer-protected gold nanoparticle films: Implications for chemical vapor sensing. *Anal Chem* 2003, *75*, 1868–1879.

30. Yang, C. Y., Li, C. L., Lu, C. J. A vapor selectivity study of microsensor arrays employing various functionalized ligand protected gold nanoclusters. *Anal Chim Acta* 2006, *565*, 17–26.

31. Joseph, Y., Besnard, I., Rosenberger, M., Guse, B., Nothofer, H. G., Wessels, J. M., Wild, U., Knop-Gericke, A., Su, D. S., Schlogl, R., Yasuda, A., Vossmeyer, T. Self-assembled gold nanoparticle/alkanedithiol films: Preparation, electron microscopy, XPS-analysis, charge transport, and vapor-sensing properties. *J Phys Chem B* 2003, *107*, 7406–7413.

32. Snow, A. W., Wohltjen, H. Size-induced metal to semiconductor transition in a stabilized gold cluster ensemble. *Chem Mater* 1998, *10*, 947–950.

33. Xu, L. P., Chen, S. W. Scanning tunneling spectroscopy of gold nanoparticles: Influences of volatile organic vapors and particle core dimensions. *Chem Phys Lett* 2009, *468*, 222–226.

34. Jadzinsky, P. D., Calero, G., Ackerson, C. J., Bushnell, D. A., Kornberg, R. D. Structure of a thiol monolayer-protected gold nanoparticle at 1.1 angstrom resolution. *Science* 2007, *318*, 430–433.

35. Deretzis, I., La Magna, A. Role of contact bonding on electronic transport in metal-carbon nanotube-metal systems. *Nanotechnology* 2006, *17*, 5063–5072.

36. Ngo, Q., Petranovic, D., Krishnan, S., Cassell, A. M., Ye, Q., Li, J., Meyyappan, M., Yang, C. Y. Electron transport through metal-multiwall carbon nanotube interfaces. *IEEE Trans Nanotechnol* 2004, *3*, 311–317.

37. Dockendorf, C. P. R., Steinlin, M., Poulikakos, D., Choi, T. Y. Individual carbon nanotube soldering with gold nanoink deposition. *Appl Phys Lett* 2007, *90*, 193116–1.

38. Chu, C. W., Na, J. S., Parsons, G. N. Conductivity in alkylamine/gold and alkanethiol/gold molecular junctions measured in molecule/nanoparticle/molecule bridges and conducting probe structures. *J Am Chem Soc* 2007, *129*, 2287–2296.

39. Vitale, V., Curioni, A., Andreoni, W. Metal-carbon nanotube contacts: The link between Schottky barrier and chemical bonding. *J Am Chem Soc* 2008, *130*, 5848–5849.

40. Devenish, R. W., Goulding, T., Heaton, B. T., Whyman, R. Preparation, characterization and properties of groups VIII and IB metal nanoparticles. *J Chem Soc-Dalton Trans* 1996, 673–679.

41. Mizukoshi, Y., Oshima, R., Maeda, Y., Nagata, Y. Preparation of platinum nanoparticles by sonochemical reduction of the Pt(II) ion. *Langmuir* 1999, *15*, 2733–2737.

42. Nemamcha, A., Rehspringer, J. L., Khatmi, D. Synthesis of palladium nanoparticles by sonochemical reduction of palladium(II) nitrate in aqueous solution. *J Phys Chem B* 2006, *110*, 383–387.

43. Cardenas, G., Munoz, C., Vera, V. Au and Pd nanoparticles prepared from non-aqueous solvents X. *Bol Soc Chil Quim* 1996, *41*, 235–241.

44. Nunomura, N., Hori, H., Teranishi, T., Miyake, M., Yamada, S. Magnetic properties of nanoparticles in Pd/Ni alloys. *Phys Lett A* 1998, *249*, 524–530.

45. Xiong, Y. J., Cai, H. G., Wiley, B. J., Wang, J. G., Kim, M. J., Xia, Y. N. Synthesis and mechanistic study of palladium nanobars and nanorods. *J Am Chem Soc* 2007, *129*, 3665–3675.

46. Mayer, A. B. R., Mark, J. E. Transition metal nanoparticles protected by amphiphilic block copolymers as tailored catalyst systems. *Coll Polym Sci* 1997, *275*, 333–340.

47. Taratula, O., Chen, A. M., Zhang, J. M., Chaudry, J., Nagahara, L., Banerjee, I., He, H. X. Highly aligned ribbon-shaped Pd nanoparticle assemblies by spontaneous organization. *J Phys Chem C* 2007, *111*, 7666–7670.

48. Chaudret, B. Organometallic approach to nanoparticles synthesis and self-organization. *C R Phys* 2005, *6*, 117–131.

49. Chen, S. W., Huang, K. Electrochemical studies of water-soluble palladium nanoparticles. *J Cluster Sci* 2000, *11*, 405–422.

50. Reetz, M. T., Winter, M., Breinbauer, R., Thurn-Albrecht, T., Vogel, W. Size-selective electrochemical preparation of surfactant-stabilized Pd-, Ni- and Pt/Pd colloids. *Chem- Eur J* 2001, *7*, 1084–1094.

51. Li, F. L., Zhang, B. L., Dong, S. J., Wang, E. K. A novel method of electrodepositing highly dispersed nano palladium particles on glassy carbon electrode. *Electrochim Acta* 1997, *42*, 2563–2568.

52. Lu, D. L., Tanaka K., Au, Cu, Ag, Ni, and Pd particles grown in solution at different electrode potentials. *J Phys Chem B* 1997, *101*, 4030–4034.

53. Schmid, G. Large clusters and colloids-metals in the embryonic state. *Chem Rev* 1992, *92*, 1709–1727.

54. Bonnemann, H., Brijoux, W., Brinkmann, R., Fretzen, R., Joussen, T., Koppler, R., Korall, B., Neiteler, P., Richter, J. Preparation, characterization, and application of fine metal particles and metal colloids using hydrotriorganoborates. *J Mol Catal* 1994, *86*, 129–177.

55. Ott, L. S., Finke, R. G. Transition-metal nanocluster stabilization for catalysis: A critical review of ranking methods and putative stabilizers. *Coord Chem Rev* 2007, *251*, 1075–1100.

56. Grzelczak, M., Perez-Juste, J., Mulvaney, P., Liz-Marzan, L. M. Shape control in gold nanoparticle synthesis. *Chem Soc Rev* 2008, *37*, 1783–1791.

57. Brust, M., Walker, M., Bethell, D., Schiffrin, D. J., Whyman, R. Synthesis of thiol-derivatized gold nanoparticles in a 2-phase liquid-liquid system. *J Chem Soc-Chem Commun* 1994, 801–802.

58. Hostetler, M. J., Wingate, J. E., Zhong, C. J., Harris, J. E., Vachet, R. W., Clark, M. R., Londono, J. D., Green, S. J., Stokes, J. J., Wignall, G. D., Glish, G. L., Porter, M. D., Evans, N. D., Murray, R. W. Alkanethiolate gold cluster molecules with core diameters from 1.5 to 5.2 nm: Core and monolayer properties as a function of core size. *Langmuir* 1998, *14*, 17–30.

59. Chen, S. W., Huang, K., Stearns, J. A. Alkanethiolate-protected palladium nanoparticles. *Chem Mater* 2000, *12*, 540–547.

60. Chen, S. W., Sommers, J. M. Alkanethiolate-protected copper nanoparticles: Spectroscopy, electrochemistry, and solid-state morphological evolution. *J Phys Chem B* 2001, *105*, 8816–8820.

61. Tong, M. C., Chen, W., Sun, J., Ghosh, D., Chen, S. W. Dithiocarbamate-capped silver nanoparticles. *J Phys Chem B* 2006, *110*, 19238–19242.

62. Chen, W., Ghosh, D., Sun, J., Tong, M. C., Deng, F. J., Chen, S. W. Dithiocarbamate-protected ruthenium nanoparticles: Synthesis, spectroscopy, electrochemistry and STM studies. *Electrochim Acta* 2007, *53*, 1150–1156.

63. Chen, S. W., Templeton, A. C., Murray, R. W. Monolayer-protected cluster growth dynamics. *Langmuir* 2000, *16,* 3543–3548.

64. Mirkhalaf, F., Paprotny, J., Schiffrin, D. J. Synthesis of metal nanoparticles stabilized by metal-carbon bonds. *J Am Chem Soc* 2006, *128*, 7400–7401.

65. Ghosh, D., Chen, S. W. Palladium nanoparticles passivated by metal-carbon covalent linkages. *J Mater Chem* 2008, *18*, 755–762.

66. Ghosh, D., Pradhan, S., Chen, W., Chen, S. W. Titanium nanoparticles stabilized by Ti-C covalent bonds. *Chem Mater* 2008, *20*, 1248–1250.

67. Ghosh, D., Chen, S. W. Solid-state electronic conductivity of ruthenium nanoparticles passivated by metal-carbon covalent bonds. *Chem Phys Lett* 2008, *465*, 115–119.

68. Chen, W., Davies, J. R., Ghosh, D., Tong, M. C., Konopelski, J. P., Chen, S. W. Carbene-functionalized ruthenium nanoparticles. *Chem Mater* 2006, *18*, 5253–5259.

69. Chen, W., Chen, S. W., Ding, F. Z., Wang, H. B., Brown, L. E., Konopelski, J. P. Nanoparticle-mediated intervalence transfer. *J Am Chem Soc* 2008, *130*, 12156–12162.

70. Lide, D. R. *CRC Handbook of Chemistry and Physics : A Ready-Reference Book of Chemical and Physical Data*. 85th ed., CRC Press: Boca Raton, Fla., 2004.

71. Clarke, L., Wybourne, M. N., Brown, L. O., Hutchison, J. E., Yan, M., Cai, S. X., Keana, J. F. W. Room-temperature Coulomb-blockade-dominated transport in gold nanocluster structures. *Semiconductor Sci Technol* 1998, *13*, A111–A114.

72. Doty, R. C., Yu, H. B., Shih, C. K., Korgel, B. A. Temperature-dependent electron transport through silver nanocrystal superlattices. *J Phys Chem B* 2001, *105*, 8291–8296.

73. Wang, L. Y., Shi, X. J., Kariuki, N. N., Schadt, M., Wang, G. R., Rendeng, Q., Choi, J., Luo, J., Lu, S., Zhong, C. J. Array of molecularly mediated thin film assemblies of nanoparticles: Correlation of vapor sensing with interparticle spatial properties. *J Am Chem Soc* 2007, *129*, 2161–2170.

74. Mann, D., Javey, A., Kong, J., Wang, Q., Dai, H. J. Ballistic transport in metallic nanotubes with reliable Pd ohmic contacts. *Nano Lett* 2003, *3*, 1541–1544.

75. Tarakeshwar, P., Kim, D. M. Modulation of the electronic structure of semiconducting nanotubes resulting from different metal contacts. *J Phys Chem B* 2005, *109*, 7601–7604.

76. Woo, Y., Duesberg, G. S., Roth, S. Reduced contact resistance between an individual single-walled carbon nanotube and a metal electrode by a local point annealing. *Nanotechnology* 2007, *18*, 095203.

77. Sampaio, J. F., Beverly, K. C., Heath, J. R. DC transport in self-assembled 2D layers of Ag nanoparticles. *J Phys Chem B* 2001, *105*, 8797–8800.

78. Gebhard, F. *The Mott Metal-Insulator Transition : Models and Methods*. Springer: New York, 1997.

79. Remacle, F., Levine, R. D. Current-voltage-temperature characteristics for 2D arrays of metallic quantum dots. *Israel J Chem* 2002, *42*, 269–280.

80. Zabet-Khosousi, A., Trudeau, P. E., Suganuma, Y., Dhirani, A. A., Statt, B. Metal to insulator transition in films of molecularly linked gold nanoparticles. *Phys Rev Lett* 2006, *96*, 156403-1.

81. Collier, C. P., Saykally, R. J., Shiang, J. J., Henrichs, S. E., Heath, J. R. Reversible tuning of silver quantum dot monolayers through the metal-insulator transition. *Science* 1997, *277*, 1978–1981.

82. Choi, J. P., Coble, M. M., Branham, M. R., DeSimone, J. M., Murray, R. W. Dynamics of CO2-Plasticized electron transport in au nanoparticle films: Opposing effects of tunneling distance and local site mobility. *J Phys Chem C* 2007, *111*, 3778–3785.

83. Terrill, R. H., Postlethwaite, T. A., Chen, C. H., Poon, C. D., Terzis, A., Chen, A. D., Hutchison, J. E., Clark, M. R., Wignall, G., Londono, J. D., Superfine, R., Falvo, M., Johnson, C. S., Samulski, E. T., Murray, R. W. Monolayers in three dimensions: NMR, SAXS, thermal, and electron hopping studies of alkanethiol stabilized gold clusters. *J Am Chem Soc* 1995, *117*, 12537–12548.

84. Chen, S. W. Discrete charge transfer in nanoparticle solid films. *J Mater Chem* 2007, *17*, 4115–4121.

85. Sachs, S. B., Dudek, S. P., Hsung, R. P., Sita, L. R., Smalley, J. F., Newton, M. D., Feldberg, S. W., Chidsey, C. E. D. Rates of interfacial electron transfer through π-conjugated spacers. *J Am Chem Soc* 1997, *119*, 10563–10564.

86. Creager, S., Yu, C. J., Bamdad, C., O'Connor, S., MacLean, T., Lam, E., Chong, Y., Olsen, G. T., Luo, J. Y., Gozin, M., Kayyem, J. F. Electron transfer at electrodes through conjugated "molecular wire" bridges. *J Am Chem Soc* 1999, *121*, 1059–1064.

87. Holmlin, R. E., Haag, R., Chabinyc, M. L., Ismagilov, R. F., Cohen, A. E., Terfort, A., Rampi, M. A., Whitesides, G. M. Electron transport through thin organic films in metal-insulator-metal junctions based on self-assembled monolayers. *J Am Chem Soc* 2001, *123*, 5075–5085.

88. Tulevski, G. S., Myers, M. B., Hybertsen, M. S., Steigerwald, M. L., Nuckolls, C. Formation of catalytic metal-molecule contacts. *Science* 2005, *309*, 591–594.

89. Cowan, D. O., Levanda, C., Park, J., Kaufman, F. Organic solid-state. VIII. Mixed-valence ferrocene chemistry. *Acc Chem Res* 1973, *6*, 1–7.

90. Day, P., Hush, N. S., Clark, R. J. H. Mixed valence: origins and developments. *Philos Trans Roy Soc A-Math Phys Eng Sci* 2008, *366*, 5–14.

91. Gamelin, D. R., Bominaar, E. L., Mathoniere, C., Kirk, M. L., Wieghardt, K., Girerd, J. J., Solomon, E. I. Excited-state distortions and electron delocalization in mixed-valence dimers: Vibronic analysis of the near-IR absorption and resonance Raman profiles of [Fe-2(OH)(3)(tmtacn)(2)](2+). *Inorg Chem* 1996, *35*, 4323–4335.

92. Williams, R. D., Petrov, V. I., Lu, H. P., Hupp, J. T. Intramolecular electron transfer in biferrocene monocation: Evaluation of Franck-Condon effects via a time-dependent analysis of resonance Raman scattering in the extended near-infrared. *J Phys Chem A* 1997, *101*, 8070–8076.

93. Brunschwig, B. S., Creutz, C., Sutin, N. Optical transitions of symmetrical mixed-valence systems in the Class II-III transition regime. *Chem Soc Rev* 2002, *31*, 168–184.

94. Sun, H., Steeb, J., Kaifer, A. E. Efficient electronic communication between two identical ferrocene centers in a hydrogen-bonded dimer. *J Am Chem Soc* 2006, *128*, 2820–2821.

95. Concepcion, J. J., Dattelbaum, D. M., Meyer, T. J., Rocha, R. C. Probing the localized-to-delocalized transition. *Philos Trans Roy Soc A-Math Phys Eng Sci* 2008, *366*, 163–175.

96. Robin, M. B., Day, P. Mixed-valence chemistry—a survey and classification. *Adv Inorg Chem Radiochem* 1967, *10*, 247–422.

97. Nishihara, H. Redox and optical properties of conjugated ferrocene oligomers. *Bull Chem Soc Jpn* 2001, *74*, 19–29.

98. Chakroune, N., Viau, G., Ammar, S., Poul, L., Veautier, D., Chehimi, M. M., Mangeney, C., Villain, F., Fievet, F. Acetate- and thiol-capped monodisperse ruthenium nanoparticles: XPS, XAS, and HRTEM studies. *Langmuir* 2005, *21*, 6788–6796.

99. Ribou, A. C., Launay, J. P., Sachtleben, M. L., Li, H., Spangler, C. W. Intervalence electron transfer in mixed valence diferrocenylpolyenes. Decay law of the metal-metal coupling with distance. *Inorg Chem* 1996, *35*, 3735–3740.

100. Levanda, C., Bechgaard, K., Cowan, D. O. Mixed-Valence Cations - Chemistry of Pi-Bridged Analogs of Biferrocene and Biferrocenylene. *J Org Chem* 1976, *41*, 2700–2704.

101. Reimers, J. R., Cai, Z. L., Hush, N. S. A priori evaluation of the solvent contribution to the reorganization energy accompanying intramolecular electron transfer: Predicting the nature of the Creutz-Taube ion. *Chem Phys* 2005, *319*, 39–51.

102. Brown, G. M., Meyer, T. J., Cowan, D. O., Levanda, C., Kaufman, F., Roling, P. V., Rausch, M. D. Oxidation-state and electron-transfer properties of mixed-valence 1,1'-polyferrocene ions. *Inorg Chem* 1975, *14*, 506–511.

103. Levanda, C., Cowan, D. O., Bechgaard, K. Electronic and Stereoelectronic Effects on Intervalence Transfer Transition in Biferrocene Cations. *J Am Chem Soc* 1975, *97*, 1980–1981.

104. Kadish, K. M., Xu, Q. Y., Barbe, J. M. Electrochemistry of a metalloporphyrin-bridging biferrocene complex - reactions of Fc-(Oep)Ge-Fc. *Inorg Chem* 1987, *26*, 2565–2566.

105. Morrison, W. H., Krogsrud, S., Hendrick, D. N. Polarographic and magnetic susceptibility study of various biferrocene compounds. *Inorg Chem* 1973, *12*, 1998–2004.

106. Yamada, M., Quiros, I., Mizutani, J., Kubo, K., Nishihara, H. Preparation of palladium nanoparticles functionalized with biferrocene thiol derivatives and their electro-oxidative deposition. *Phys Chem Chem Phys* 2001, *3*, 3377–3381.

107. Horikoshi, T., Itoh, M., Kurihara, M., Kubo, K., Nishihara, H. Synthesis, redox behavior and electrodeposition of biferrocene-modified gold clusters. *J Electroanal Chem* 1999, *473*, 113–116.

108. Li, D., Zhang, Y. J., Jiang, J. G., Li, J. H. Electroactive gold nanoparticles protected by 4-ferrocene thiophenol monolayer. *J Coll Interface Sci* 2003, *264*, 109–113.

109. Wolfe, R. L., Balasubramanian, R., Tracy, J. B., Murray, R. W. Fully ferrocenated hexanethiolate monolayer-protected gold clusters. *Langmuir* 2007, *23*, 2247–2254.

110. Chidsey, C. E. D., Bertozzi, C. R., Putvinski, T. M., Mujsce, A. M. Coadsorption of ferrocene-terminated and unsubstituted alkanethiols on gold - electroactive self-assembled monolayers. *J Am Chem Soc* 1990, *112*, 4301–4306.

111. Uosaki, K., Sato, Y., Kita, H. Electrochemical characteristics of a gold electrode modified with a self-assembled monolayer of ferrocenylalkanethiols. *Langmuir* 1991, *7*, 1510–1514.

112. Cuadrado, I., Moran, M., Casado, C. M., Alonso, B., Lobete, F., Garcia, B., Ibisate, M., Losada, J. Ferrocenyl-functionalized poly(propylenimine) dendrimers. *Organometallics* 1996, *15*, 5278–5280.

113. Zamora, M., Herrero, S., Losada, J., Cuadrado, I., Casado, C. M., Alonso, B. Synthesis and electrochemistry of octamethylferrocenyl-functionalized dendrimers. *Organometallics* 2007, *26*, 2688–2693.

114. Pietron, J. J., Hicks, J. F., Murray, R. W. Using electrons stored on quantized capacitors in electron transfer reactions. *J Am Chem Soc* 1999, *121*, 5565–5570.

115. van Herrikhuyzen, J., Portale, G., Gielen, J. C., Christianen, P. C. M., Sommerdijk, N. A. J. M., Meskers, S. C. J., Schenning, A. P. H. J. Disk micelles from amphiphilic Janus gold nanoparticles. *Chem Commun* 2008, 697–699.

116. Wu, Q., Van Voorhis, T. Extracting electron transfer coupling elements from constrained density functional theory. *J Chem Phys* 2006, *125*, 164105.

117. Bylaska, E. J., de Jong, W.A., Kowalski, K., Straatsma, T.P., Valiev, M., Wang, D., Aprà, E., Windus, T.L., Hirata, S., Hackler, M.T., Zhao, Y., Fan, P.-D., Harrison, R.J., Dupuis, M., Smith, D.M.A., Nieplocha, J., Tipparaju, V., Krishnan, M., Auer, A.A., Nooijen, M., Brown, E., Cisneros, G., Fann, G.I., Früchtl, H., Garza, J., Hirao, K., Kendall, R., Nichols, J., Tsemekhman, K., Wolinski, K., Anchell, J., Bernholdt, D., Borowski, P., Clark, T., Clerc, D., Dachsel, H., Deegan, M., Dyall, K., Elwood, D., Glendening, E., Gutowski, M., Hess, A., Jaffe, J., Johnson, B., Ju, J., Kobayashi, R., Kutteh, R., Lin, Z., Littlefield, R., Long, X., Meng, B., Nakajima, T., Niu, S., Rosing, M., Sandrone, G., Stave, M., Taylor, H., Thomas, G., van Lenthe, J., Wong, A., Zhang, Z. *NWChem, A Computational Chemistry Package for Parallel Computers, Version 5.0*. Pacific Northwest National Laboratory, Richland, Washington, 2006.

118. Chen, W., Brown, L. E., Konopelski, J. P., Chen, S. W. Intervalence transfer of ferrocene moieties adsorbed on electrode surfaces by a conjugated linkage. *Chem Phy Lett* 2009, *471*, 283–285.

119. Hush, N. S. Homogeneous and heterogeneous optical and thermal electron transfer. *Electrochim Acta* 1968, *13*, 1005–1023.

4 Sol-Gel Electrochemistry
Silica and Silicates

Ovadia Lev and Srinivasan Sampath

CONTENTS

4.1　INTRODUCTION

Sol-gel technology has undergone a major transformation during the past two decades. New discoveries, particularly those pertaining to mesoporous silicates, hybrid nanoparticle building blocks, and new biological opportunities opened up by sol-gel hybrids and composites, have also been made. Currently, sol-gel technology is considered a versatile and useful method of modifying electrodes and membranes in widely different electrochemical fields.

Several comprehensive reviews on sol-gel electrochemistry [1–10] and sol-gel materials for analytical chemistry [11–16] have appeared over the past decade. These allowed us to concentrate here only on emerging trends in sol-gel science for electrochemical applications rather than addressing all of the many proposed applications.

We present an introduction to sol-gel processing with an emphasis on silicate formation. Dip- and spin-coatings, the two most useful thin-film processing techniques for electrochemical applications, are briefly described. The section on thin films also includes an expanded account of recent developments in sol-gel electrodeposition of functional silicates, a field that has recently attracted considerable scientific attention. A classification of ways to electrodeposit thin films is provided.

Silicates, more than any other metal oxide, provide versatile ways to bring inorganic and organic functionalities within the same network. A dedicated section provides a general overview of the methods for the formation of single-phase organic–inorganic hybrids. Recent activities in sol-gel composites, in which the silicate grains are embedded in a host material or vice versa, is provided. Fuel cell membranes and organically modified electrolyte, two fast-growing areas, demonstrate how simple sol-gel methods can improve electrochemical performance.

Structure-forming technologies to direct sol-gel polymerization and achieve high analytical specificity, controlled micro and meso-porosity, and even hierarchical networks with multimodal pore size distribution, are reviewed in the last section of this account.

The chapter emphasizes the application of emerging new sol-gel derived materials, rather than the electroanalytical perspective. Thus, there are no dedicated sections on sensors, biosensors, membranes, and fuel cell applications, and they are addressed only within the discussions of properties and demonstrating examples of the different classes of sol-gel materials.

Although sol-gel is generic for metal oxides, we focus attention on silica and silicate processing. Electrochemical activities in other sol-gel derived metal oxides are described when a particular application is also relevant to silica and silicates.

4.1.1 INTRODUCTORY REMARKS ON SILICON CHEMISTRY

For many years, the limited similarity between silicon and carbon excited the scientific community. Carbon and silicon share the same outer shell electronic structure, s^2p^2, which permits sp^3 hybridization and dominant tetrahedral coordination, as well as dominance of the tetravalent oxidation state. Nevertheless, silicon chemistry is markedly poorer compared to that of carbon. Double silicon bonds and silicon catenation are scarce, and crystalline silicon, which is so widely used in the electronics industry, is never encountered in nature. Instead, silicon-oxygen bonds dominate natural silicon chemistry, and solid silica and silicates have no common physicochemical features with carbon dioxide and carbonates. The silicon atom is larger than carbon, it is less electronegative, has lower nuclear electric charge shielding and, perhaps most importantly, it has vacant d-orbitals in its outer shell; all these dictate the reactivity of silicon. Several consequences of these differences are especially significant, and they are also relevant to sol-gel electrochemistry.

Prevalence of Si-O in Nature and the Stability of Si-C Bonds: After oxygen, silicon is the second most abundant element on Earth's upper continental crust. About 95% of Earth's crust is composed of silica, silicates, or aluminosilicates. In all these forms, silicon is predominantly coordinated to oxygen. The prevalence of Si-O bonds in nature, in marked dissimilarity with carbon chemistry, is one of the most distinctive characteristics related to contemporary sol-gel chemistry. Another distinct characteristic is the stability of organosilicon compounds, which, in fact, are seldom seen in natural systems. Indeed, all the silicon compounds that are mentioned in this chapter share a common feature: the silicon participates either in Si-O, Si-N, or in Si-C bonds. Unlike aluminum (the next most abundant element on Earth's crust), the Si-C bond is not hydrolyzed in aqueous solution and, once formed, is stable over a large range of ambient conditions and even withstands elevated temperatures (this is indeed the reason for the interest in methyl siloxanes for thermal insulation). Table 4.1 delineates the energetic reasons for the dominance of silicates in natural systems and the relative stability of the organosilane compounds. The enthalpy of formation of the Si-O bond is far larger than all possible alternative silicon bonds, particularly catenation (involving silicon-silicon bonds) or silicon hydride formation. Table 4.1 shows that the Si-C bond is more energetic than the Si-Si bond and also less polar ($\Delta\chi_p = 0.65$) compared to the Si-Cl ($\Delta\chi_p = 1.26$), Al-C ($\Delta\chi_p = 0.95$), and other metal-carbon bonds. This explains the relative hydrolysis resistance of the former as compared to the latter.

TABLE 4.1
Mean (Single) Bond Enthalpies (kJ/mol) and Pauling Electronegativity

Element	Si	C	Pauling Electronegativity χ_p
Si	226	301	1.9
C	301	348	2.55
O	466	360	3.44
H	318	412	2.7
Cl	401	338	3.16

Source: Shriver, D. F., P. W. Atkins, and C. H. Langford, 1994. *Inorganic Chemistry*, 2nd edition. Oxford University Press: Oxford.

Brönsted–Lowry Acidity of Surface Silanols: Silanols ($\equiv$SiOH) are much stronger acids than alcohols due to the larger polarity of the Si-O bond as compared to Si-C bonds. Silicic acid has a fairly high (first) proton dissociation constant ($pK_a = 9.8$), but the presence of nearby siloxane bonds increases the polarity of surface silanols and increases the deprotonation propensity. The isoelectric point of silicate is in the range of pH 2–4; above this point, the surface silanols are increasingly deprotonated. This is the reason for the natural prevalence of silicates (in which hydrogen is replaced by another cation) and for the important electrostatic interactions and ion exchange properties of natural silicates as well as sol-gel products. The negative charges of inorganic sol-gel silicate films determine the permselectivity of modified electrodes, though the permeation selectivity can be reversed by working with positively charged organically modified silicate films.

Coordination Expansion: Silicon can undergo $3sp^3d^2$ and $3sp^3d^1$ hybridizations, and thus the coordination of the silicon can expand to penta- and hexacoordination, which lowers the activation energy for substitution reactions and ligand exchange. Transitions involving hexacoordinated and pentacoordinated intermediates are common ways to substitute silicon ligands, pathways that are less obtainable for carbon atoms, whose smaller size and exclusive 4-coordination restrict such substitution reactions. The hydrolysis, condensation and dissolution of silicate, which are described in the next section, exemplify the importance of this attribute.

4.1.2 THE SOL-GEL PROCESS OF SILICA AND SILICATES

Sol is defined as small particles dispersed in a medium, while gel is a relatively rigid interconnected network containing different-sized pores connected by polymeric chains. The wet gel can result in a xerogel if the volatile components are removed under ambient conditions, whereas it will become an aerogel when the volatile constituents are removed under supercritical conditions of a solvent. The sol-gel process, the parameters associated with it, and the consequent structures have all been extensively reported in comprehensive books [18–21]. The formation of a gel from alkoxysilanes may be thought of as formed by hydrolysis of a precursor, followed by condensation to give polymers, and subsequent network formation or the growth of discrete small particles that connect to give a solid network.

In the first step of hydrolysis, a precursor is mixed with water, a solvent that can dissolve the monomer (mostly an alkoxide). Often, a catalyst is added to produce a homogeneous solution yielding at the first stage hydroxosilane species or even silicic acid. Further condensation results in three-dimensional siloxane, Si–O–Si linkages and the hydrolysis and condensation by-products trapped in the pores of the matrix. The sol particles are formed when sufficient numbers of Si–O–Si linkages are generated and the particles subsequently interact with other particles (leading to densification of the matrix). An oversimplified scheme presenting the rate of the coinciding chemical reactions as a function of pH is presented in Figure 4.1. Not only do the three different mechanisms compete, they influence each other's rate and give rise to different forms of sol that lead to different types of gels. Nevertheless, a description of the different chemical steps and the way they influence the respective rates may become useful.

4.1.2.1 Hydrolysis

The rate of hydrolysis increases linearly with the employed acid or base concentration. Scheme 4.1 describes the hydrolysis mechanism of tetramethoxysilane in acid and base aqueous solutions. In both cases, the first stage of the S_N2 (second-order nucleophilic substitution) mechanism involves nucleophilic attack on the positively charged silicon followed by expansion of the coordination of

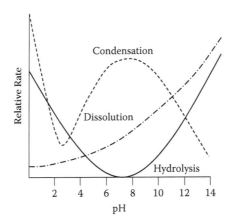

FIGURE 4.1 Schematic trends of the rate of hydrolysis, condensation, and dissolution of alkoxysilane and its products.

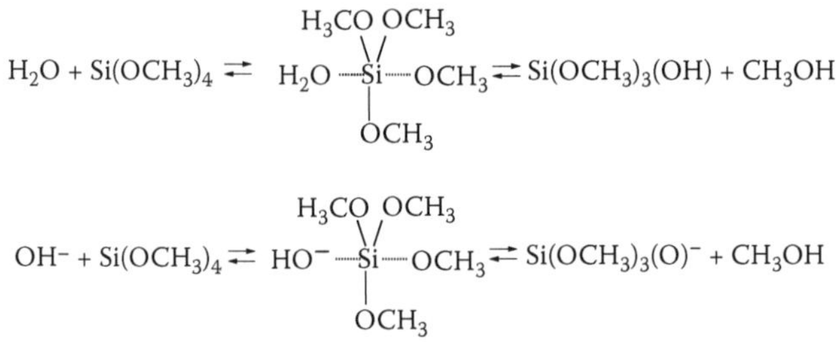

$$H_2O + Si(OCH_3)_4 \rightleftarrows H_2O\overset{\overset{\displaystyle H_3CO\;\;OCH_3}{\diagdown\diagup}}{\underset{\underset{\displaystyle OCH_3}{|}}{-Si-}}OCH_3 \rightleftarrows Si(OCH_3)_3(OH) + CH_3OH$$

$$OH^- + Si(OCH_3)_4 \rightleftarrows HO\overset{\overset{\displaystyle H_3CO\;\;OCH_3}{\diagdown\diagup}}{\underset{\underset{\displaystyle OCH_3}{|}}{-Si-}}OCH_3 \rightleftarrows Si(OCH_3)_3(O)^- + CH_3OH$$

SCHEME 4.1   Hydrolysis of tetramethoxysilane by nucleophilic substitution (S_N2) under acidic and basic conditions.

the silicon, formation of distorted bipyramidal structure and, finally, cleavage of the $Si\text{-}OCH_3$ bond, which is (mostly) accompanied by inversion of the pyramidal structure.

Important evidence in support of this mechanism was provided by carrying out hydrolysis with oxygen-isotope-labeled water ($H_2^{18}O$). The isotope test showed that the alkoxysilane cleavage is carried out by breaking the $Si\text{-}OC$ bond rather than by cleavage of the $SiO\text{-}C$ bond, resulting in the formation of the heavy silanol $\equiv Si^{18}OH$ and not the light $\equiv Si^{16}OH$.

$$Si(OR)_4 + H_2^{18}O \rightleftarrows Si(^{18}OH)_{4-x}(OR)_x + \text{Condensation Products} \quad (4.1)$$

Protic and polar solvents that can participate in hydrogen bonding may activate the alkoxide, make it a better leaving group, and accelerate the rate of the hydrolysis reaction. Likewise, polar solvents lower the energy involved with charge separation and accelerate hydrolysis. Indeed, it was found that the rate of hydrolysis of silicon alkoxide monomers varies in the order of acetonitrile, methanol, dimethylformamide, dioxane, and formamide [22]. A larger alkyl group in the alkoxy function is known to affect the rate of hydrolysis negatively [23,24].

In a series of papers, Sommer and coworkers [25] used enantiomerically pure methyl phenyl napthyl alkoxysilane monomers to exemplify that hydrolysis is accompanied by an L-to-D tetrahedron inversion. The stability of the pentacoordinated intermediates sometimes allows true substitution with retention of the original structure. In this case, the high-coordination intermediate is sufficiently stable to allow time for repositioning of the leaving group from the trans-position (with respect to the nucleophile) to a cis-position that prevents inversion. This mechanism predominates when the leaving group is hydrogen or alkoxide, which are both poor leaving groups.

4.1.2.2   Condensation

The condensation step also follows an S_N2 mechanism enabled by coordination expansion, but in this case, the nucleophilic attack is conducted by deprotonated

$$\equiv SiO^- + \equiv SiOH \rightleftharpoons \equiv SiOSi \equiv + OH^-$$

$$\equiv SiO^- + \equiv SiOH_2^+ \rightleftharpoons \equiv SiOSi \equiv + H_2O$$

SCHEME 4.2 Condensation reactions under near-neutral (up) and acidic conditions.

silanol moieties. Above the isoelectric point of the silicon moieties, the condensation involves reactions (Scheme 4.2) between deprotonated silanols and protonated ones, and therefore it is faster under near-neutral conditions. However, a second increase of the condensation rate constant exists at pH 2–4 due to the formation of $SiOH_2^+$ moieties, increasing the positive charge on the silicon and further accelerating the condensation reactions.

4.1.2.3 Dissolution

Silicate dissolution at high pH (as well as at high fluoride concentration in acidic solution) stems from a similar mechanism, that is, nucleophilic attack of the hydroxide (or fluoride ion) on the positively charged silicon and coordination expansion followed by cleavage of a siloxane bond. Therefore, the rate of siloxane cleavage and dissolution of silicic acid and other silicon monomers increases monotonically as the pH is raised above the isoelectric point. In fact, at very high pH values, the dissolution rate exceeds the condensation rate, and the sol-gel process does not take place at all, unless the pH is lowered. This is, for example, the reason for the success of high-pH sodium silicate solution (also called waterglass) as a binder. At high pH, the waterglass is fully dissolved because the rate of siloxane breakdown exceeds the rate of the condensation reactions. As the pH is lowered by CO_2 dissolution, the polycondensation rate exceeds the dissolution rate, and a film is formed.

4.1.2.4 Sol and Particle Formation

The fast rate of hydrolysis with slow condensation, which takes place at low pH, prefers the formation of linear polymeric siloxane chains. These polymers cross-link to form polymeric, dense microporous gels. At high-pH conditions, condensation is faster than hydrolysis, and the higher dissolution rate provides a constant source of small clusters and monomers. Under such conditions, particles can grow considerably before aggregation takes place. These conditions lead to the formation of a more porous silicate gel [26]. An extreme case of this process is called the Stöber process, which is usually carried out with an ammonia catalyst to guarantee high-pH conditions and excess of water conditions that lower aggregation. Large and spherical silicate particles are formed under these conditions by Ostwald ripening mechanism. Small particles are less stable than large ones, and therefore they dissolve more quickly, thus supplying monomers that support the growth of the large ones. Moreover, the dissolved monomers tend to precipitate in the negative curvature region of the silicate gel,

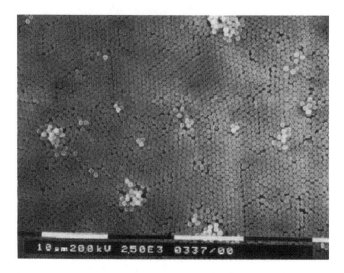

FIGURE 4.2   Hydrophobic micron-dimension Stöber silica monolayer deposited by the Langmuir Blodgett (LB) method. (From Szekeres, M., O. Kamalin, P. G. Grobet, R. A. Schoonheydt, K. Wostyn, K. Clays, A. Persoons, and I. Dékány, 2003. *Coll Surf A: Physicochem Eng Aspects* 227:77–83. Used with permission.)

resulting in small pores getting filled. Hence, the average pore size increases and surface area decreases. The process results in micron-dimension-dense particles such as those shown in Figure 4.2.

4.1.2.5   Gelation and Drying

The gel point is defined by the presence of a drastic change in viscosity. It was also pointed out by Brinker that a frozen structure is formed at the point of gelation and that the structure depends on many factors such as temperature, solvent, pH, and the rate of removal of the volatile components. The pH dependence of the gelation time was reported to be bell shaped. The gelation time was at a maximum at an intermediate pH value, while highly acidic and highly alkaline conditions favored immediate gelation. The amount of water is known to affect the gelation time. When the mole ratio of water to alkoxide is 2, the gelation time was determined to be around 7 h, whereas it was reduced to 10 min when the ratio of water to alkoxide was 8 (given that all other conditions were the same). The role of temperature was also studied, revealing that the polymerization process is generally thermally activated [28].

Aging of the gel before it gets dried dictates the final structure of the matrix. Iler had studied this process extensively, and Scherer carried out some theoretical calculations [29]. The newly formed gel is still not highly cross-linked and, therefore, it is very flexible and allows further movement of silicate segments and further condensation reactions between nearby silanols. In addition, under high-pH conditions, dissolution can take place and Ostwald ripening can proceed. These lead to increased rigidity of the gel, a process that may persist long after the gelation point and may lead to a considerable change of mobility of electroanalytes through the gel, and even to gradual blocking of accessibility to electrode surfaces.

The drying process involves three stages. The first stage is related to the decrease in volume of the gel, which is exactly the same as that of the liquid expelled. The rate of the initial stage of drying is not constant for gels with pore sizes less than 20 nm. Stage 2 begins when a critical point of highest capillary pressure is reached. The compression of the gel cannot be sustained, and the pores are emptied when liquid transport occurs by flow and subsequent evaporation takes place. When the pores are emptied, the third stage begins, in which the diffusion of the remaining liquid to the surface takes place before evaporation. Subsequently, densification takes place. The critical step from an electrochemical film production point of view is the second stage, especially when some of the pores (usually the larger ones) are already empty and others are still filled with the solvent. Surface tension then creates large pressure differences and, because at this stage the pore walls are already stiff, the process may result in cracks. The tendency to fracture increases as the film thickness grows, and it is very difficult to receive crack-free, thick (>1 µm) silicate monoliths without special measures. Several ways to overcome this problem have been demonstrated:

1. Introduction of compounds that tend to reduce surface tension and thus lower the differential pressure over the pore walls. The most important are formamide and cationic surface-active agents such as cetylperidinium bromide or cetyltrimethylammonium bromide.
2. Incorporation of silane moieties having only two or three hydrolyzable groups (e.g., methyltrimethoxysilane), which have only three or fewer cross-linking degrees of freedom and thus form a more flexible siloxane skeleton.
3. Application of composite materials, as discussed in Section 4.3.2. Here, the foreign entity, be it polymer, particle, or even fiber or nanotube, provides ample degrees of freedom for stress relaxation.
4. Sequential coating of the electrode by thin films of around 0.1 µm or less is another viable route to increase film thickness, though this technique is labor intensive when thick films are desired.
5. Drying under supercritical or freeze-drying conditions yielding aerogels. Aerogels are gels whose void fraction exceeds 90% and whose density may reach 1–2 mg L^{-1}. When the aqueous solvent is replaced by a liquid whose critical temperature is low enough, it is possible to remove the solvent without phase transformation and the formation of a liquid–gas interface within the silicate, and thus avoid fractures [30].

4.2 SOL-GEL PROCESSING OF THIN FILMS AND SOL-GEL ELECTRODEPOSITION

4.2.1 Nonelectrochemical Film Formation

The most useful configuration of sol-gel materials for electrochemical applications is the thin film, which is also the strongest aspect of sol-gel processing. This is because the sol-to-gel stage development, as described earlier, goes through a change in viscosity from a liquid state of low viscosity to a gel state of very high viscosity. This has given rise to the use of various methods of film formation such as dip coating, spin coating, and electrophoretic and electrochemical deposition and their electrochemical applications. The advantages of sol-gel processing to prepare thin films include the use of low temperature, ease of thickness variation ranging from single monolayers which come prepared in aprotic dry solvent, to thick films and high optical quality.

4.2.1.1 Dip Coating

In the dip coating process, the substrate is immersed into a sol containing the precursor, catalyst, and the solvent and removed in a very controlled manner, resulting in a film that gels as the substrate is taken out. Subsequent heat treatment yields the desired product. If the withdrawal speed is such that the shear rate–shear stress dependence obeys Newtonian flow conditions, the coating thickness follows the Landau–Levich equation:

$$h = 0.94 \; [(\eta \cdot \nu^{2/3}) \, / \, \gamma^{1/6} \, (\rho g)^{1/2}] \tag{4.2}$$

where h is the coating thickness, η is the viscosity, ν is the withdrawal speed, γ is the liquid–vapor surface tension, ρ is the density, and g is the acceleration of free fall [18,31]. The thickness attained in the experimental conditions matched well with the expected values for the acid-catalyzed silicate sol [32,33]. An angle-dependent dip coating process that results in precise, uniform coatings was developed as well [34]. Dip coating processes were developed for curved surfaces too. In such a procedure, the thickness achieved is generally submicron and is good insofar as adhesion properties are concerned. When the thickness is small, the bonding force between the film and the substrate leads to shrinkage of the film only in the direction normal to the substrate and, hence, the coating is intact. On the other hand, when the thickness is large, the coherent force within the film would lead to shrinkage parallel to the substrate during the drying process, resulting in nonuniformity as well as peeling off of the film. All four methods for crack prevention that were discussed at the end of Section 4.1 can assist in film peel-off prevention as well. One of the instances where a thick film of 2 μm was reported to be coated was based on the use of trimethylethoxysilane and TEOS to produce silica hybrid coatings [35]. The viscosity of the sol is sometimes controlled using additional polymers such as hydroxypropylcellulose along with the sol-gel components [36]. Figure 4.3 shows the variation in thickness of the films formed in the presence and in the absence of hyroxypropylcellulose. If one

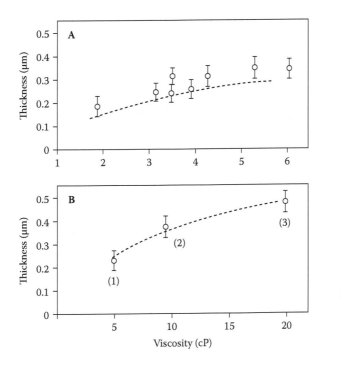

FIGURE 4.3 Variation in thickness of the film prepared from TEOS in the (A) absence and (B) presence of hyroxypropylcellulose. (From Sakka, S., K. Kamiya, K. Makita, and Y. Yamamoto, 1984. *J Non-Cryst Solids* 63:223–225. Used with permission.)

wishes to attain a large thickness, it is sometimes advisable to repeat the dip procedure rather than increase the thickness in one dip itself.

The complicated structural variations in the coating process were followed by Brinker and Scherer [18]. Various parameters such as viscosity of the solution, withdrawing speed of the substrate, oxide concentration in the sol, and final heat treatment temperature were taken into account. The following relationship that was derived for styrene-methacrylate copolymers [37] was used for sol-gel coatings as well:

$$t = J\xi \left(\frac{\eta - \eta_s}{\eta_0} \right)^{0.84} \left(\frac{\eta v}{\rho_s g} \right)^{\frac{1}{2}} \qquad (4.3)$$

where J is a proportionality constant, ρ_s, the solvent densities, η (viscosity of the solution) is the effective sum of η_s (viscosity of the solvent) and η_0 (coefficient), ξ is the ratio of densities of solvent and copolymer, g is the acceleration due to gravity, t is the thickness of the coating, and v is the substrate-withdrawing speed. The expected slope of log v versus log t was ½ and was reported to be the same for silica films from TEOS [32,33], while Dislich and Hussman [38] obtained a value of 2/3 for sols of metal alkoxides. Guglielmi and Zenezini [39] analyzed the t versus v^n relationship data by various researchers and reported that the n value varied between 0.1 and 1 while most of the studies reported close to 1/2. The thickness variation as a function of viscosity (replotted in logarithmic scales) yielded a slope of around 1/2.

4.2.1.2 Spin Coating

In spin coating technology, the substrate spins around an axis that is perpendicular to the coating area. Spin coating techniques require tuning of the viscosity of the sol. The parameters that control the process are spin speed and the temperature of the substrate as well as the sol used. The centrifugal driving force, viscous rheology, and solvent evaporation kinetics all affect the thickness as well as the uniformity of the film. Meyerhofer [40] discussed the dependence of the film thickness of a spin-coated film and the angular viscosity and evaporation rate of the solvent, using an empirical relationship,

$$h = (1 - \rho_a/\rho_0)\ (3\eta\ m/2\ \rho_0 \cdot \omega^2)^{1/3} \qquad (4.4)$$

where η is the viscosity, ω is the angular velocity, m is the evaporation rate of the solvent, and ρ_a and ρ_0 are densities corresponding to r_a and r_0 that are the mass of volatile solvent per unit volume and initial value of r_a, respectively. Because m is an empirical quantity, the equation can be written as $h = A\ \omega^B$, where h is the thickness of the film, ω is the angular velocity, and A and B are empirically determined constants. This equation was verified for various angular velocities, and the parameter B was found to lie between 0.4 and 0.7. The sol-gel spin coating process was also shown to yield preferred orientation of ceramic films such as $LiNbO_3$ and other perovskites [41]. The epitaxial precipitation from an amorphous sol indeed opens up a controlled way of preparing oriented thin films under ambient conditions. The formation of mesoporous materials through dip coating and other evaporation-induced self-assembly has been attracting considerable attention recently and will be discussed in Section 4.4.2.

A recipe that was shown to yield very uniform silicate coatings on a glass substrate is given in the following text [42]. The molar composition of the sol used was TEOS:water:nitric acid:ethanol as 1:2:0.01:1, and 1:r:0.01:2, where r is 4 or 10. Water present in the acid was included in r. Spin coating was carried out using 0.5 mL of the sol, spread with a syringe in 3 s on a soda lime silicate glass with dimensions $26 \times 76 \times 1$ mm rotating at a speed of 3440 rpm with the substrates kept rotating for 1 min after dispensation of the sol.

4.2.2 SOL-GEL ELECTRODEPOSITION OF THIN FILMS

4.2.2.1 Principle of Electrophoresis and the Advantages of Electrodeposition

Sol-gel electrodeposition deserves a special section in this chapter because it is attracting significant scientific interest, its special characteristics are slowly being unraveled, and its mechanistic aspects are still not fully understood. Therefore, although sol-gel electrodeposition is probably much less useful compared to the more mature spin, dip, and spray coating techniques, a more detailed account of recent developments is presented here.

The principle of electrophoresis, the most straightforward method for electrodeposition is based on the electric field-driven charged particles (silicate, organosilicate, metal oxides, micelles, or polymer composite particles) to an electrode at an electrophoretic velocity, v, which is determined by Stoke's law.

$$v = \frac{qE}{6\pi\mu r}$$

(4.5)

where q is the particle charge, E is the applied electric field, μ is the viscosity, and r is the effective radius of the particle. When the concentration of the particles near the electrode (anode or cathode, depending on charge) exceeds the flocculation concentration, the particles adhere and further build up on the support, probably by van der Waals interaction or hydrogen bonding and subsequently by condensation reactions with the surface silanols or other M-OH moieties. Often, charge neutralization is employed to increase the particle settling rate. When the precursor is a metal salt, which is converted to particulates by a faradaic process (mostly acid generation at the anode or formation of base at the cathode), the process is termed *electrolytic deposition* (ELD) [43,44,353]. ELD is a mature technology that works with or without binders, and it usually yields compact films comprising small, nanosized building blocks. Particulates that are formed near the electrodes deposit onto it before they have sufficient time to agglomerate.

Electrophoretic and electrolytic deposition of silicates and other metal oxides have many advantages and, of course, a drawback as well. The advantages include the following:

1. The deposition is very uniform compared to other deposition techniques, in which dense and large particles settle first. Again employing Stokes' law under gravity driving force

$$v = \frac{(\rho_p - \rho_l)g}{18\mu} D_p^2$$

(4.6)

   where D_p is the particle diameter, $(\rho_p - \rho_l)$ is the difference between the density of the particle and the deposition solution, and g is the acceleration of free fall; large and dense particles settle first, and then small and voluminous particles hit the surface. In electrophoretic deposition because gravity is not the driving force, density gradients are not observed.
2. It is easy to deposit films over inhomogeneous and curved surfaces. A striking example of this is the deposition of silicate over carbon nanotubes as described in Section 4.3.2.3.3.
3. The coating thickness can be easily controlled by the deposition period and the applied voltage or current.

4. Coating thickness is uniform because there is a large autohealing effect. Wherever a defect or a thin section is formed, the ohmic resistance near the defect becomes low; redistribution of the current stream lines near the defect takes place, additional particulates are driven to the defect, and thus it is cured. This negative feedback for the deposition of mixed conductive-insulating composites helps in achieving a uniform coating. However, in some cases, such as in the deposition of conductive particles, positive feedback takes place. Conductive particles tend to electrode-posit faster in high-conductivity locations, that is, on those regions that are already covered by conductive particles. An example of this effect is given in Section 4.3.2.3.2 (Figure 4.4).

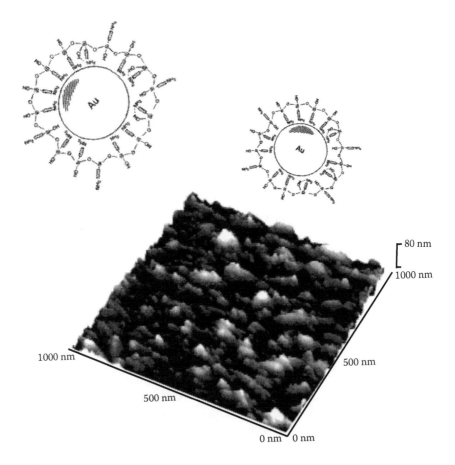

FIGURE 4.4 A scheme of aminosilicate-capped nanoparticles and an AFM micrograph of gold nanoparticles-silicate electrodeposit. (From Bharathi, S., J. Joseph, and O. Lev, 1999. *Electrochem Solid State Lett* 2:284–87. Used with permission.)

5. Very often, the applied electric field induces reorientation of anisotropic particles and buildup of aligned particulates. This was elegantly demonstrated by Walcarius et al. [45] for the electrodeposition of perpendicular orientation of mesoporous electrodeposit on electrodes.
6. The electrodeposition processes are soft and can accommodate delicate moieties, including biomolecules that are very important in contemporary electrochemical sensing.
7. Electrodeposition processes are environmentally friendly and they do not require aggressive deposition conditions. For the most part, they are carried out under close to ambient conditions and minimal solvent wastes are generated.

The most restrictive drawback is that the deposition should be carried out on a conducting or a doped semiconducting surface. This is why it is not yet popular for most high-tech electro-optic applications, whereas it is extremely useful for coatings in the automotive industry, for the fabrication of batteries, fuel cells, and electrochromic windows, all of which involve films on conductive supports. Electrophoretic deposition conditions involve the presence of colloidal suspensions, but methods for in situ formation of particulates by faradaic acidification or base formation were amply described as well [44]. Finally, gas evolution (H_2) may be considered a technicality, but it adversely affects the film quality and may pose a safety hazard unless appropriately handled.

Both potentiostatic and galvanostatic deposition techniques are used, though the latter is often preferred, because the galvanostatic mode implies that the deposition rate is almost constant, thus resulting in uniform films. Modulated current and pulses may be useful to build up and dislodge concentrated solutions from the electrodes and hence help in confining micropatterned coatings.

Several other techniques have been developed over the past few years to accelerate electrophoretic deposition or use faradaic processes to obtain films of functional materials. The general topic of electrophoretic and electrolytic deposition of inorganic materials has been dealt with by others [43,44].

4.2.2.2 Examples of Sol-Gel Deposition of Functional Silicates

4.2.2.2.1 Electrophoretic Driving Force

The simplest and most well-known approach is to deposit silicate or other oxides by an electrophoretic driving force. For example, we used this approach to deposit a naphthoquinone redox group appended to a trimethoxy silane (Scheme 4.3), 1-propanaminium, N-[2[(3-chloro-1,4-dihydro-1,4-dioxo-2-naphthalenyl)amino] ethyl]-N, N-dimethyl-3-(trimethoxysilyl)-, bromide (NPQ), on a glassy carbon electrode (GC) [46].

Deposition was achieved after surface activation of glassy carbon to generate oxygenated moieties on the electrode surface. The oxygenated species provided

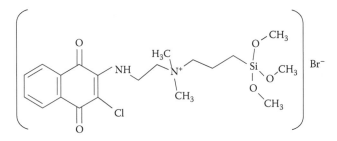

SCHEME 4.3 1-Propanaminium, N-[2[(3-chloro-1,4-dihydro-1,4-dioxo-2-naphthalenyl) amino]ethyl]-N, N-dimethyl-3-(trimethoxysilyl)-, bromide (NPQ).

chemical anchors for condensation reactions with the silanols of the hydrolyzed NPQ. In this case, the GC electrode was held for 2 min at +2.0 V versus the Ag/AgCl reference followed by a cathodic reduction at −1.0 V for 2 min. At this stage, the electrode turned purple, and a pair of redox peaks developed at about −0.1 V. The deposited film was used for oxygen sensing and for hydrogen peroxide formation. Carbon ceramic electrodes (CCEs) and GC electrodes were coated by this method and used for oxygen gas sensing [46].

Electrophoretic deposition combined with an oxidation process. A somewhat complex electrophoretic deposition technique involves coupling the electrophoretic driving force and faradaic oxidation to yield composite metal–silicate films. First, aminosilane-coated gold sol was produced by the addition of borohydride to a solution of $HAuCl_4$ and hydrolyzed aminopropylsilane. This procedure yielded nanoparticles that were protected by aminosilane coatings. It was proved that the amine was oriented toward the gold surface [48], and thus the nanoparticles bear a negative zeta potential. Deposition was carried out by cycling the electrode potential between −0.4 and 1.0 V versus Ag/AgCl at a scan rate of 100 mV/s at pH 4.5 [48]. A conductive, irregular (see Figure 4.4) deposit was obtained, and the deposition process was found to be selective for the capped gold nanoparticles over free aminosilane monomers and oligomers.

The amine moieties, which were oriented toward the gold surface, could not get protonated and thus the capped nanoparticles were more negatively charged as compared to the aminosilane moieties that enhanced their electrophoresis. Another important mechanism that facilitated gold electrodeposition was anodic oxidation of the gold nanoparticle surface, which removed the gold capping and thus increased its tendency to participate in film formation.

The general deposition scheme is demonstrated in Scheme 4.4 [48]. First, the gold-capped nanoparticles were attached to the surface through condensation of the outer shell, SiOH moieties with surface SnOH or SbOH groups present on the ITO surface. Then oxidation of the gold surface took place and desorbed some of the aminosilicate shells, because the amine had much less affinity for gold oxide than the bare, uncoated gold surface. Next, another layer of gold particles became

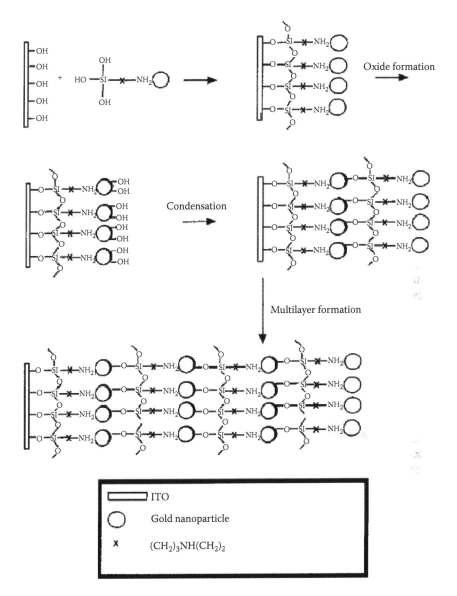

SCHEME 4.4 Mechanism of oxidation-assisted electrodeposition of aminosilane-capped gold nanoparticles. (From Barathi, S., M. Nogami, and O. Lev, 2001. *Langmuir* 17: 2602–2609. Used with permission.)

attached to the oxide-coated gold deposit by reaction of the silanols on the outer capping shell of nanoparticles from the solution with the gold oxide surface. Layer formation could be repeated several times. Glucose oxidase entrapment and glucose sensing were also demonstrated, thus showing the biocompatibility of this process [354]. Figure 4.4 shows that the tendency of gold nanoparticles to prefer gold-coated locations results in a very rough surface area.

4.2.2.2.2 Faradaic Neutralization Leading to Deposition of Silane Monomers

A third electrodeposition technique involves a charge transfer step to convert the charge-stabilized sol to uncharged moieties that can agglomerate and precipitate on the electrode surface. An example of this technique was given by Leventis and Chen [49], who demonstrated the polymerization of methylene blue appended trimethoxysilane moieties on the surface of ITO. The mechanism of electrodeposition is given in Scheme 4.5. First, a sol of methylene blue functionalized trimethoxysilane (MB-Si(OMe)$_3$) was introduced to a slightly basic starting solution. The monomers hydrolyzed and partly condensed to yield oligomeric charged species. At this stage, the moieties were positively charged and tended to dissolve in the solution. Then the methylene blue redox functionality became reduced to leucomethylene blue, which was electrically neutral. The reduction promotes sedimentation of uncharged species, and subsequent cross-linking with the film commenced in the usual manner.

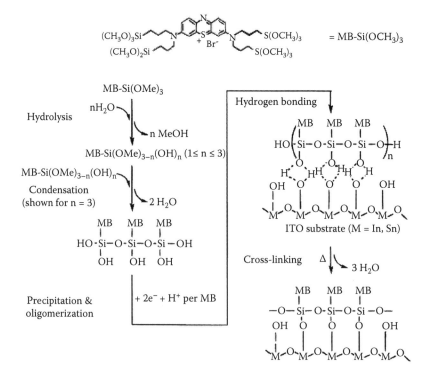

SCHEME 4.5 Electrodeposition of MB-Si(OCH3)3. (From Leventis, N., and M. Chen, 1997. *Chem Mater* 9:2621–2631. Used with permission.)

4.2.2.2.3   Electrodeposition by pH Change Induced by Faradaic Reaction

Recently, Shacham and coworkers [50] proposed using acidification of the solution by the electrochemical process to promote electrochemically driven deposition of alkoxysilanes. Faradaic reactions were used to increase the pH near the electrode surface that catalyzed the gelation process. This report triggered substantial electrochemical activity, which was well covered by recent reviews [51,52]. Deepa and coworkers further showed that dissolved oxygen reduction can increase the pH of the solution very close to the electrode and promote the deposition step [53], whereas acidification by anodic water splitting might sufficiently reduce the solution pH. If the pH was initially low enough, it would promote acidic deposition of silicates [54]. Collinson also conducted a detailed study illuminating the role of the solvent and supporting electrolyte in cathodic or anodic deposition of silicates [55]. It was further shown that incorporation of neutral supporting electrolytes (e.g., KCl and KNO_3) was favorable for cathodic deposition, whereas acidic supporting electrolytes (e.g., KH_2PO_4 and H_2SO_4) promoted anodic deposition. As expected, the addition of the buffer hindered the anodic as well as the cathodic deposition of silicates due to the buffering of the electrochemically induced pH change near the electrode surface. Perhaps the most remarkable demonstration of the versatility of electrodeposition was the ability to electrodeposit ordered 2-D hexagonal mesoporous materials with cylindrical open structures that were oriented predominantly perpendicular to the conductive support [111]. This process was further described within the context of mesoporous material formation. Jia and coworkers [56] showed that the electrochemical deposition process was soft enough to allow the encapsulation of enzymes such as glucose oxidase in the silicate film, and they further demonstrated the fabrication of an amperometric glucose sensor. Kanungo et al. [57] electrodeposited a silicate film over single-wall carbon nanotubes. The film thickness could be controlled by either sol concentration or by applied potential. The authors pointed out the following advantages for the electrodeposition of carbon nanotubes: (1) the coating was noncovalent and therefore nondestructive; (2) the deposition was environmentally friendly, requiring only a minimum amount of reactants, and mild pH and temperature conditions; and (3) the deposition time was low. Other notable electroanalytical applications include the electrochemical sensing of mercury ions by TEOS and mercaptosilane codeposition [58], Cr(VI) detection by concentration of the analyte over pyridine-silicate film [59], and cystidine recognition by 3-(aminopropyl)trimethoxysilane (APS)-derived Ormosil film [60]. Herlem et al. [61] used APS film to graft α-lactalbumin and Tian et al. [62] electrodeposited silicate film on ruthenium purple (an analogue of Prussian Blue), which was first deposited on a microelectrode, to create ATP and hypoxanthine microbiosensors. Nadzhafova and coworkers [63] reported direct charge transfer from electrochemically deposited hemoglobin and glucose oxidase to a glassy carbon electrode.

The striking feature of these past few examples is that most of them were reported within the past couple of years, which highlights how important this field could become in the years to come.

4.3 ORGANIC–INORGANIC HYBRIDS
AND SOL-GEL COMPOSITES

Silica gel and sol-gel porous films are chemically inert materials. Except for the deprotonation capability of silanols, these materials are highly stable, do not participate in faradaic and redox transformations, resist chemical and biological degradation, and also exhibit very low catalytic power. From an electrochemical point of view, however, these advantages may become too limiting because silica and silicates have a narrow vocabulary of specific functionalities and redox transformations. Accordingly, their electrochemistry is dull.

Various characteristics such as special electrocatalytic and catalytic effects, porosity control, flexibility, pore-wall strain relaxation modes, tailored dielectric properties, ionic or electronic conductivities, water retention as well as many other properties that are required in order to optimize specific applications cannot be achieved by a single inorganic phase and require combining two types of materials-i.e. a composite. The length scale of the guest phase in composites can vary from the molecular level to composite materials in which distinct two phases are visible. In some cases, molecular doping or Ormosil–Ormocer formation is sufficient and, in other cases, a two-phase—either entangled (i.e., a blend) or isolated (i.e., a composite)—exhibits better properties. In this section, we refer to any combination of organic and inorganic moieties as hybrids, and we reserve the term composite for two-phase systems.

The versatility of sol-gel technology and its film-forming capabilities are fully exploited in the formation of hybrids and composites. In contrast to other silicate technologies, there is no need for high-temperature processing, and the entire process can often be maintained even well under the decomposition temperature of most of the active components, including even delicate biomolecules. Although complex multiple-step preparation technologies are frequently used for specific sol-gel applications, they are not always necessary. A one-pot preparation of aerogel and xerogel films is often amply sufficient for the incorporation of many functionalities in close proximity within sol-gel films. Currently, over one-third of the publications on sol-gel electrochemistry deal with hybrids and composite materials [64], which underscores the large electrochemical advantages entailed in the incorporation of a second phase in sol-gel films. In the following paragraphs, we provide a general classification of hybrids and composites that are currently used for electrochemistry and then address the incentives for using two-phase silicate composites in electrochemistry by a few illustrative examples.

4.3.1 Organic–Inorganic Hybrids

The functional advantages of organic moieties and biomolecules can be combined with the benefits of sol-gel silicates by forming organic–inorganic hybrids, in which mixing of the organic and inorganic materials is performed at the molecular level. Currently, there are over 10 million registered organic compounds in the chemical abstract register; thus, formation of organic–inorganic hybrids becomes an attractive pathway to benefit from the inertness, rigidity, porosity, high surface area, processability, and other attractive features of sol-gel silicates, and at the same time to take advantage of the enormous versatility inherent in organic chemistry [65–72].

Two scientific developments that took place in the early 1980s signaled the evolution of hybrid organic–inorganic hybrids. Schmidt and coworkers exploited the fact that Si-C bonds are stable and do not hydrolyze during sol-gel processing in order to develop organically modified ceramics (Ormocers) and silicates (Ormosils), using organofunctional silane precursors such as methyltrimethoxysilane, MTMOS [73]. In this class of materials, the organic moiety is covalently bonded to the siloxane backbone. These materials containing Si-C covalent bonds are now classified as class I organic–inorganic hybrids. The covalent bonding between the organofunctional group and the cross-linked siloxane backbone is very similar and almost indistinguishable from materials made of silylation of large-surface silica gels. However, the Si-containing monomers can form a maximum of three siloxane bonds, which influence the rigidity and other physical properties of the end product. Silylation products are also classified as class I materials, because the covalent bond should be broken to release the organofunctional group. Avnir et al. [74] developed class II organic–inorganic hybrids by the so-called sol-gel doping process. The first dopants were dye molecules that were introduced into the silicate during sol-gel polymerization. Avnir et al. [75–78] further showed that it is possible to encapsulate colorimetric reagents that change color in response to chemicals in the silicate environment. Doped sol-gel materials and impregnated or adsorbed organic compounds on sol-gel silicates are now termed class II hybrids because some of the organic moieties can leach out.

Sol-gel hybrids can be prepared by the following methods:

1. **Impregnation**: Physical adsorption of an organic material on an inorganic support is simply done by first preparing a neat sol-gel silicate film and then immersing the film in a concentrated solution of the organic compounds. Evaporation of the external solvent leaves a silicate impregnated by the organic molecules. The interaction between the organic and inorganic moieties may be direct (e.g., by hydrogen bonding to silanols) or indirect (e.g., by adsorption of a cluster of molecules to the silicate surface).

2. **Sol-gel doping**: An important variant of the last approach is termed sol-gel doping. Here the guest molecule, be it an organic or inorganic moiety, is introduced along with the sol-gel precursors and is entrapped in the sol-gel matrix by the polymerization of the silicate around it. The intimate contact between the guest molecule and the polymerizing host provides ample routes for optimized interaction between the inorganic host and the guest molecule. In addition, caging of the organic moiety within the silicate porous structure can take place. This caging mechanism is particularly interesting from an electrochemical point of view when the silicate pores form a morphology of bottlenecks that hinder the mobility of large

encapsulates. These materials have interesting functionalities while still allowing the movement of electrochemical reactants and products and thus rapid interaction of modified electrodes with the environment.

3. **Polymerization of Ormosils carrying desirable functionality**. This procedure involves the synthesis of hybrid materials from silicon monomers containing hydrolysable groups that can participate in sol-gel polymerization reactions and also carry desirable functionalities that remain bound by Si-C bonds to the silicate network after gelation. A combination of several functionalities by mixing different monomers is a viable and useful possibility. Silicon monomers containing double bonds, epoxides, thiols, aldehydes, and amines (see several examples in Table 4.2) are frequently used for Ormosil formation. Carboxylates containing monomers are less useful because they can participate in intramolecule reactions with the silanols. Redox-active functionalities such as metallocenes (ferrocene), methylene blue, and quinones were also proposed.

4. **Grafting on preprepared Ormosils or inorganic silicate**. In this technique, the organic moiety is bonded to silicate silanol by a covalent bond. A variant of the same method is to produce organically modified silicate containing a pendant functionality (e.g., by sol-gel polymerization of epoxy-, vinyl-, thio-, or aminosilicate) that can be used to anchor the desirable organic moiety after gel formation.

Impregnated and doped sol-gel hybrids sometimes result in permanent immobilization of the guest molecule with zero observed leaching of the organics. However, because no covalent bonding is involved, some or even the entire organic modifier is removed from the silicate in most cases, particularly when an appropriate solvent is used for extraction.

4.3.1.1 Encapsulation of Biomolecules

Biomolecules constitute an important class of organic compounds whose immobilization in silicate and other inorganic matrices has attracted considerable attention. The power of this technique, which was first presented over half a century ago by Dickey [79,80], was unraveled by a series of publications starting with the work of Braun et al. [81,82]. The versatility and the synergy between the bioentities and the silicate hosts are still being unraveled, yielding surprising revelations [13]. Delicate, heat-sensitive biological entities capable of direct recognition of small ligands or large bioentities (e.g., aptamers, oligonucleotides, and antibodies) and highly specific biocatalysts (e.g., enzymes and catalytic antibodies) can be encapsulated in sol-gel matrices by sol-gel doping or by linkage to pre-prepared silicate or Ormosil matrices. Not only can these delicate entities survive the harsh conditions involved in sol-gel polymerization, but at times they can also exhibit improved performance as compared to the native bioentities. Remarkable changes of the binding affinity, turnover numbers, and redox potential of encapsulated enzymes were reported by a mere change of the encapsulation environment. Increased stability under far-from-neutral pH conditions or elevated temperature and the ability to

TABLE 4.2
Examples of Some Useful Commercially Available Silicon Monomers for Polymerization of Ormosils

Monomer	Structure	Abbreviation	Usage
n-octadecyltrimethoxysilane		ODTMOS	Surface hydrophobization
Methyltrimethoxysilane	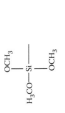	MTMOS	Surface hydrophobization
Octyltrimethoxysilane	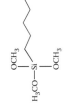	OHMOS	Surface hydrophobization
Phenyltrimethoxysilane		PhTMOS	Surface hydrophobization
Perfluoroheptylethyltriethoxysilane	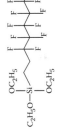		Surface hydrophobization

(*continued*)

TABLE 4.2 (CONTINUED)
Examples of Some Useful Commercially Available Silicon Monomers for Polymerization of Ormosils

Monomer	Structure	Abbreviation	Usage
3-Cyanopropyltrimethoxysilane			Nonpolar surface hydrophilization
3-Mercaptopropyltrimethoxysilane		MPTS	Metal coordination, binding nanoparticles
3-Aminopropyltrimethoxysilane		APTS	Cationic. Nanoparticles binder, binding proteins
3-Aethacryloxypropyltrimethoxysilane		MEMO	Curing agent for polyaddition
3-Glycidoxypropyltrimethoxysilane		GLYMO	Curing agent for polyaddition

(continued)

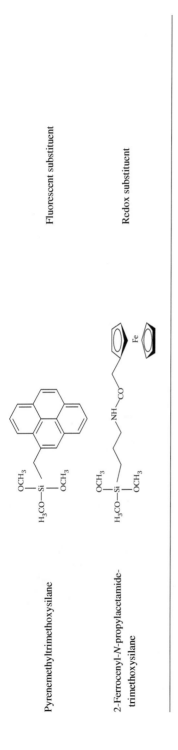

Pyrenemethyltrimethoxysilane

Fluorescent substituent

2-Ferrocenyl-*N*-propylacetamide-
trimethoxysilane

Redox substituent

function in organic solvents and in the encapsulated form extremely hydrophobic conditions were reported. For example, alkaline phosphatase exhibits its optimum activity at pH 9.5 and remains active at a pH as low as 0.9 [83]. Entraped creatine kinase retains 50% of its activity ten times longer at 47°C than the free enzyme at the same temperature [84]. Encapsulated lipases that can function in organic as well as inorganic solvents are commercially available from Fluka [85].

Bioencapsulation is not limited to biomolecules but can be extended to live microorganisms as well. Viruses, bacteria, yeasts, algae, and even mammalian tissues were successfully immobilized within silicate shells and retained their viability [13]. Two inspiring examples include the encapsulation of bioartificial organs and bioengineered microorganisms. Langerhans islets were encapsulated in silicate–alginate composites, which were successfully implanted in diabetic mice and kept the mice alive for a lengthy period [86]. While the electrochemical application of this approach is rather far-fetched, though it can be envisioned in the future, electrochemistry is likely to play an increasing role in our second example, the encapsulation of recombinant microorganisms [87–89]. Figure 4.5 (after work conducted by Permkumar et al. [87]) demonstrates the dynamic response of a single recombinant *E. coli* bacterium that was engineered to express green fluorescent protein (gfp) in response to specific stress conditions. Similar engineering with the expression of β-galactosidase or glucose oxidase will enable

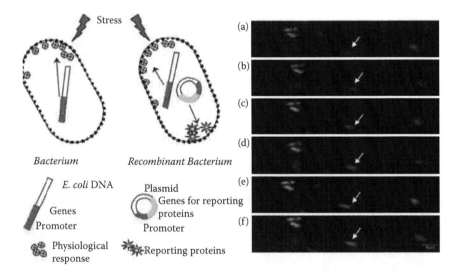

FIGURE 4.5   Left: Scheme of stress-induced response of native and recombinant bacteria. The recombinant bacterium senses the stress and expresses green fluorescence protein in addition to the physiological defense mechanism of the native bacterium. Right: Single *E. coli* cell fluorescent response after induction by 1.2 mmolar mitomycin C (an antibiotic). Confocal microscope images were taken after incubation times of 0, 140, 260, 340, 420, and 480 min with the antibiotic. (From Avnir, D., T. Coradin, O. Lev, and Livage, J., 2006. *J Mater Chem* 16:1013–30. Used with permission.)

electrochemical sensing as well. Needless to say, the expression of insulin in response to glucose deficiency as well as the expression of green fluorescent protein in response to stress conditions imply that the encapsulated cells were viable, and that nutrients and analytes could be delivered through the hybrid matrix.

4.3.2 COMPOSITES IN SOL-GEL ELECTROCHEMISTRY

Composites are materials made up of two or more phases. From the materials chemistry point of view, the straightforward way to classify composites is based on the degree of interpenetration between the two phases and the plasticity or elasticity of the components. Because the degree of interpenetration is not a binary quantity, and it may span a large range of dimensions of the embedding and guest domains, one would prefer to use an alternative classification based on how the composite is formed, which bears also on the degree of interpenetration and wettability of the silicate by the foreign phase.

4.3.2.1 Organic Polymer–Inorganic Host and Organic Host–Inorganic Polymers

Five different methods are currently being used to design organic polymer–inorganic component composites, and most of them have already found electrochemical usefulness, though practical applications are slow in coming.

4.3.2.1.1 Encapsulation of Preprepared Organic Moieties in Sol-Gel Silicates

This method of obtaining inorganic–organic composites is by far the most popular and one of the simplest to implement. All that the experimenter has to do is carry out conventional sol-gel processing in the presence of a foreign organic polymer in a compatible cosolvent that will allow dissolution of the precursors as well as the polymer. The method is generic, and it is compatible with all sol-gel processing techniques, including the conventional dip coating, spray coating, and spread coating as well as electrophoretic and electrochemical sol-gel deposition techniques. Numerous applications were reported, including modified working electrodes, sensors and biosensors, solid electrolytes, stand-alone membranes for humidity sensors, and fuel cell electrodes.

The list of polymer dopants that have already been used for electrochemical applications is quite extensive and includes hydrophilic polymers such as polysaccharides and the positively charged chitosan [90–97] as well as exceedingly hydrophobic neutral dopants (e.g., poly(dimethylsiloxane) and DMDS [98,99]). Several examples are given in Scheme 4.6. The guest polymer can be tailored to retain water (if the application is in the area of polyelectrolyte membrane fuel cells (PEMFCs)) or reject water (for gas-phase sensors), depending on the choice of the end application. Likewise, many of the polymer dopants are natural polymers, whereas others are synthetic and may span a large range of price and complexity, including relatively inexpensive poly(ethylene glycol) and expensive, carefully designed redox polymers such as poly(neutral red) [100] or osmium bipyridine [101].

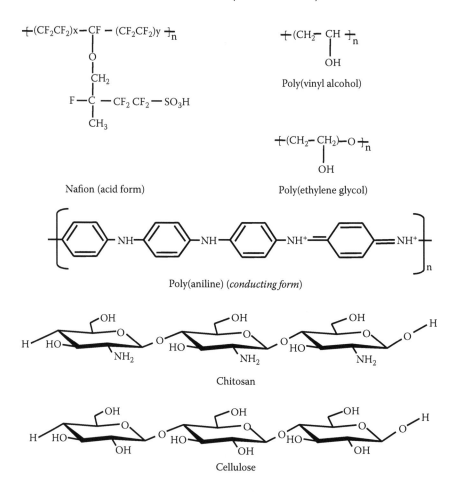

SCHEME 4.6   Structure of common polymer dopants mentioned in this chapter.

For most electrochemical applications, the mechanical strength of the composites is quite sufficient, although for certain applications such as fuel cell membranes, the loss of strength may necessitate the addition of rigid support that would decrease the power per weight output of the fuel cell. Nevertheless, it was noted that better connectivity between the phases (and smaller domain size) increases the strength and the Young modulus of the composite. Interpenetrating and segregated phases are not usually differentiated by electrochemical means, and therefore electrochemists do not perceive the degree of interpenetration unless clear optical or structural changes accompany the phase segregation.

Novak [102] provided important guidelines for the preparation of alloys and interpenetrated polymers relative to conditions leading to the formation of segregated phases. These general guidelines may be used also as a qualitative way to control the domain size of partly interpenetrating polymer composites.

Conditions that are close to those forming interpenetrated polymers may yield smaller domains of the guest phase. These include the following:

1. The cosolvent should be compatible with the precursors as well as with the hydrolyzed intermediates and oligomers that are formed during polymerization, otherwise phase separation will occur. This is not always simple because most sol-gel precursors are hydrophilic or become hydrophilic after the initial hydrolysis step (which replaces the hydrophobic silicon alkoxide by a more hydrophilic silanol), whereas many of the desirable polymer dopants are hydrophobic in nature. Hence, successful cosolvation for the first stage of polymerization does not necessarily guarantee that the inorganic phase will not precipitate at an advanced stage of the sol-gel process.
2. To enhance polymer interpenetration conditions and to form small domain sizes, compatibility with the solvent should be maintained even after water or alcohol is released as a polymerization by-product.
3. Better interpenetration of the two phases may be encountered when they have high affinity toward each other. For example, this may be the reason for obtaining clear, transparent composite from poly(methyl methacrylate), PMMA, poly(vinyl alcohol) (PVA), and poly(vinyl acetate) and poly(vinyl alcohol) in silicate and methyl silicate gels [102,103].
4. As an extension of the last criterion, hydrogen bonding between the two phases enhances the interpenetration and yields small-size domains. This makes polyols, for example, especially attractive as sol-gel guests.
5. Chemical bonding between the two components may enhance polymer interpenetration (though this class of grafted composites belongs to our next classification). Trimethylsilane terminated polyols and poly(ethylene oxide) are especially attractive due to the combination of labile hydrogen bonds and facile formation of covalent linkage to the siloxane backbone.

Indeed, polysaccharide-silica composites are frequently used for the formation of electrochemical films, because there is a large affinity between cellulose and the silicate oligomers in addition to the compatibility in the nature of solvents that can be used. The opposite is also true, and silicates (particularly sodium silicate) are useful additives in the paper industry. Chitosan is quite appealing, at least from the interpenetration point of view, because in addition to all these benefits, it is also positively charged and thus undergoes electrostatic interactions with the deprotonated silanols and can also assist in the attachment of biomolecules that are often negatively charged.

The example of Ormolytes (organically-modified electrolytes). Sol-gel-derived silica/alumina/zirconia/titania are poor conductors, and one of the ways to increase their ambient temperature ionic conductivity is to add organic components that incorporates ionic salts. When the processing is carried out under ambient conditions, the organic moiety will be intact. The molecular-level

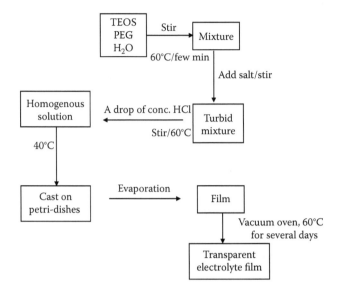

SCHEME 4.7   Preparation of hybrid, ionically conducting matrix. (From Judeinstein, P., and C. Sanchez, 1996. Hybrid organic–inorganic materials: A land of Multidisciplinarity. *J. Mater. Chem* 6:511–525. Used with permission.)

mixing of the organic and inorganic precursors ensures homogeneity at the microscopic level. The literature contains a large number of studies in which various combinations of polymeric species with different electrolytes are reported to yield highly conducting matrices. They are referred to as organicly-modified electrolytes (Ormolytes). The conducting hybrids that are generally used as electrolyte materials are treated as "biphasic" at the microscopic level [104–106]. The formation of hybrids may be due to weak interactions based on van der Waals forces, hydrogen bonding, or due to strong covalent bonding [107]. The electrical conductivity depends on the intimate mixing of the inorganic and organic phases and also the ionic component present. The basic recipe reported by Judeinstein is used by many with little modification, depending on the components.

A schematic of the process is given in Scheme 4.7, where the use of TEOS and poly(ethylene glycol) are shown, and the resultant Ormocer with an ionic component was reported to have very good ionic conductivity, on the order of 10^{-6}–10^{-4} S/cm at 25°C. Salts based on lithium, magnesium, calcium, zinc, and other trivalent metal ions such as europium, dysprosium, and ytterbium have been used to prepare sol-gel-derived hybrid electrolytes [108–114]. Ravine et al. [115] used lithium nitrate and lithium perchlorate salts to achieve ambient temperature ionic conductivities of $\approx 7 \times 10^{-5}$ S cm^{-1} at 25°C using SiO$_2$–PEG gels. The metal ion would be solvated by the ethylene oxide units present in the matrix. Hybrids of tetraethoxysilane and tetraethylene glycol were used as matrices for lithium salts to achieve high lithium ion conductivity [116,117]. In spite of the conductivities achieved using hybrid materials, there is still a long way to go before they can be used in actual devices such as batteries. Other considerations, including a solid polymer interface (SPE) that generally occurs with a polymer electrolyte–electrode interface, thermal cyclability with retention of ionic conductivity, transport number, etc., are to be addressed further. Further research is required to optimize the gel properties so that the contact resistance with the electrode surface is minimized. This is an essential requirement for application in the area of electrochemical energy systems, particularly rechargeable batteries. The use of silicates alone would not give rise to high conductivities because a plasticizer is required in the form of an organic component to improve ionic conductivity.

4.3.2.1.2   Grafted Organic Polymer–Silicate Network

Criterion 5 (chemical bonding between the two components) for the enhancement of polymer–silicate interpenetration brings us to the next class of composites in electrochemistry. Although the preparation conditions are similar to the previous protocol, a covalent linkage between the two phases provides strong attachment of the pre-prepared polymer to the siloxane backbone, and intimate mixing of the two components is achieved. Covalent bonding is important when otherwise leachable, linear polymers are processed, or when interpenetrating polymer formation is desirable, that is, especially when optical transparency is preferred. The covalent bond between the two polymers is often created by condensation reactions between the silane monomer and polyols. This is indeed the case in the formation of PVA–poly(vinyl pyridine) (PVA-g-PVP), poly(ethylene glycol), and dextran sulfate-silica sol composites [118,119].

Limited interpenetration of the inorganic and organic polymers is usually achieved by preparation of the silicate sol prior to its mixing with the organic polymer and other additives. This procedure is especially preferred over the mixing of the organic polymer with the silane monomer in order to minimize contact of the evolving methanol and the starting solution with delicate bioentities that are used for the production of biohybrids. Another synthesis procedure involves linear polymers containing reactive end groups such as alkoxysilane or hydroxyl-terminated linear polymers. For example, hydroxyl-terminated poly(dimethyl siloxane), $[OH [Si(CH_3)_2 O]_n Si(CH_3)_2 OH]$ [120], and triethoxysilane-terminated poly(ethylene oxide) are frequently used for composite preparation, depending on the desirable hydrophobicity. When the embedded phase is highly hydrophobic, and this is certainly the case for dimethylsiloxanes, the covalent bonding to the silicate is an absolute must in order to prevent phase separation during sol-gel processing.

4.3.2.1.2.1   Interpenetrating Composite Aerogels by Supercritical Solvent Evacuation   A related procedure was reported for the preparation of highly porous, low-density composites. This process relies on the selection of a cosolvent that is suitable for the preparation of interpenetrating organic–inorganic composites, and can also be replaced by carbon dioxide under supercritical

conditions. This process yields interpenetrating composite aerogel at the end. Novak et al. [121] demonstrated the formation of interpenetrating composites films starting with poly(2-vinylpyridine) (PVP), poly[methyl methacrylate-co-(3-trimethoxysilyl)-propyl methacrylate)], silanol-terminated poly(dimethylsiloxane) (PDMS), or cross-linked poly(N,N-dimethylarylamide) (PDMA), which can all form hydrogen bonds with silicate oligomers, TMOS, and a compatible solvent (e.g., methanol, THF, or formamide). The solvent is exchanged by carbon dioxide, and the latter is evacuated under supercritical conditions, leaving porous transparent material. Although, to the best of our knowledge, this procedure has never been used for the preparation of electrochemical materials, it is a promising route for the production of transparent and highly voluminous electrode materials.

4.3.2.1.2.2 Concurrent Organic–Inorganic Polymerization   Another procedure to produce truly interpenetrating polymer–silicate composites is by incorporating vinyl containing polymers that can undergo radical polymerization and concurrently cross-link with the polymerizing silicate. Novak et al. [121] used this method to produce poly(acrylamide)–silicate gels by TMOS hydrolysis and polycondensation, which occurred simultaneously with the free-radical polymerization of *N,N*-dimethylacrylamide. The process is carried out in a methanol–formaldehyde solution with a small percentage of *N,N'*-methylenebisacrylamide cross-linking agent, ammonium persulfate initiator, and *N,N,N',N'*-tetramethylethylenediamine (TMEDA). The following reaction scheme demonstrates this procedure:

$$CH_2C(O)N(CH_3)_2 + CH_2(NHC(O)CH_2)_2 + Si(OCH_3)_4$$

$$\xrightarrow{\text{(NH}_4)_2S_2O_8, H_2O, TMEDA, CH_3OH, HCO_2H} \begin{bmatrix} \text{Interpenetrating -Gel} \\ \text{Silica/Polyacrylamide} \end{bmatrix} \quad (4.7)$$

4.3.2.1.3   Encapsulation of Oxides in Organic Materials and the Relevance to PEMFC and Lithium-Conducting Polymers

The incorporation of small titania, silica, zirconia, and other metal oxide grains within Nafion and other anionic polyelectrolytes is gaining interest in electrochemistry. The first article [122] in this area described the favorable physical characteristics of sol-gel-derived silica particles doped within Nafion membranes. It was soon noted that a dispersion of 10% fumed silica or titania particles in Nafion membranes increases the water retention capabilities of the membrane [123] and thus allows its operation at high temperatures. The real breakthrough was achieved when Bocarsly and his colleagues [124,125] showed that Nafion–sol-gel composite membranes could operate as proton exchange membranes in polyelectrolyte membrane fuel cells (PEMFCs) above 100°C and still retain superior performance compared to Nafion membranes. Their membranes were prepared by acidic TEOS polymerization of the silicate nanoparticles within the network of the wet precast Nafion membrane or by casting a Nafion gel concurrent with the TEOS polymerization. Both membranes exhibited better water retention properties, better strength, and higher temperature durability than pure Nafion, the industry reference material.

A few words on the factors that currently impede commercialization of PEMFCs are due here. PEMFCs, one of the most promising emerging clean energy technologies, generates a potential drop between the hydrogen (or some other organic fuel such as methanol) oxidation and the oxygen reduction electrodes. Hydrogen–air is the most advanced fuel cell of the category. Because there is no natural high-capacity source of hydrogen, it is produced by reforming, a catalytic process to convert methane and other hydrocarbons to CO_2 and hydrogen. However, this process produces carbon monoxide as a by-product, which is retained in the hydrogen feed to the fuel cell and poisons the electrocatalysts. Increasing the temperature of the cell shifts the relative adsorption equilibria in favor of hydrogen adsorption, and thus reduces CO catalyst poisoning. Additionally, high operating temperatures give good kinetics at both electrodes, and improve heat dissipation, thus improving the PEMFC performance. However, at high temperatures, sulfonated polymers, including Nafions, lose water, which results in low elasticity, and even worse, a decrease in the ion exchange capacity and proton transport rate. Thus, the ability to increase water retention at elevated temperatures is a desirable property, and overcoming such barriers is considered to be a major breakthrough that will have practical implications. It not surprising therefore that Bocarsly's research stirred considerable interest in sol-gel doping of silicate and other micro- and nanoparticle oxides in Nafion and other sulfonated membranes. Many variants of the originally proposed idea were put forward during the past six years. Recently, Hartmann-Thompson and coworkers [126] reported functionalized oligosilsesquioxane (POSS) as hygroscopic dopants for sulfonated polyphenylsulfone proton-conducting membranes. The POSS were derivatized with sulfonic acid groups, mixed sulfonic acid and alkyl groups, and phosphonic acid groups to improve the conductivity of the proton-conducting membranes. Though conductivity was improved, the authors reported that the mechanical stability of these membranes was slightly inferior. This example demonstrates a rather general contemporary trend to improve the inorganic filler by grafting a cation exchange ionomer onto it. Another example of the same approach is provided by Tay and coworkers [127], who incorporated ionomer coated with 10-nm-silicate nanoparticles in Nafion membranes. Additional recent publications testifying to the wide interest in adding sol-gel silicate additives to proton-conducting membranes include references [128–133]. This activity is by no means confined to silicates. Other porous metal oxides were also studied. For example, titania fillers [134,135], zirconium phosphate matrices [136], ruthenium oxide [137], tungsten oxide [138], zirconia [139], and MxOy (where M = Ti, Zr, Hf, Ta and W) [140] have been reported.

Another important sol-gel filler additive to PEMFC membranes is mesoporous silicate nanoparticles. Valle et al. [141] reported a general pathway for the production of hierarchically structured transparent composites starting with a mixture of tetraalkoxysilane, as silica precursor, a surfactant as template-forming agent, a solvent, water, and a hydrophobic polymer (e.g., PVDF). The precursors force the silicate polymerization to occur in isolated mesoporous spheres entrapped within

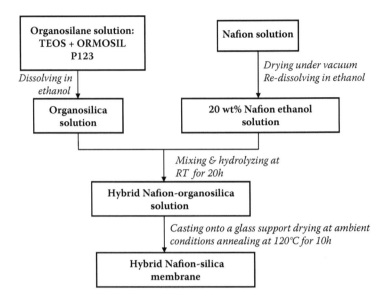

SCHEME 4.8 Protocol for the preparation of mesoscopic silicate–Nafion composite membranes. (From Pereira, F., K. Valle, P. Belleville, A. Morin, S. Lambert, and C. Sanchez, 2008. *Chem Mater* 20:1710–18. Used with permission.)

the hydrophobic nonporous organic polymers. Surprisingly, conductivity studies showed that a continuous inorganic phase was produced, at least when a large concentration of silica-forming agents was used. Surfactant-templated silica in a PVDF network that was created by this route exhibited 4×10^{-4} S cm^{-1} (at 100% RH and 20°C), over two orders of magnitude higher than pure PVDF. An interesting and highly successful variant of this method was recently reported [111]. In this preparation, protocol (shown in Scheme 4.8) mesoporous silicate–block polymer composite was prepared. The authors used Pluronic P123, a poly(ethylene glycol)-poly(propylene glycol)-poly(ethylene glycol), $(HO(CH_2CH_2O)_{20}(CH_2CH(CH_3)O)_{70}(CH_2CH_2O)_{20}H)$ block polymer, for templating of mesoporous particles and then mixed them with Nafion or with another sulfonated polymer precursor. The authors found that high silicate contents (>13%) tend to retain more water at high temperature but lower the proton density in the composite. The authors reported that the membrane exhibited a fivefold improved proton conductivity at 95°C compared to Nafion throughout the range of studied ambient humidity, RH = 50%–100%, attributed to the highly accessible mesostructure of the silicate filler. This important article demonstrated that good control of pore dimensions of the silicate can be obtained by mesoporous templating and further demonstrated that advances in mesoporous materials production—one of the hot contemporary materials science fields—can improve PEM specifications.

Lithium-conducting polymers with inorganic additives. An additional energy-related field in which silicate additives to continuous organic membrane is gaining importance is in "solid organic electrolytes" (Ormolyte) for secondary lithium batteries. The activity in this area is centered on poly(ethylene oxide) (PEO) polymers that can coordinate lithium and thus allow lithium transport between the two electrodes. However, amorphous PEO exists only at elevated temperatures (about 60°C), and its crystalline structure at room temperature shows poor conductivity. So the lithium conductivity of PEO electrolytes is about 10^{-4} S cm^{-1} only at above 60°C. The group of Croce [142] has demonstrated that addition of a few percent alumina or titania nanoparticles hinders crystallinity and increases room temperature conductivity. It was shown that the addition of a few percent of hectorite clay platelets of hydrophobic fumed silica increases the conductivity, decreases the crystallinity of the PEO membranes, and increases their mechanical stability [143]. Conductivity of up to 10^{-4} S/cm^{-1} was reported at room temperature. The use of sol-gel technology for the introduction of such fillers is to be anticipated. Kweon and Noh [72] synthesized and studied the conductivity and stability of triethoxysilane-terminated linear PEO polymer membranes that were prepared under acidic conditions in the presence of tetraethoxysilane and LiClO$_4$ salt. The favorable impact of silicate formation was noted even in this early report. Mello et al. [144], by NMR studies, showed that lithium motion in the PEO–silica–LiClO$_4$ matrix is accompanied by segmental motion of the organic PEO polymer, which underscores the importance of the open, noncrystalline nature of the polymer for efficient lithium transport. A similar synthesis protocol for obtaining silica incorporating poly(ethylene glycol) was used [145]. Liu et al. [146] used copolymerization of TEOS- and UV-initiated free radical polymerization of poly(ethylene glycol), dimethacrylate, PEGDMA, and poly(ethylene glycol) monomethacrylate (PEGMA) to give an amorphous PEO-silicate network (Figure 4.6 and Scheme 4.9). The ratio between the PEGDMA cross-linker to the PEGMA determined the flexibility of the membrane and its conductivity. The pure PEGDMA–silica film yielded highly cross-linked brittle material, whereas increasing the amount of PEGMA resulted in a conductive and very flexible material, though for PEGDMA:PEGMA (1:4) the material was too weak.

4.3.2.1.4 Polymerization of Organic Polymers within Preformed Silicate Films

In a simple approach, monomers of the guest polymer are first impregnated into a preformed porous silicate film, and subsequently polymerization is initiated either by a free radical initiator (UV, peroxides), by thermal treatment, or by electrochemical means. On the positive side, this procedure is favorable for the incorporation of linear polymers in silicates, but on the downside, some of the major advantages of composites, such as biocompatibility and adhesion to the substrate, are lost. The polymerization of organic moieties within the interconnected pores of the organic material can be carried out by two general methods that are addressed in the following text.

4.3.2.1.4.1 Chemical Polymerization

Chemical polymerization of an organic monomer can be held concurrent with or after the polymerization of the inorganic

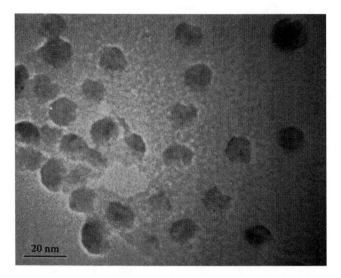

FIGURE 4.6 Transmission electron microscopy (TEM) image of composite electrolyte prepared by in situ polymerized PEGDMA and silica. (From Liu, Y., J. Y. Lee, and L. Hong, 2004. *J. Power Sources* 129:303–11. Used with permission.)

oxide. Three (organic) polymerization methods are widely used: UV curing, thermal initiation, and chemical oxidation. Ultraviolet-initiated polymerization is often carried out by incorporation of monomers containing epoxide groups such as 3-methacryloxypropyl trimethoxysilane, which yields siloxane backbone linkage to the organic polymer [147]. The use of methylmethacrylate (MMA), PEGMA, and PEDGMA, which form the organic polymer by UV curing, was

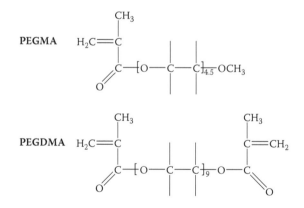

SCHEME 4.9 Structure of UV-curable PEO precursors (Liu, Y., J. Y. Lee, and L. Hong, 2004. *J. Power Sources* 129:303–11).

already discussed in the context of the encapsulation of silicates in organic polymers. It may be useful to increase the extent of cross-linking between the organic and silane moieties using an interaction between pendant alcohol groups and the inorganic silanols. For example, 2-hydroxyethylmethylmethacrylate (HEMA) was used as a copolymer that bonds strongly hydrolyzed alkoxysilane and polymerize (or copolymerizes with MMA) upon the introduction of free radical initiator (e.g., benzoyl peroxide) [148]. The authors noted that vacuum evaporation of the condensation products prevented matrix shrinkage, yielding a transparent membrane. Thermally initiated polymerization was reported, for example, for the copolymerization of 3-glycidoxypropyl-trimethoxysilane monomers [149]. Chemical oxidation (for example, by peroxodisulfate) of aniline and thiophene polymers was also initiated either within the dry silicate gel or in the sol state [148]. Interestingly, better homogeneity was obtained in this research by polymerization of aniline that was introduced into the wet inorganic gel rather than by copolymerization. However, most researchers prefer the concurrent polymerization methods relative to the in-series processes involving wet silicate gel production and subsequent incorporation and polymerization of the organic monomer within the porous structure. Diffusion time and maximal organic polymer loading considerations seem to favor the former processes.

4.3.2.1.4.2 Electrochemical Polymerization The polymerization of conductive polymer can be carried out within preformed silicate or other inorganic films [149–152]. Polyaniline, polypyrrole, and polythiophene are widely used by electrochemists, and they were readily polymerized by electrooxidation within the interconnected pores of silicate films. This mode guarantees electronic connectivity between the support electrode and every segment of the polymer dopant. In fact, once electron percolation between the conductive film–liquid interface and the solid electrode is terminated, the electropolymerization is no longer sustained and the doped film growth is terminated. However, because long-chain polymers are essential for high electric conductivity, the composite exhibits somewhat lower conductivity compared to pure conductive polymers. Simultaneous electropolymerization of aniline and electrodeposition of the silicate was recently reported [153].

4.3.2.1.5 Layered Composites
Composites can also be formed by sequential, layer-by-layer construction of multilayered materials. The silicate or an Ormosil thin film is used to bridge the different layers and immobilize the enzyme. In many cases (e.g., [100,154,155]), a mediator layer is first deposited on the electrode, and then layers containing the enzyme, the silicate, and sometimes another layer of permselective Nafion are deposited to minimize anionic interferences.

Multiple layer deposition is especially useful for the preparation of composites containing nanoparticle layers, where the capped gold nanoparticles are deposited on thiolsilicate films and in some cases covered by additional layers [156–160]. Enzyme electrodes can be constructed by incorporation of enzymes either within the layers or "sandwiched" in between different layers. If the silicate layers are sufficiently thin, perpendicular electron percolation can be maintained, and in some cases direct charge transfer from the gold to the enzyme active center can also be attained. A related assembly was proposed [161] that comprised a positively charged boron-doped diamond substrate, coated with (the negatively charged) sol-gel silicate, on which a layer of cytochrome c was deposited and held in place by another layer of sol-gel silicate. Electrochemically generated doubly charged cytochrome c could oxidize nitrite to nitrate, and it could be regenerated by the diamond nanparticle support.

4.3.2.2   Inorganic Nonmetallic Particulates: Inorganic Binder

Prussian blue, polyoxometalates, and vanadium dioxide as well as titania, zirconia, and even aluminosilicate clays are often incorporated in silicates by the sol-gel process. These composites are prepared for two reasons: The first involves obtaining desirable dielectric and optical characteristics of the inorganic film. This can be performed by the production of a single phase with molecular blending (alloying) of different oxides. This is indeed the situation for most cases where titania, zirconia, or alumina–silicate composites are produced. The second class, which is more relevant to a section dealing with composites, involves the production of redox inorganic polymers that can be used for catalysis, electrochromic windows, and charge mediation in electroenzymatic devices. It is beneficial if the second component forms a separate continuous phase, which will allow electric connectivity with the conductive electrode support. Indeed, Prussian blue, a selective hydrogen peroxide reduction electrocatalyst, and its nickel and copper analogues are often used to enhance electronic communication between enzymes and electrodes. The inorganic redox polymer can be either deposited as a single layer along with the silicate [162–164], as a part of carbon ceramic electrode [165], or as a separate layer underneath the silicate phase (e.g., [166–168]). A tungsten oxide electrochromic phase within sol-gel silicate windows [169,170], polyoxometalate–silica proton conductors [171], and electrocatalysts for bromate, nitrate, and chlorate reduction [172] were also reported. An inverse approach in which sol-gel derived titania and silica grains were incorporated in WO_3 films was reported to increase counterion transport to support electroswitching of tungsten oxide layer [173].

4.3.2.3   Graphite and Metal–Silicate Hybrids

CCEs and the closely related metal–silicate composite ceramic electrodes were first reported by the Tsionsky group [174]. The electrode material and its many offspring are increasingly used in different areas of amperometric sensing and biosensing, and some potentiometric sensors and biosensors and energy-storage-related applications. The originally proposed electrode comprised a conductive network of graphite particles embedded in sol-gel-derived ceramic or Ormocer binding material. The metal oxide, which is predominantly a silicate or organically modified silicate, serves as a porous binder for the conductive network. The conductive component can be in the form of graphitic powders (e.g., activated carbon,

carbon black, and glassy carbon particles) or fibers. The originally proposed carbon filler was sometimes replaced by metal powders, conductive platelets of coated aluminosilicates, and increasingly by carbon nanotubes. Sol-gel-coated porous macro-reticulated carbon [175] and carbon cloth ceramic electrodes were also proposed [176]. The latter is especially useful for improving the contact between sol-gel solid electrolyte and the carbon electrode support, and hence it becomes very useful for all–sol-gel electrochemical fuel cells and lithium batteries, where both electrodes as well as the electrolyte are solids made by sol-gel processing.

CCEs became so useful because the material provides high electric conductivity by the interconnected conductive powder of its surrogates, catalytic reactivity is guaranteed by the addition of metals or catalytic fillers, it is compatible with enzymelectrodes, and it is possible to control the thickness of the wetted section of the electrodes in aqueous electrolyte.

Four different classes of composite ceramic electrodes were reported: (1) ceramic carbon electrodes, (2) metal powder–ceramic composite electrodes, (3) carbon nanotube–silicate composites, and (4) coated aluminosilicate–ceramic electrodes. The first two fillers are basically three dimensional, whereas the third and fourth classes represent 1-D and 2-D anisotropic fillers.

4.3.2.3.1 Carbon Ceramic Electrodes (CCEs)

Carbon ceramic composites are prepared by mixing graphitic material, silica precursor, acid catalyst, water, and a cosolvent, and letting the material gel and dry under ambient conditions [174,177,178]. Acid catalysis or addition of another catalyst, and high temperatures may be used to enhance gelation. The composite electrodes benefit from the mechanical rigidity and deformation resistance of the silicate backbone, from electron conductivity through the interconnected carbon powder, as well as from the ability to incorporate different hydrophilic or hydrophobic silicate precursors, sol-gel dopants and biodopants, metallic and organometallic catalysts, and templating agents within the same matrix.

4.3.2.3.1.1 Composition and Configuration of CCEs The preparation protocols of CCEs entail a large number of degrees of freedom, including versatility in choosing every ingredient of the CCE composition. Most publications to date focus on silicate and Ormosil binders. Ormosil binders allow easy control over the thickness of the wetted electrode in aqueous electrolytes. Appropriate selection of the silane monomer precursor changes the hygroscopic nature of the CCE. Hydrophobic binders such as alkyl- or phenyl-tetraalkoxysilane yield highly hydrophobic porous structure that rejects water, so that only the external surface of the electrode is wetted by the electrolyte, whereas hydrophilic silicate made of tetraethoxysilane or aminopropyltrimethylsilane is completely wetted by aqueous solutions. Selection of appropriate organofunctional groups is a useful way to bond entities of desirable function. Thus, for example, aminopropylsilane functionalities provide a useful anchor for aldehyde and carboxylate reagents and, as such, they can be used to readily bond bioentities. Thioalkylsilanes are useful tethers to bond gold and silver nanoparticles

to silicates. Finally, the silane monomer may contain organofunctional ligands, charge mediators, electrocatalysts, or redox groups of desirable electroanalytical or electrocatalytic function.

Although most applications relate to silicate and Ormosil-based composites, other inorganic oxides such as zirconium and titanium oxides were also reported [47,174,180]. Sol-gel ruthenium oxide–carbon composites are used for supercapacitors [181–183]. Ruthenium is much more expensive than carbon and it is therefore commonly used in low concentrations that do not allow efficient binding of the high surface area carbon material in supercapacitors. Additionally, ruthenium oxide morphology is altered by the charging process, which may lead to undesirable volume changes during the charging/discharging operation if the binder comprises ruthenium oxide alone. Therefore, in most supercapacitor applications, an added polymer binder (e.g., carboxymethylcellulose (CMC), polytetrafluoroethylene (PTFE), Nafion) is employed. Because the inorganic phase does not form a binder in these applications, these promising sol-gel composites will not be discussed further within the context of composite ceramic electrodes.

The choice of carbon powder affects the properties of the CCEs significantly [184,185]. The conductivity percolation threshold ranges between a few weight percent for carbon black and approximately 30% for 40 μm graphite powders. The maximum attainable carbon loading ranges between approximately 15% by weight for carbon black electrodes and over 90% for 40 μm graphite particles. The pore size distribution of CCEs can be influenced by changes in the water:silicon ratio and the pH [186]. The choice of carbon may also be used to control wettability of the electrode. Recently, Macdonald et al. [187] reported the incorporation of highly hydrophilic, sulfonated carbon nanoparticles in CCEs. Naturally, the choice of carbon affects the electrocatalysis and is best illustrated by the high sensitivity of carbon nanotubes to dopamine and other analytes [188].

It is possible to cast carbon ceramic electrodes in different electrode configurations. Figure 4.7 demonstrates that it is readily possible to cast supported and

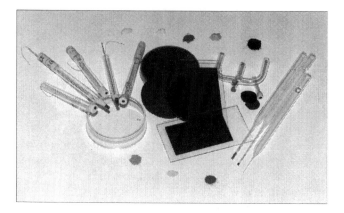

FIGURE 4.7 Different configurations of carbon ceramic electrodes (CCEs).

unsupported thick films, rods, disks, and even microelectrodes by simple one-pot sol-gel processing. Large-grain carbon powder CCEs do not shrink during gelation, probably due to the dense packing of the carbon grains, and because the silicate binder adheres very well to metal oxide supports, it is possible to cast electrodes and microelectrodes in glass tubes and use them with no further sealing treatment. Additional measures to prevent electrolyte wetting through the electrode–glass support interface is not necessary. Screen-printed CCEs are widely used [189–192], and the ability to cast the electrode for specialized configurations was demonstrated by Hua and Tan [193], who reported an on–capillary CCE electrochemical detector for capillary electrophoresis.

A remarkable property of graphite CCE is the ability to control the configuration of the wetted section of the electrode in aqueous electrolytes by changing the hydrophobicity of the graphitic filler, and even more readily, by using different silicate precursors. Hydrophobic silicate binders formed from methyl, alkyl, or phenyltrimethoxy silane reject water, and thus only small islands of carbon at the outermost surface of the electrode are exposed to the electrolyte and participate in the electrochemical process. On the other hand, hydrophilic silicate binders such as unmodified silicate or aminopropyl-modified silicate are completely wetted by the electrolyte, and practically all the surface area of the carbon grains is exposed to the electrolyte. Hydrophobic CCEs exhibit characteristic voltammetry of an array of microelectrodes, because only some carbon dots at the outer surface of the CCEs remain exposed to the electrolyte. The background current in electrochemical measurements is proportional to the water-wetted conductive surface. The faradaic signal is determined by the diffusion of the analytes to the whole cross section of the electrode; hydrophobic CCEs provide a favorable balance between low background current under dynamic electrochemical studies and high accessibility of solution analytes. The ratio of the faradaic signal to the background current of carbon black CCEs is several orders of magnitude superior to the response of glassy carbon electrodes [178], as can be readily seen in Figure 4.8. In a way, the closest electrode configuration to the carbon ceramic electrodes is the carbon paste electrode, which comprises a paste made by mixing graphite powder and hydrophobic solvent. Both the carbon paste and the CCE exhibit an electrochemical response typical of an ensemble of microelectrodes, but the former exhibits inferior operational stability [184,185]. Additionally, CCEs are porous and rigid, and thus it is possible to synthesize CCEs with controlled thickness of their wet section, whereas the active section of carbon paste electrodes is confined to the outermost paste surface. A controlled wetted porous section of up to 0.1 mm was demonstrated. This, of course, has direct implications for the maximal amount of catalyst or chelating functionalities that can be loaded and remain within electron-hopping distance from the wet, active section of the electrode. Only functionalities that are held at hopping distance from the electron-conductive support can contribute to the faradaic current and the observed signal. Several research groups have demonstrated that the active section of the CCEs does not clog upon repeated polishing due to the deformation resistance of silicates, and thus the electroactive section can be renewed by mechanical polishing

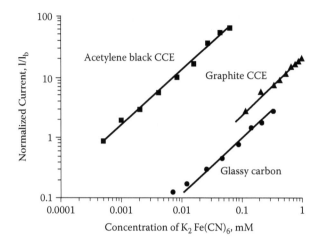

FIGURE 4.8 A comparison of the signal to background (current density divided by the background current density, I/Ib) obtained for CCE (made of small-grain acetylene black graphite, large-grain graphite, and glassy carbon electrodes). The signal-to-noise ratio is considerably enhanced by the formation of the ensemble of microelectrodes configuration.

after each measurement. The reported relative standard deviations of surface renewability [47,194–196] or sensor-to-sensor reproducibility by screen printing are usually only a few percent [47,197]. A comprehensive survey of different areas in which CCEs are currently used is given by Rabinovich et al. [47].

4.3.2.3.2 Metal Ceramic Composite Electrodes (Metal CCEs)

Metal powders and nanoparticle dispersions can be used instead of graphite fillers in CCEs. The specific density of gold is 19.3; thus, while 15–20 weight percent of carbon is sufficient for electron percolation, over 80% of the weight of gold–CCE should be composed of the metal in order to exceed the electron percolation threshold level and obtain reasonable conductivity. This introduces a large cost burden on thick-film metal CCE electrodes and limits their practical application to very thin films or microelectrodes. On the other hand, metal ceramic electrodes have several advantages compared to graphite CCEs: (1) the metal provides better electrocatalysis, (2) it is easier to prepare metal nanoparticles of controlled dimension, and (3) added functionalities can be readily tethered to the metal nanoparticles by self-assembly techniques.

The simplest reported way to obtain metal CCEs was introduced by Wang et al. [198]. They mixed water, 0.5-micrometer-sized gold particles, tetramethoxysilane monomer, and glucose oxidase to obtain a composite glucose biosensor. A more elaborate method was needed to prepare gold nanoparticle–CCEs by Bharathi et al. [46]. First, aminosilane-stabilized gold nanoparticles (about 3–10 nm) were prepared (see Figure 4.4). Dip-coating on ITO slides provided semitransparent thin gold–silicate composite films [199,200]. Lack of transparency is a serious drawback of metal CCEs, prohibiting their use in optical applications. The metal nanoparticle–silicate films provide sufficient transparency to overcome this problem, though this is limited to thin-film electrodes. We used the aminosilane-stabilized gold nanoparticles to prepare glucose biosensors by mixing glucose oxidase with the gold sol. However, in this construction, the active concentration of gold particles was less than 2%. It was possible to increase the active gold concentration in the film and even to exceed the percolation threshold by electrochemical deposition of gold on conductive substrates as described in Section 4.2.2.2 [46].

Several other methods for the preparation of metal CCEs, mostly using stabilized nanoparticles, were reported. For example, Chen et al. [201] demonstrated improved accessibility of gold nanoparticles to 3-mercaptopropyltrimethoxysilane-based aerogel film, which also provided better ascorbate and hydrogen peroxide electrocatalysis compared to simple xerogel film. Other examples of gold and silver CCE films include reports on potentiometric diphtherotoxin sensors by electrochemical ELISA [202,203]. Amperometric immunosensors were reported first by Wang et al. [204]. Cai et al. [205] reported a gold-$CaCO_3$–silicate matrix for enzymatic sensors. Xu et al. [206] reported direct electron transfer from a conductive surface to horseradish peroxidase, (HRP) enzyme based on biocompatible carboxymethyl chitosan–gold nanoparticle–silicate composite. More recent reports involve mainly biosensor fabrications (e.g., [191,192,207,208]). Less recent biosensing applications of CCEs were reviewed by Rabinovich et al. [47].

4.3.2.3.3 Carbon Nanotube CCEs

Although noted before by carbon scientists, Iijima's paper [209] on the preparation of carbon nanotubes revolutionized the area of composite science and technology. CNTs are a hundred times stronger than steel and considerably lighter; besides, they are more flexible than carbon fibers and much stronger as well. No wonder that they are already used for reinforcement of plastics. Carbon nanotubes in the form of single wall (SWCNT) or the less expensive multiwall CNTs (MWCNT) are increasingly being used for the development of electrochemical sensors and biosensors. The important benefits of using CNTs in combination with sol-gel silicates lie in their high surface area, high aspect ratio, essentially unidimensional configuration, and above all, their electrocatalytic activity. Britto et al. [210] showed superior electrocatalytic oxidation of dopamine by multiwall carbon nanotubes compared to other forms of carbon, and soon enough, the superior electrocatalytic properties of carbon nanotubes were established by others (e.g., [210–212]). Gong and coworkers [213] were the first to realize the relevance of these findings to sol-gel electrochemistry and that carbon nanotubes are compatible with sol-gel processing. They treated MWCNT by a nitric acid–hydrogen fluoride mixture, which is known to introduce defects and shorten the CNT length. Sodium dodecylsulfate was used as a dispersing agent. The solution was sonicated, and the CNT was filtered and washed with caustic soda solution. Hydrophobic sol-gel electrodes, which the authors called ceramic-carbon nanotube nanocomposite electrodes (CCNNEs), were then produced by the addition of the treated MWCNT to acidic MTEOS

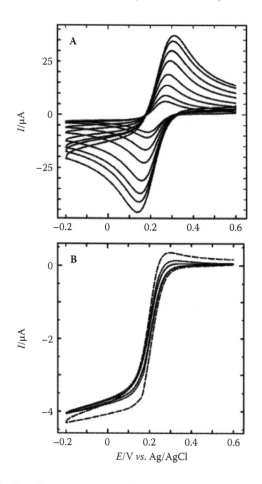

FIGURE 4.9 Cyclic voltammetry curves for 1.0 mM $K_3Fe(CN)_6$ in 0.10 M PBS at the CCNNEs prepared by dispersing (A) 2.0 and (B) 0.08 mg/mL MWCNTs into the silica sol. Potential scan rates (A) (from inner to outer) 10, 20, 50, 100, 200, 300, 400, and 500 mV s^{-1} and (B) 10 (solid line), 50 (dotted line), and 200 mV s^{-1} (dashed line). (From Gong, K. P. M. N. Zhang, Y. M. Yan, L. Su, L. Q. Mao, Z. X. Xiong, and Y. Chen, 2004. *Anal Chem* 76:6500–6505. Used with permission.)

sol-gel precursors and drop-coating of the homogeneous suspension on glassy carbon electrodes. Indeed, the authors showed that the new electrodes benefit from the hydrophobicity of the methyl silicate as well as from the electrocatalytic activity of the MWCNT. The authors showed that the electrochemical response of CCNNEs could be tuned by the level of CNT loading. At low loadings, the authors obtained cyclic voltammetric response similar to that from an ensemble of microelectrodes, and this was converted into conventional response of a planar electrode by increasing the MWCNT level (see Figure 4.9) [213]. The procedure was later extended to incorporate metal-modified CCNNE [214].

The CCNNEs received considerable attention, and several other reports of CCNNEs soon followed, mostly focusing on biosensing applications [93,215–219]. Shi et al. [219] demonstrated the first sol-gel CNT biosensor. They constructed a cholesterol detector based on the sol-gel immobilization of cholesterol oxidase on top of platinum catalyst-modified MWCNT. Gooding et al. [220] demonstrated direct charge transfer between a peroxide mimic, microperoxidase MP-11, and MWCNT array electrode. Although MP-11 is a small enzyme that requires only a short hopping distance between the heme center and the electrode, this publication showed the feasibility of direct enzyme–CNT charge transfer and opened the road for devising direct charge transfer devices involving larger enzymes and, indeed, similar sol-gel-based applications soon followed. Recently, Chen and Dong [221] demonstrated direct, mediatorless electron transfer from the heme center of HRP, an about 40 kDa enzyme, to MWCNT sol-gel electrode.

Nonsilicate–carbon nanotube composites were also reported by the same group [221]. They reported that a SnO–CNT anode showed improved lithium intercalation properties compared to the respective CNT and tin oxide electrodes. Wang et al. showed that Nafion addition can solubilize CNTs in neutral-pH phosphate-buffered aqueous and alcohol solutions [222]. Soon afterward, titania–MWCNT and mesoporous titania–Nafion–SWCNT and MWCNT electrodes were reported [223,224].

4.3.2.3.4 Modified Clay–Silicate Ceramic Electrodes

There is a practical need for transparent, porous, and electrically conductive electrode materials that can be prepared at or near room temperature. Such electrodes benefit from all the advantages of carbon CCEs, and their transparency still allows optical and dual electrochemical–optic applications. The available options for achieving such electrodes are currently not many. Conductive polymers such as polythiophene and polyaniline have narrow electrochemical windows. Very thin films of metallic nanoparticles–silicate composites may be both transparent and conductive, but the reagent loading level is very limited. The inherent anisotropy of CNT may reduce the threshold loading for electron percolation. Some 1% MWCNT renders CNT-sol-gel electrodes with sufficient conductivity for most electrosensing applications. However, even at this very low loading level, the electrodes are not transparent. Efficient ways to disperse the nanoparticles within the electrode material and thus reduce the percolation threshold are needed in order to exploit the advantages of using CNT fillers. Natural platelets such as clays provide alternative anisotropic fillers, and they can be easily made conductive by coating with transparent and conductive oxide such as antimony-doped tin oxides or indium tin oxide. Moskovite (a type of mica) platelets are especially attractive because their grain size is relatively large, on the order of 10 micrometers, and their thickness, even without a special delamination process, is on of the order of 0.5 micrometer and less. We have demonstrated that it is possible to disperse antimony-doped tin oxide (ATO)-coated mica in silicate and methylsilicate sol-gel films. It was demonstrated that the platelet clays predominantly self-align in a layered configuration parallel to the electrode (see Figure 4.10) [225]. This special configuration decreases

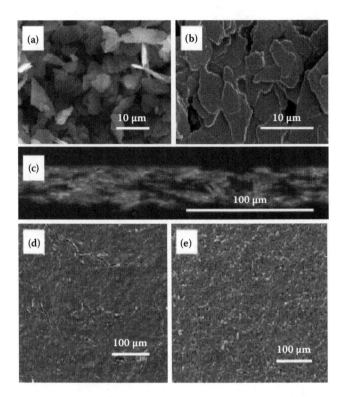

FIGURE 4.10   (a) Antimony-doped tin oxide (ATO)-modified clay powder. (b) 40% ATO-modified mica platelets in methyl silicate film deposited on a glass slide. (c) Cross section of a film of the same composition as (b), but this film is deposited on PET and covered by epoxy resin. (d) 40% ATO-modified mica platelets in silicate. (e) 40% ATO-modified mica platelets in methyl silicate matrix. (From Sadeh, A., S. Sladkevich, F. Gelman, P. Prikhodchenko, I. Baumberg, O. Berezin, and O. Lev, 2007. *Anal Chem* 70:5188–5195. Used with permission.)

the electron percolation threshold, and reasonable conductivity is attained using almost two orders of magnitude lower loading of ATO compared to the experimental percolation threshold of ATO in a clay-free, homogeneous sol-gel silicate matrix. Typical resistivity values of 100 kΩ/sq and 1.5 OD for a 20-μm-thick film were reported. The transparency is lower than that for sputtered ATO glasses, but this is probably the best method for the low-temperature preparation of transparent, porous, and electrically conductive electrode materials. It was demonstrated that similar to carbon CCEs and sol-gel silicate films, the electrode material provides permselectivity and improved detection of the positively charged methyl viologen compared to negatively charged ferricyanide. Prussian blue modified ATO-coated platelets dispersed in sol-gel-derived silicate were used to demonstrate the feasibility of a transparent and electrically conductive porous electrochromic material.

4.3.3   ELECTROCHEMICAL PROPERTIES OF COMPOSITES

A significant part of contemporary sol-gel activity is based on the exploitation of the versatility of sol-gel science and the ability of sol-gel synthesis to produce different composite materials and nanomaterials. The favorable properties of heterogeneous materials are not intuitively explicit, and therefore we feel that it is appropriate to reiterate the electrochemical incentives for the widespread use of sol-gel composites in electrochemistry.

Composite materials based on sol-gel technology are used by material scientists to achieve improved strength, stiffness and toughness, and improve optical and dielectric properties, or gain temperature, and corrosion resistance [202,226]. However, in electrochemistry, and particularly in analytical electrochemistry, these attributes are important only for some specific applications, and electrochemists resort to composites for different reasons. In this section, we describe and exemplify some electrochemical properties that may be gained by electrode and membrane design based on heterogeneous multiphase materials.

4.3.3.1   Sensitivity and Selectivity of Composite Amperometric Sensors

Amperometric sensors as well as most electrocatalytic and battery electrodes operate under diffusion-controlled conditions, and thus their response depends on the flux of electroactive solutes (or their transformation products) to the conductive electrode surface, be it a film-coated electrode or the three-dimensional conductive network of porous electrodes. In order to examine the qualitative factors that influence the flux to such electrodes, it is valuable to examine a simple conceptual model. We examine here the diffusion-limited current to a test case planar electrode of area, A, coated by a thin film of thickness, d. Diffusion-limited conditions frequently accompany a high overpotential operation such that the concentration of the key analyte is negligible at the surface of the electrode. Under these conditions, the concentration drop across the membrane determines the faradaic current:

$$i = FAD_{eff} \frac{\Delta c}{d} \qquad (4.8)$$

where F is the Faraday constant (96,500 Coulombs/equivalent), D_{eff} is the effective diffusion coefficient of a key solute in the film, and Δc represents the concentration difference between the two ends of the membrane: near the solution interface and in the close vicinity of the solid conductor. However, because the concentration near the solid electrode under diffusion-limited conditions is practically zero, and the concentration on the outer side is given by the product of the partition coefficient, K, and the solution concentration of the analyte, C_0 (assuming negligible external resistance), then the current depends on the product of the partition coefficient and the effective diffusion, on geometrical parameters, and on the concentration of the key analyte in the solution.

$$i = F(KD_{eff})\left(\frac{A}{d}\right)Co \qquad (4.9)$$

Equation 4.9 represents a somewhat oversimplistic picture of homogeneous, unidimensional diffusion in a single rate-limiting film, a situation that is almost never encountered in practice. Nevertheless, it provides a framework for the analysis of more complicated real systems involving heterogeneous, two-phase film-modified electrodes that exhibit nonlinear diffusion to the electrode. Equation 4.9 shows that the ratio between the electrode response to the analyte concentration, that is, the signal sensitivity, is determined by several effective parameters that appear either directly or indirectly in Equation 4.9. These parameters can be influenced to a large extent by the incorporation of a second phase in the sol-gel membrane or electrode.

4.3.3.1.1   Electrode Surface Area (A)

For the simple planar homogeneous film-coated electrode such as the one described in our model, the surface area term, A, cannot be manipulated at all. However, for a composite electrode, increasing the surface area by the incorporation of a conductive network such as metal nanoparticles, carbon powder, or carbon nanotubes and fibers increases the effective surface area for charge transport without a corresponding increase of the cross-sectional area of the electrode.

Another way to enhance the electrode's performance is to increase the signal-to-noise ratio. Noise level and background current in electrochemical systems are often contributed by the wetted electron-conductive area of the electrode. For example, consider the background current for an electrode that is subjected to linear voltage increase, a condition encountered during cyclic voltammetry. The background current, assuming no faradaic reactions, is given by

$$i = Sc_{dl}v \tag{4.10}$$

where c_{dl} stands for the double-layer capacitance of the electrode, S is the surface area of the conductive surface at the electrolyte interface, and v is the rate of potential change (V/s). The noise level in electroanalytical systems often depends on the background current or is contributed by faradaic reactions that are not kinetically limited and thus depends linearly on the exposed surface area of the electrode, S.

Equations 4.9 and 4.10 show that the signal-to-noise ratio of the electrode can be enhanced by influencing the ratio between S and A, that is, by increasing the effective surface area for diffusion and at the same time decreasing the charge transfer interfacial area. The simplest way to do so is to use nonplanar electrode geometry. Cylindrical or spherical film geometries will increase the ratio A/S. A microelectrode configuration is perhaps the ultimate way to increase the effective A/S ratio, because the diffusion to the electrode occurs across a sphere of much larger average diameter than the actual diameter of the microelectrode. A way to approach the favorable diffusion configuration of a microelectrode is by the formation of a configuration of a so-called ensemble of microelectrodes or partially blocked electrode. Here, the surface of a planar electrode is mostly blocked by an insulating layer, and only some small scattered dots on the electrode surface are conductive and available for faradaic charge transfer and capacitive currents

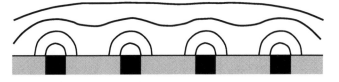

FIGURE 4.11 A scheme of the equiconcentration lines in the vicinity of partially blocked electrode, such as hydrophobic carbon ceramic electrodes (CCEs).

(Figure 4.11). In an ensemble of microelectrodes configuration, only the array of microelectrodes at the outer surface of the electrode is exposed to the electrolyte, whereas the remaining surface is isolated from the electrolyte by the insulating surface. We have demonstrated that incorporation of an intercalated conductive phase in hydrophobic silicate (such as those formed in hydrophobic CCEs, carbon nanotubes composite electrodes, or metal–methyl silicate electrodes) exposes only the outer surface of the CCE to the surrounding electrolyte and prevents the wetting of the bulk of the composite by the electrolyte (e.g., [47]). Thus, hydrophobic carbon ceramic electrodes provide an ideal balance between low background and high accessibility of solutes. The ratio of the faradaic signal to the background current of carbon black CCEs is several orders of magnitude superior to the relative response of glassy carbon electrodes, as observed in Figure 4.8 (compare acetylene black CCE with glassy carbon electrode).

4.3.3.1.2 Film Thickness (d)

The determining barrier film length is often the average distance between the outer gel–liquid interface and the interface on which charge transfer takes place. For a planar film-coated electrode, d is simply the film thickness. Clearly, the thinner the film, the smaller the diffusion barrier. Therefore, it is favorable to reduce the thickness of the film as long as the outer diffusion barrier does not become the rate-determining step. The external, or outer diffusion barrier, is influenced by the hydrodynamic conditions and, unless controlled hydrodynamic conditions prevail (e.g., rotating disk electrode or thin channel flow), external diffusion control is undesirable for electroanalytical applications. Additionally, decreasing the film thickness also implies a decreased capacity for loading coordinating ligand and catalyst and biocatalyst electrode modifiers. Introducing a percolating conductive network into the film, for example, by CCE configuration, allows a decrease in the thickness of the inner film surrounding each carbon or metal grain without compromising the ability to load the electrode with a large amount of catalysts or capturing ligands [191,227–229]. A similar effect can be achieved by coating reticulated vitreous carbon with sol-gel film [230].

4.3.3.1.3 The Effective Diffusion Coefficient and the Electrode Void Fraction

Equation 4.8 reveals the importance of the effective diffusion coefficient for attaining high sensitivity. In most membranes, the surface diffusion is negligibly small compared with the diffusion of electrolytes in the liquid-filled pores of the membrane, that is, the void fraction. The effective diffusion parameter can be expressed as the product of the void fraction and the diffusion coefficient in the electrolyte-filled pores. Thus, the larger the void fraction, the larger the flux to the electrode, and the larger the resulting current signal. Thin silicate films that are produced by most sol-gel protocols (with the notable exception of mesoporous silicate films, to be discussed in the next section) are more compact than monoliths that are formed under similar precursors and acidity conditions, and thus possess relatively smaller void fractions. This limits the diffusion rate inside the porous network, which in turn reduces the attainable current and the response time in amperometric applications. Incorporation of a guest phase may improve diffusion in the composite electrode in two ways:

1. Incorporation of a second polymeric phase and subsequently burning or leaching it out vacates large voids in the sol-gel silicate [5,6,231] and other films and monoliths (e.g., titania, [232] and tungsten oxide, [233]). This concept was first used for sol-gel thin-layer chromatographic plates [234] and monolithic columns [235].

   The Nakanishi, Soga, and Tanaka groups paved this way by the incorporation of poly(ethylene oxide), a water-soluble polymer that undergoes phase separation during sol-gel processing of monoliths into sol-gel silicates. They used this technique to form monolithic chromatography columns with exceptionally high accessibility [235]. The chromatographic monoliths are now marketed by Merck Ltd. under the trade name Chromlith™. Template formation by introduction of porogens or surfactants is another attractive and popular method of forming ordered mesostructured materials and thus increasing the void fraction in sol-gel silicates [236] and films [237]. These methods have been used intensively by electrochemists over the past decade and will be discussed further in a separate section devoted to mesoporous structured electrodes (Section 4.4.2).

2. Incorporation of a guest filler into the sol-gel starting solution provides yet another way to increase the rate of mass transfer through composite electrodes. Leddy [238] performed a detailed analysis of several exemplary cases of the mass flux in structured composites. The flux increases with the increase of the interfacial grain boundary surface per volume (i.e., the surface area between the two phases divided by the volume of the film or membrane). The manner in which the diffusion is accelerated is not at all intuitive, because the diffusion pathway along such interfaces is tortuous and does not follow the shortest line from the solution to the electrode. It may be postulated that rapid polymerization of the silicate occurs near the guest phase, which creates a third phase of macroporous voids or low-density film through which accelerated diffusion is facilitated. It is also possible that the adherence between the two segregated phases is low, and a third phase with constrained geometry is formed near their interface. The diffusion can be considerably enhanced

in such phases. The large nanostructured graphite powder–silicate interface is probably one of the reasons for the favorable electrochemical characteristics of carbon and metal-doped silicates. It is also reasonable to hypothesize that at least part of the faster conductivity exhibited by PEMFC and DMFC membranes incorporating sol-gel silicate grains (e.g., [124,125,130]; Section 4.3.2.1.1) is due to the fast H^+ diffusion rate along grain boundaries between the negatively charged filler and the sulfonated polymer support.

4.3.3.1.4 Partition Coefficient (K)

Larger K values imply larger uptakes of the analyte by the film and thus larger flux to the electrode. Not only can the sensitivity be enhanced by larger K values, but Equation 4.9 shows that the selectivity is determined by the ratio of the respective product $D_{eff}K$ of the key analyte and $D_{eff}K$ of interfering solutes. The partition coefficient, K can be influenced by the incorporation of specific ligands that have high affinity to the analyte but sufficiently high lability to allow transport of the analyte across the membrane. Acid dissociation of surface silanol groups ($\equiv SiO^-$) takes place already at very low pH (despite the fact that the first acid dissociation constant of silicic acid is $pK_a = 9.9$ at 30°C). At pH > 2, sol-gel-derived silicates are negatively charged in aqueous solutions. Thus, unmodified silica gel films exhibit large K values for positively charged species. Indeed, electrodes coated by unmodified silicate films show preferential uptake and faster diffusion of anionic redox probes such as ferricyanide/ferrocyanide, while the mobility of cationic redox probes such as ruthenium(II)tris(bipyridine) and ruthenium(II)hexaammine is significantly hindered [225,239,240]. This preferential uptake and mobility can be easily restored by incorporation of 3-aminopropylmethyldiethoxysilane into the starting solution, which gives the silicate anion exchange properties and accelerates the uptake of ferricyanide/ferrocyanide ($Fe(CN)_6^{4-}/Fe(CN)_6^{3-}$) redox pair compared to, for example, ruthenium(II)hexaammine probes [241]. Of course, what can be done with the silicate binder can also be done by a guest polymer phase. Incorporation of appropriately charged guest polymer is often used to alter the selectivity of silicates. For example, dramatic preferred permeability of positively charged moieties is obtained by incorporation of Nafion and other negatively charged polysulfonated polymers that facilitate H^+, cations (e.g., methylviologen), and positively charged colloids transport [242,243]. Amine- and ammonium-containing polymers, such as poly(ethylene imine), (PEI) or natural anion exchange biopolymers, such as chitosan [244], are used to enhance the transport of small anionic species. Obviously, the polymer film should provide connectivity throughout the thickness of the composite film, which is usually guaranteed by incorporation of more than 16%–20% polymer by volume. Not surprisingly, most of the recently reported sol-gel-derived fuel cell membranes for direct methanol and high-temperature polymer electrolyte fuel cells utilize negatively charged polysulfonated membranes and negatively charged nanoparticle additives [170,112–119].

4.3.3.2   Other Advantages of Composite Sol-Gel Silicates

Equations 4.8–4.10 provide a general framework for understanding the electro-chemical driving force for enhanced mass transport and selectivity in composite sol-gel electrodes. However, somewhat less obvious practical incentives for the use of composite materials are sometimes encountered. In the following section, we briefly address some of these incentives. We leave the discussion of templated sol-gel systems and mesoscopic silicates to Section 4.4 although their method of formation goes through composite silicate intermediates.

4.3.3.2.1   Size Exclusion and Templates

Silicate composites can provide an efficient size-exclusion membrane. The pore size distribution can be controlled, at least to some extent, by incorporation of water-soluble polymers such as poly(ethylene glycol) and their leaching after sol-gel formation or by pH control. On the low-pore-size side, low-pH silicate processing may yield silicate membranes that exhibit Knudsen flow and can separate nitrogen from air [245]. The upper limit for porous silicates probably lies in living cell templated materials [246,247] or latex sphere templates [248], and in sol-gel-derived ruthenium and vanadium oxide aerogels for supercapacitor applications [249,250].

4.3.3.2.2   Specific Recognition

Silicates have low specific affinity to organic compounds, but, although they attract positively charged molecules, this interaction is not specific. Incorporation of organo-functional groups or the addition of chemical or biochemical entities can be used to attract, with high specificity, certain compounds or classes of compounds. The most striking example of specific recognition is through the encapsulation of enzymes or antibodies, but specific recognition can also be achieved by low-molecular-weight dopants [1,11] or by molecular imprinting, as discussed in Section 4.4.1.2 [251,252]. Single-molecule spectroscopic methods were used [253] to study specific interactions between Nile red dopant and organically modified silicates. It was found that the more rigid fractions tend also to be more hydrophilic. The heterogeneity of hybrids, which visually seem homogeneous, is accentuated by this method.

4.3.3.2.3   Improved Biocompatibility

As far back as the first publication of Braun and his coworkers on the successful bioencapsulation of enzymes in silicates [82], it was noted that sol-gel encapsulation stabilizes the proteins and provides protection against thermal denaturation. This was confirmed by a large number of authors for a wide range of bioentities (e.g., [84,85,254–255]). Notably, Heller's group reported a 200-fold stability increase of glucose oxidase by encapsulation in monolithic silica [256].

However, the encapsulation of proteins in thin silicate films is less straightforward as compared to the encapsulation in monoliths. Very few reports on successful immobilization of proteins in thin (noncomposite) sol-gel silicate films exist to date (e.g., [63,257–260]). The successful encapsulation protocols in thin films can, in most cases, be attributed to the incorporation of the enzyme in-between layers [219] or in a sandwich construction between the silicate layer and the conductive matrix

[133]. Indeed, many of the reports regarding noncomposite sol-gel biosensors, where the enzyme was mixed with the sol-gel precursors prior to the immobilization, also report on direct charge transfer from the enzyme to the electrode surface, which may indicate that the viable enzyme is attached to the electrode surface (e.g., [63, 258]).

However, successful encapsulation of enzymes was reported for many composite films. Presumably, the incorporation of the organic polymer phase provides modes for stress relaxation and reduces the physical constraints on the encapsulated proteins, probably by creation of flexible microdomains within the silicate matrix. In a way, a similar mechanism can explain, at least in some cases, the stabilization of the proteins in sandwich constructions. Additionally, it seems that the adsorption of the proteins onto solid grains within the silicate films stabilizes the globular 3-D structure of the biochemicals. These two mechanisms are probably responsible for the success of guest fillers such as glycerol, which was introduced by Gill and Ballesteros [261]. Various fillers such as positively charged polymers such as Chitosan (e.g., [155,262–265]), poly(N-vinyl imidazole) and poly(ethylene imine) [256], negatively charged polymers such as Nafion [242,266], and neutral polymers such as cellulose acetate [267] have been used to stabilize biomolecules. Inorganic filers based on graphite powder [268,269,195,196,270–280], metallic particles [198], nanoparticles [46,208,279], carbon nanotube–silicates [93,278], and Prussian blue [155,162–164,167] have also been used to maintain the enzyme activity in sol-gel composite films.

4.3.3.2.4 Composites as Reagent Nanocontainers

This interesting approach was originally proposed by Opallo [279,280], who used carbon ceramic electrodes loaded with hydrophobic charge mediators such as t-butyloferrocene and Co(II)tris((bipyridine) that could be slowly released from the electrode. Rozniecka et al. [281] incorporated ionic liquids in CCEs and other sol-gel electrodes. The approach was further developed by the covalent immobilization of ionic liquid in silicate electrodes by the same group [282].

Recently, encapsulation of benzotriazole [283], a corrosion inhibitor in silica–zirconia anticorrosion film on aluminum alloy substrates, has been reported. The benzotriazole could be slowly released from the hybrid film deposit on the substrate and provide long-term corrosion protection. This approach is especially attractive because the dissolution and redeposition of the benzotriazole is pH dependent and increases when local corrosion takes place. This type of self-healing material is likely to play an increasing role in "futuristic" active smart materials or corrosion protection as well as other applications.

4.3.3.2.5 Decreasing Ohmic Drop by Improving Electric and Ionic Conductivity

Electrochemistry involves electric charge transport in an electron conductor and ionic conductivity in an adjacent phase, be it the silicate coating or the nearby electrolyte. One way to decrease the ohmic drop is by increasing the pathway of electron percolation through the conductive part and decreasing it in the silicate phase. A good way of doing so is by CCE construction, and another is by using electrically conductive organic polymers such as polyaniline and polypyrrole in CCEs [149–151].

Electric conduction can also be carried out by a hopping mechanism between adjacent redox sites. This mode of self-exchange charge transfer within silicate composite was exploited by the incorporation of redox polymers such as the osmium redox polymer, [Os(bpy)(2)(PVP)(10)Cl]Cl in the silicate [101]. A similar approach based on the creation of naphthoquinone-modified silicate CCE was also demonstrated by Rabinovich et al. [47]. Here, the binder itself functions as a charge transfer polymer due to the presence of pendant quinone functionalities.

Improving the ionic conductivity in silicate sensors is another way to reduce undesirable ohmic drop in amperometric applications. The conductivity of silicate networks is dependent on the embedded salts and is amply discussed in the context of ionic membranes [284]. Conductivity of silicates can be improved by surface modification so as to increase the concentration of the surface charges within the porous network, or by incorporation of cationic or anionic polymers such as Nafion and chitosan.

4.3.3.2.6 Improved Adhesion

Composite sol-gel silicate-organic compounds can be tailored to have extremely good adhesion to oxide substrates due to the $\equiv$Si-O-M$\equiv$ bonding (where M = metal). For example, polyaniline–silica composite coatings have much higher operational stability as compared to polyaniline films [285]. The stability of thick composite films is an important property, because it is difficult to construct sol-gel films thicker than about 10 μm. The high surface tension that builds up in partially filled pores during the drying stage fractures sol-gel matrices and limits their usefulness. The inherent heterogeneity of composite electrodes provides stress relaxation pathways and prevents fractures. Indeed, thick well-adherent polymer–silicate, carbon–silicate, and metal–silicate supported films can easily be prepared.

4.3.3.2.7 Water Retention

Silicates, titanium oxide, zirconium oxide, zirconium sulfate, and other metal oxides adsorb water, and therefore incorporation of a few percent of silicates or other metal oxides in polysulfonated membranes improved their performance at elevated temperatures and low relative humidity (e.g., [124,125]).

4.4 MOLECULAR IMPRINTING AND FORMATION OF MESOPOROUS AND HIERARCHICAL STRUCTURES

Nanoscale design of silicates and other metal oxides is an intriguing subject that can be approached in a number of ways, for different length scales, and for various classes of materials suitable for the desired end application. At least three different approaches, all based on the incorporation of pore-forming agents and their subsequent removal after gel formation, have already reached electrochemical implementation or will probably come under electrochemical scrutiny in the near future and are therefore mentioned below:

1. **Molecular imprinting**, largely originated by the work of Dickey [80], which involves the incorporation of a guest molecule or a protein and its subsequent removal after gel formation. This should leave the fingerprints of the entrapped molecule embedded in the silicate or metal oxide. A subclass of this technology is the formation of a macroporous structure by sol-gel film formation in the presence of large entities, as large as latex beads or living cells, and their subsequent removal after gel formation.
2. **Formation of mesoporous materials by surfactant and block polymer templates**. Here, a long-range periodic order is attained by the formation of a surfactant template during sol-gel polycondensation. This approach was initiated by Kresge et al. [236], and its electrochemical offspring are becoming widely used.
3. **Gels with hierarchical, two-mode pore size distribution**. This approach involves the formation of interconnected porous material by introduction of a guest polymer and its subsequent removal after gel formation. This technology, which was initially proposed by Nakanishi et al. [286], has had many spin-offs that make it especially useful for chromatography, even though its electrochemical implications are only now emerging.

These three different approaches are distinguished by the type of pore formers that are introduced in each case: leaving particulates, molecules or functional groups in the former, self-organized entities (mainly micelle and lamellar structure formers) in the second approach, and a continuous polymeric phase in the latter approach. The three different approaches also yield, respectively, very different gel morphologies: microporous or macroporous material; mesoporous materials; and hierarchical pore structures with macro- or mesoporosity as well as nanoscale pores within the same material domain.

The end applications of these materials are also rather different: The first approach is probably most useful for molecular recognition and electrosensing or other sensing applications, whereas the interconnected large pores in the third approach are useful for flow-through applications, particularly for chromatography and packed bed catalysis. Mesoporous materials have a much broader range of applications, including template formation for casting different inorganic or carbon materials, large accessible surface area adsorbents, and numerous catalytic and electrocatalytic applications.

Sol-gel processing is especially attractive to achieve successful structuring by the incorporation of a guest molecule and its subsequent removal. Template structuring and inclusion of a guest molecule in solid materials are but different facets of the same approach [287–289], and they both benefit from the advantages offered by sol-gel processing:

1. Sol-gel polymerization is a soft process, and it does not deform large globular imprints, let alone destroy small-size organic dopants.
2. The silicate is sufficiently cross-linked, which entails rigidity, and thus it may, to a large extent, retain the shape of the imprint or template long after its removal from the polymer. This said, one must also acknowledge that gradual changes of pore dimensions by aging (i.e., long immersion in the electrolyte) are also known for silicate gels.
3. Organically modified silicates, particularly Ormosils, may be designed to contain mixed functionalities within close proximity, and these can complement matching groups on the imprint, thus obtaining delicate chemical imprint rather than physical shaping alone.
4. Silicates are porous and allow evacuation of the imprinting molecule. Inorganic silicates are not damaged by calcinations and thus allow thermal removal of organic guest molecules.
5. Sol-gel processing allows formation of thin films, which are preferable sensor materials. Additionally sol-gel powders and increasingly monoliths are suitable for chromatographic and fixed bed catalytic applications.

If for the sake of balance we must name the less advantageous properties of silicates, then we may mention the slow aging of high-surface-area silicates—driven by the higher stability of siloxanes relative to silanols—and the nonspecific adsorption on silicates due to interaction with uncapped silanols or surface hydroxyl groups. The first leads to changes of binding capacity over long-term use, and the second may lead to somewhat lower specificity and to interference by uptake of positively charged groups. Wise use of tailored Ormosils and Ormocers may probably remedy both. It is perhaps appropriate to start this section with a brief description of the progress in the area most closely related to Dickey's original target: silicate artificial antibodies.

4.4.1 IMPRINTING BY MOLECULES AND LARGE MOIETIES

A comprehensive and readable review of molecular imprinting, which, however, preceded most of the sol-gel developments discussed here, was written by Haupt and Mosbach [290]. Additional relevant reviews are given by references 291–294. Three general modes of molecular imprinting in sol-gel silicates are currently used to retain the memory of the imprinted molecule within the inorganic hosts.

4.4.1.1 Imprinting in a Monofunctional Host Matrix

This is the most straightforward method of imprinting. Sol-gel polymerization of the silicate or other inorganic oxide takes place in the presence of an inert molecule that does not participate directly in the polymerization process. Removal of the guest molecule by extraction, thermal evaporation, or calcinations yields empty pores with the shape of the imprint. This is essentially a direct descendent of an approach that was introduced by Dickey [79,80] for the preparation of chromatographic media by imprinted methyl orange and its alkyl homologues in sodium-silicate-derived sol-gel powders. The resulting matrix often lacks specificity, presumably due to distortions during imprint removal and slow polycondensation, which may proceed long after imprint removal. Specificity to the guest imprint is also limited because the recognition is solely by size and shape, and thus smaller molecules can also compete for the same site. Nevertheless, this straightforward approach was rather successful for the preparation of macroporous materials, and it is especially useful for the formation of ordered macroporous films.

Macroporous material. Tsionsky et al. [234] incorporated a large concentration of small molecules (e.g., bromocresol) in the sol-gel precursors to form macroporous chromatographic materials by spread coating. The dopant, say, bromocresol, segregated out as crystals during the evaporation of the solvent and gel formation. After the dopant was extracted out, macroporous materials with controlled average pore dimension were formed. Thin-layer chromatographic TLC plates were demonstrated. The imprint was used here to obtain large void fractions, and no attempt was made to obtain specificity to the target molecule.

Macroporous structures were obtained by introduction of micron dimension particles, such as latex beads [248,251,295]. This is not unique for silicates and many oxides (Si, Ti, Zr, Al, W, Fe, Sb, and a Zr–Y mixture) were templated by a similar way [296]. Live or dead *Saccharomyces-cerevisiae* yeast cells (of 3–5 micrometer diameter) were also used to form highly ordered 2-D hexagonal templates [246] (Figure 4.12). Liu's group [297] linked paramagnetic nanoparticles to *Bacillus megaterium* bacteria via carbodiimide coupling; then they directed the bacteria to predetermined locations on a solid support by magnetic force,

FIGURE 4.12   Ordered macropatterned thin film produced by dip coating of a mixture of dead *Saccharomyces-cerevisiae* yeast templates and TEOS-derived sol. (From Chia, S. Y., J. Urano, F. Tamanoi, B. Dunn, and J. I. Zink, 2000. *J Am Chem Soc* 122:6488–89. Used with permission.)

and, finally, they froze the structure by sol-gel processing to obtain aligned wire structures containing the microorganisms.

The encapsulation of small molecules or organometallic complexes in single-component inorganic silicate or Ormosil with the aim of achieving electrochemical specific recognition of the target compound was also attempted. Shustak et al. [298] embedded tris(2,2'-bipyridine)iron(II) (Fe(bpy)$_3^{2+}$), in TMOS:phenyltrimethoxy silane-derived thin silicate films. The authors showed that diffusion can be remarkably controlled by PEG addition, though specificity was not achieved. Even more successful specific recognition was reported [299] with silica-alumina sol-gel film. The authors reported high specificity for catecholamines by an electrochemical sensor made of TEOS and AlCl$_3$ precursors. Probably the combination of the two elements provided sufficient multifunctional affinity to the imprint. Despite Ling's exceptional results. Dickey's [299] and Shustak's [298] studies underscore the need for additional means to obtain specific recognition, and the fact that structure alone is mostly insufficient to obtain molecular imprinting. Multifunctionality of the host polymer is desirable for achieving high specificity.

4.4.1.2   Multifunctional Molecular Imprinting Polymers (MIPs)

Here, the host polymer is polymerized from monomers having different groups with complementing functionalities on the guest imprint (Figure 4.13). Multifunctional monomers, tailored to wrap around the target molecule and interact with the complementing sites on the target, were devised but have not yet been introduced to sol-gel science, let alone sol-gel electrochemistry. It is hypothesized that the different silane monomers are first attached to the target molecule or larger entity already in the sol state, and the target molecule along with its complementing sol-gel monomers is further set in place within the film by the polymerization step. Removal of the guest molecule leaves behind a cavity that not only has the shape of the originally introduced template but also has a three-dimensional arrangement of functionalities that can accommodate the guest molecule and reject compounds having either different functionalities or different structural configurations of the matching functionalities. Most of the current activities in this field [290] and most of the

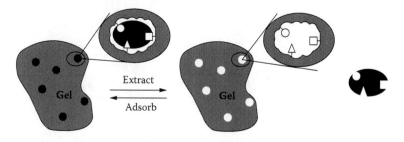

FIGURE 4.13   Imprinting by the incorporation of multifunctional target analyte in multifunctional silicate and subsequent removal of the encapsulated analyte. (From Srebnik, S., and O. Lev, 2003. *J Sol-Gel Sci Technol* 26:107–13. Used with permission.)

electrochemical activity [300] are still directed to acryl and vinyl polymers due to the enormous number of different monomers with these polymer backbone forming functionalities, and also because of the very small backbone of these polymers.

Perhaps the first to follow Dickey's footsteps in the silicate arena was Mosbach's team. Glad et al. [301] reported selective rhodanile blue and safranine O chromatographic media by Dickey's procedure, with a substantial modification, and phenyl and amine silicates were used to aid the imprinting of the target molecules in the silicate. Recently, successful examples of electrochemical and other sensors based on sol-gel processing were reported, though the range of functionalities that were employed thus far is rather limited compared to the range of available organosilane functionalities. The current vocabulary for sol-gel imprinting is mostly based on silanols, methyl, and phenyl functionalities (with titania, zirconia, or alumina additions), a dull selection compared to the rich silane chemistry.

The method is already widely used for obtaining sol-gel electrochemical sensors. Target molecules used include molecules such as the catecholamines, dopamine [251,295], and epinephrine [302], a beta blocker, propranolol [303], and parathion and paraoxon pesticides [304]. Marx and Lion [304] reported that the selectivity for binding of parathion compared to paraoxon increased from 1.7 to 30 by the sol-gel imprinting. The insert of Figure 4.14 shows that paraoxon is merely a thio-parathion that is, the P=S bond in parathion is replaced by a P=O bond in paraoxon, but this meager difference was sufficient to drive very large molecular selectivity.

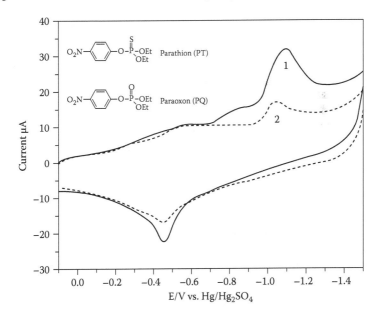

FIGURE 4.14 CV of parathion-imprinted film after incubation in (1) 0.1 M parathion and (2) 0.1 mM Paraoxon for 10 min. (0.1 M phosphate buffer, scan rate 100 mV s). (From Marx, S., A. Zaltsman, I. Turyan, and D. Mandler, 2004. *Anal Chem* 76:120–26. Used with permission.)

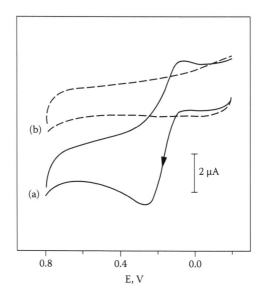

FIGURE 4.15 Cyclic voltammetry of 0.1 mM dopamine at (a) dopamine MIP and (b) nontemplated film (scan rate = 100 mV/s; 0.1 phosphate buffer, pH 7.4). (From Makote, R., and M. M. Collinson, 1998. *Chem Mater* 10:2440–45. Used with permission.)

The work of Makote and Collinson [295] is described here because of its ingenuity, and partly because it was the first convincing demonstration of sol-gel MIPs in electrochemistry. A sol-gel dopamine MIP was fabricated from ethoxy-ethanol solution of phenyltrimethoxysilane (PTMOS), methyltrimethoxysilane, and tetramethoxysilane. The solvent was selected because of the high solubility of all the sol-gel precursors and their hydrolysis products as well as the dop-amine imprint. The phenyl pendant group complements the aryl group of the catechol. The uncapped silanols formed hydrogen bonding with the hydroxyls. Methyltrimethoxysilane contributed hydrophobicity and increased the adhesion to the solid support. Multifunctionality was essential for obtaining a specific sen-sor for dopamine. In fact, Makote and Collinsion demonstrated that removal of any of the ingredients led to loss of selectivity. Electrode preparation was simple. Dopamine was added to a sol containing all three ingredients, and a glassy car-bon electrode was dip-coated by the sol. The electrode showed high dopamine accessibility (Figure 4.15) [295], with no interference by negatively charged com-pounds such as ascorbic acid, (dihydroxyphenyl)alanine, and (dihydroxyphenyl) acetic acid and low accessibility for positively charged interfering catecholamines such as epinephrine and norepinephrine.

4.4.1.3 Imprint Formation by the Removal of a Pendant Group from the MIP

Figure 4.16 demonstrates yet another approach involving polycondensation of a suitable sol-gel precursor to obtain R-silicate gel (where R represents

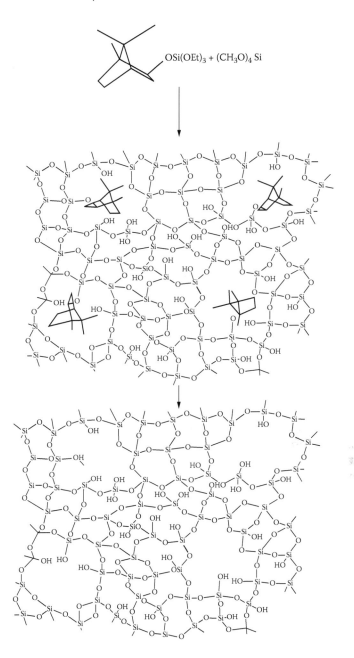

FIGURE 4.16 MIPs made by incorporation of borneol-appended triethoxysilane and TMOS and subsequent removal of the borneol functionality. (From Srebnik, S., and O. Lev, 2003. *J Sol-Gel Sci Technol* 26:107–13.) (The original work was conducted by Hunnius, M., A. Rufinska, and W. F. Maier, 1999. *Microporous Mesoporous Mater* 29:389–403. Used with permission.)

a desirable organofunctional group). Hydrolysis of R and its extraction then induces pore formation. Although successful electrochemical imprinting by this mode was never demonstrated, significant steps were made toward achieving successful imprinting in silicates. Hunnius [305] used bornyloxytriethoxysilane or fenchyloxytriethoxysilane and methyltriethoxysilane sol-gel precursors. Although good selectivity was obtained in this case, the authors refrain from attributing it to successful imprinting. A similar approach was used for the preparation of inorganic porous silicate networks by removal of an organic bridge from organic silsesquioxanes (of the general structure $(OCH_3)_3Si(CH_2)_nSi(OCH_3)_3$, $n = 2$–4) using a mild ammonium fluoride treatment [306]. Again, controlled pore formation was achieved, but true molecular imprinting was not demonstrated.

4.4.1.3.1 Chiral Recognition

Chiral recognition by sol-gel MIPs is another challenging field that is attracting increasing electrochemical attention, and is treated here within the context of multifunctional interactions with the imprint. Selectivity due to the mere grafting of a chiral selector on silicate media (without imprinting) was demonstrated already by Mikes et al. [307]. The technique is now widely used to obtain chiral separation of enantiomers by gas and liquid chromatography through the formation of labile diastereomers (non-mirror-image combinations of enantiomers). Obviously, incorporation of the molecular selector (or even an antibody or enzyme) as a dopant or as a comonomer in sol-gel processing yields similar results that can be useful for chromatography as well as for sensing or catalytic applications (e.g., [308,309]). For example, a generic way to achieve chiral recognition is by appending cyclodextrins on the silicate. For example, Huq et al. [310] grafted β-cyclodextrin on mesoporous silica by the procedure given in Scheme 4.10 [310]. The reaction involved the condensation of trialkoxysilane with the chloro derivative of the cyclodextrin.

True molecular imprinting to obtain chiral electrochemical recognition by sol-gel matrices was also demonstrated by several groups. Interestingly, several teams demonstrated impressive chiral recognition even without the incorporation of a multifunctional host matrix. Willner's group [311,312] reported an ion-selective field-effect transistor (ISFET) with sol-gel-derived TiO_2 film imprinted with 4-chlorophenoxyacetic acid or 2,4-dichlorophenoxyacetic acid. Similar specificity was reported for enantiomers of 2-methylferrocene carboxylic acids, 2-phenylbutanoic acid, and 2-propanoic acid imprinted by deposition of titania on ISFET (ion-selective field effect transistor) sensors. A discussion related to advantages of ISFET compared to amperometric detection is given at the end of this section.

Similar impressive results for the imprinting of l and d–dopa (3,4-dihydroxy-l,d-phenylalanine) in 5:1 TMOS–PTMOS (dissolved in ethoxyethanol) derived silicate have been reported [313]. They showed over 20 times preferred uptake of the imprinted dopant. TMOS–PTMOS mixture also gave successful chiral imprinting of *N,N'*-dimethylferrocenyl ethylamine (Figure 4.17).

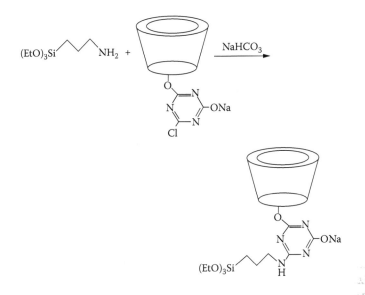

SCHEME 4.10   Preparation of trimethoxysilane-appended cyclodextrin. (From Huq, R., L. Mercier, and P. J. Kooyman, 2001. *Chem Mater* 13:4512–4519. Used with permission.)

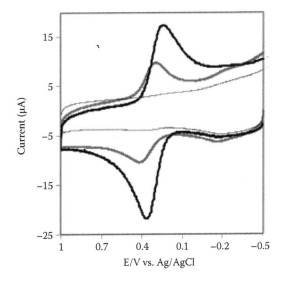

FIGURE 4.17   Cyclic voltammetry of chiral ferrocene (R-Fc) imprinted silicate film. From outer CV to inner one: Rebinding of R-Fc (30 µM); rebinding of S-Fc (30 µM); rebinding of R-Fc (30 µM) to nonimprinted films. All CV measured in a clean buffer after incubation in the FC solution. Scan rate, 100 mV s^{-1}. (From Fireman-Shoresh, S., I. Turyan, D. Mandler, D. Avnir, and S. Marx, 2005. *Langmuir* 21:7842–47. Used with permission.)

4.4.1.3.2 Prospects for Practical Sensing Applications
Based on Sol-Gel Imprinting

There are several factors that still hinder the practical application of MIPs, and sol-gel systems do not escape these constraints. System heterogeneity, due to differently configured recognition sites, leads to lower specificity. Nonspecific binding to only one of the recognizing functionalities (say, silanols) and their non-optimal combination is usually possible, particularly when excess complementing functionalities are available on the silicate. Careful tailoring of highly specific host cages that wrap the imprint either before or after film casting provides a way to approach the high homogeneity of antibodies, but this requires a lower density of sites and results in a lower mass transport rate. Indeed, slow mass transport is another obstacle faced by MIPs; higher specificity requires reduction of the imprint density in order to prevent aggregation of the imprints and overlap of the evacuated pores, and this inevitably leads to mass transport hinderance.

Obtaining imprinted silicates for amperometric sol-gel sensing is even more challenging, particularly due to the required selectivity. For most practical applications, the sensing selectivity of the target drug versus its natural interferences should be at least two or three orders of magnitude. In order to appreciate this difficulty, it may be useful to compare amperometric sensing with more successful applications of MIPs: chromatography on the one hand and photometric detection, quartz crystal microbalance, and potentiometry on the other. The required selectivity for electrochemical sensing far exceeds that required for chromatographic applications. Chromatographic media can be successful even when the selectivity (the ratio of the binding constant of the analyte to that of its interference) is very low, because multiple adsorption and desorption steps take place when the analyte flows through the chromatographic media. Because electrochemical sensors are based on thin MIP films, the selectivity factor between the target (i.e., the ratio of the binding coefficient of the analyte to that of its interference) and the interference should also be greater than 100, to guarantee interference-free sensing (at least when the interfering compound and analyte are present on the same level).

The molecular-imprinting approach is more successful for potentiometric sensing (e.g., ISFET) and photometric detection compared to amperometric applications. A low binding affinity is required in order to allow mobility of the analyte through the imprinted film, as otherwise the analyte will concentrate near the external interface of the silicate, far from electron-hopping distance from the electrode surface. Facilitated-transport sensors relying on hopping of the analyte between nearby recognition sites are possible only when the bond between the analyte and the silicate is very labile. However, when lability is high, the binding constant is usually small, and the resulting selectivity is low. High lability and high stability are both required for successful amperometric sensors, and these usually conflict. Indeed, the situation is somewhat similar to sensing with antibody-encapsulated sol-gel films. Potentiometric success is often demonstrated because bindings may change the Donnan potential, whereas amperometric applications are rather scarce due to the high antibody–antigen affinity that hinders mobility toward the electrode. Artificial antibodies are even more demanding than natural antibodies because the film thickness required for achieving efficient 3-D imprinting is usually thicker for the former and does not allow direct charge transfer from the recognition site to the conductive support. Furthermore, for electrochemical sensors, it is not sufficient to gain good transport selectivity to the targeted analyte alone; the electrochemical reaction product should also have high mobility through the film because otherwise it will accumulate in the MIP and hinder transport of the analyte to the electrode surface. In short, high affinity to the analyte is not sufficient for amperometric MIPs. Selective uptake and high transport rate of the analyte as well as its redox product are also needed in addition to the obvious need for electrochemical reactivity.

4.4.2 MESOPOROUS MATERIALS

The successful production of mesoporous silicate molecular sieves by Mobil Oil scientists [236] has inspired the electrochemical community. The transport advantages of mesoporous (2–50 nm pore size) materials, their high surface area (up to 1500 m^2/g), and their aesthetically ordered structure are attracting increasing attention. Kresge and coworkers showed that micelles and bicontinuous structures that are formed by surfactants in aqueous solutions might serve as templates for the polymerization of mesoporous silicates. The organic template can be removed by thermal or chemical treatment after the polymerization, leaving the skeleton of the inorganic phase. Soon the method was expanded to other metal oxides. Ordered mesoporous oxides of W, Mg, Al, Mn, Fe, Co, Ni, and Zn were synthesized [314] using positively or negatively charged surfactants. The rapidly reacting sodium silicate precursor was replaced by the more friendly alkoxides, and the spin-coating technique was successfully demonstrated [315]. Whereas the first examples of mesoporous materials were of little relevance to electrochemists because they were confined to powdery material, it was soon discovered that it is possible to construct the mesoporous materials on mica and other solid supports [316,317]. A breakthrough in the method of disseminating mesoporous material in electrochemistry came with the work of the sol-gel scientists from the Lu groups [237], who demonstrated mesoporous material formation on solid supports by the sol-gel dip coating process. Thus, continuous homogeneous films could be formed by this process. Starting with a few percent (relative to the silicon precursor) of positively charged surface-active agents such as cetyl-trimethyl-ammonium bromide (CTAB) and hydrolyzed TEOS-derived silica sol, they demonstrated the formation of highly ordered mesoporous structures by dip coating.

In a subsequent article, Brinker et al. [318] coined the terminology Evaporation-Induced Self-Assembly (EISA) to describe the complex processes associated with the formation of the mesoporous materials. Brinker described the processes that occur during the slow evaporation of ethanol–water–silicate sol-containing surface-active agents. The general process is depicted in Figure 4.18 [319]. The starting solution for the dip coating is below the critical micelle concentration (CMC), and the surfactant is mostly in the form of free molecules. The concentration in

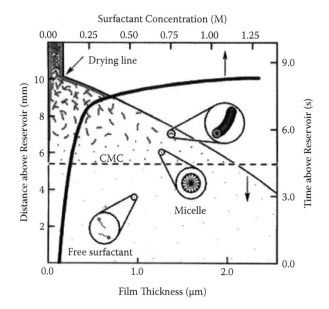

FIGURE 4.18 Film-thinning profile established during dip coating in surfactant, silica ethanol, and acidic aqueous solution. Just above the water surface, the solution near the substrate surface is identical to the precursors solution. As the substrate is raised above the water level, the solution evaporates and the CMC level is exceeded, forming micelles (some 5–6 mm above the water level [left curve] when the film thickness is 2 mm [right curve]). Further evaporation consolidates the film and orients hexagonal structures in parallel with the substrate. (From Huang, M. H., H. M. Soyez, B. S. Dunn, J. I. Zink, A. Sellinger, and C. J. Brinker, 2008. *J Sol-Gel Sci Technol* 47:300–310. Used with permission.)

the solution is also the concentration near the slide section that emanates from the solution bath during the dip-coating process. As the slide moves upward, the ethanol–water solution becomes enriched in water, and the concentration of the surface-active agents increases (and so does the acid catalyst concentration, if this is used). Eventually, the concentration of the surfactant exceeds the CMC, and micelles start to form. Further concentration leads to the formation of cylindrical micelles, and then the micelles self-organize on the surface to give the following ordered structures (Figure 4.19): hexagonal arrangement of micelles (3-D) (space group *P6₃/mmc*); 2-D hexagonal arrangement of elongated, cylindrical micelles (symmetry *P6m*); and cubic (*Pm3n*), or lamellar phase. Concurrently with these processes, the negatively charged hydrolyzed silicate precursor is attracted to the polar, positively charged ammonium-rich external surface of the self-assembled structure, and thus it undergoes polycondensation on the outer surface of the surfactant structures. The end product of the dip-coating process is a highly ordered organic–inorganic hybrid, whose structure is predominantly governed by the surfactant phase equilibria at the moment of the loss of fluidity. This is a general process, and it may take place in other sol-gel protocols that are driven by evaporation of the solute and gradual concentration of the surfactant and silicon precursors (e.g., spray drying, spin coating, spread coating, or simple dropping). Thus, despite the fact that the starting solution is poor in surfactants, and their concentration is initially far below the CMC, solvent evaporation enriches the solution and forms the template structure and the mesoporous silicate. Needless to say, the affinity of the silicate precursors and the surfactant template is essential for the formation of ordered mesoporous material. Shear forces tend to align the cylindrical phase in parallel to the substrate, which has important electrochemical consequences as discussed later. Initially, mesoporous material formation by dip coating was considered to be of low reproducibility, but this was largely overcome by better control of dip-coating parameters and especially the humidity in the coating chamber.

Evaporation-induced self-assembly is not limited to low-molecular-weight surfactants, and equally rich meso-structured oxides can be formed by template formation on amphiphilic triblock poly(alkylene oxide) and other polymers. Amphiphilic block polymers, as their name implies, have dual (amphi) affinity (philic) character. For example, the Pluronic family of BASF includes hydrophilic poly(ethylene oxide) and hydrophobic poly(propylene oxide) segmented polymers (e.g., the structure of Pluronic P 123, which is widely used, is $HO(CH_2CH_2O)_{20}(CH_2CH(CH_3)O)_{70}(CH_2CH_2O)_{20}H)$, denoted also as $PEO_{20}PPO_{70}PEO_{20}$). These allow the formation of micelle and lamellar structures, much like low-molecular-weight surface-active agents but with higher predictability. There is a growing interest in the ability of these compounds to serve as templating agents for the formation of mesoporous silicates and other oxides. Although originally employed to produce large-pore-size silicates [321], it was soon enough demonstrated for other metal oxides. Yang et al. [322] demonstrated the formation of mesoporous metal oxides for a large range of elements (Ti, Zr, Nb, Ta, Al, Si, Sn, W, Hf) and numerous mixed metal oxides, using an amphiphilic poly(alkylene oxide) block copolymer as a templating agent.

4.4.2.1 Selected Electrochemical Examples of Mesoporous Material Electrochemistry

Mesoporous materials have attracted considerable electrochemical attention, and several comprehensive reviews have recently been devoted to the subject [323,324]. There are currently three general ways to make use of ordered mesoporous electrode materials in sol-gel electrochemistry:

1. Preprepared mesoporous particulates in a silicate or nonsilicate binder
2. Evaporation-induced self-assembled thin films
3. Electrochemical deposition of mesoporous materials

A short description of each pathway with typical examples follows. In all cases, organic modification of the mesoporous material is feasible either by adsorption of organic moieties onto the mesoporous material or by grafting of the organic material onto the preprepared mesoporous oxide. From a surface manipulation point of view, mesoporous materials are no different from amorphous silicates.

4.4.2.1.1   Incorporation of Preprepared Ordered Mesoporous Materials

The early reports of ordered mesoporous materials described the response of mesoporous silicate (or other oxides) in carbon paste electrodes. For example, thiol and amine grafting [252,325], and ferrocenyl-grafted MCM 41 particulates [326] have been reported. MCM41 is a hexagonally ordered powder prepared according to a Mobil Oil preparation protocol. In all of these cases, the process is straightforward. Pre-prepared (or commercially available) mesoporous powder is introduced to the starting solution of the electrode modifiers, and the electrode is then cast in the usual manner. Simple pressing of the powder into carbon paste electrodes also provides reasonable results with minimal synthesis efforts. Carbon ceramic electrodes incorporating mesoporous silicates have also been reported. For example, mesoporous silicate–graphite CCE [327] was prepared by sonication and was used for catechol analysis. Inorganic silicate [328] or organic binders such as Nafion [97] or PVA [329] were also used to bind the modified mesoporous particulates to the surface of electrodes. The polymer–mesoporous precursors were simply dropped on the electrode surface by a micropipette to produce thin-film-modified electrodes.

The applications of these materials span an equally wide range of analytes. Several articles have dealt with the fundamental aspects regarding charge and mass transport in mesoporous particulates modified by adsorption or grafting of redox species. Copper and mercury ion concentration on thiol and amine grafted ordered mesoporous materials were studied by the group of Walcarius and showed promise for accumulation-stripping electroanalysis of the target analytes (e.g., [252,330]). This account of available electrochemical studies and the many electrochemical sensors that were produced by ordered mesoporous materials is far from being exhausted, and the reader may also consult relevant recent reviews for more examples of mesoporous hybrids [10,52]. In general, however, the range of analytes and the range of polymers used are similar to silicate–polymer composites.

Several biosensing elements based on mesoporous silicate and other metal oxides have recently been reported. Unlike amorphous silicates and composites, mesoporous materials offer ordered structure and uniform pore size. Pore size homogeneity allows one to study the effect of the pore size of the host matrix on immobilized enzymes in a methodical way. Lee et al. [331] studied the stability and specific activity of cytochrome c immobilized in several ordered mesoporous materials such as MAS-9 (pore size 90 Å), MCM-48-S (2.7 nm), MCM-41-S (2.5 nm), and Y zeolites (0.74 nm). They checked the residual activity of the cytochrome c after thermal treatment by exposure to boiling water for 12–24 h. The best thermal stability and the best specific catalytic activities were obtained for MCM-48 and MCM-41, whose pore dimension are barely sufficient to contain cytochrome c (having dimensions 2.5 mm × 2.5 mm × 3.7 nm). The authors attributed the thermal stability to good interaction between the enzyme and the MCM silanols that prevent protein unfolding. In addition, the authors demonstrated a change of axial ligation of the Fe(III) center of the cytochrome c heme after entrapment in MCM 41 and MCM 48 hosts. Both low and high spin states were observed—by EPR studies—for the immobilized cytochrome c, whereas the native enzyme predominantly showed the low spin state. The authors showed improved activity toward the oxidative bleaching of pinacyanol dye, which was attributed to the replacement of methionine axial ligand of the heme. Its replacement by a water molecule (or as the authors suggest, also by a surface silanol) improved the accessibility of the dye to the active center. This work shows that it is possible to manipulate the electronic state of the entrapped enzyme by appropriate selection of the pore dimension of the host. Mass transport through the substrate becomes important for the oxidation of large substrates such as polycyclic aromatic hydrocarbon, and then MAS 9 indeed exhibited the highest activity. Change of reactivity of bioentities due to confinement by silicate cages was observed several times earlier (e.g., [84,85,254–256,332]), but it seems that the uniformity of mesoporous materials allows better control of the confinement and thus better manipulation of the electronic state of enzymatic active centers. This opens the door for studies that could unequivocally prove delicate bioentity–host interactions by electrochemistry, where delicate shifts of the formal potential of the enzyme itself may be used to probe changes in the electronic state. This, of course, is limited to rather small enzymes that can manifest mediatorless charge transfer to the supporting electrode. Indeed, Ganesan and Viswanathan [333] demonstrated that the redox potential of cytochrome c was positively shifted by some 70 mV upon immobilization in MCM-41, which is similar to the improved reactivity of cytochrome c shown earlier [331]. This can be attributed to porewalls-induced conformational change of the enzyme.

Improved activity upon immobilization in mesoporous silicates was also observed by electrochemical studies. It was demonstrated that the specific activity of organophosphorus hydrolase entrapped in mesoporous amine functionalized ordered-silicate powder could reach more than twice as high as that of the free enzyme, and in a subsequent article it was reported that enzymatic sensor for paraoxon based on the immobilization of enzyme-grafted mesoporous silicate particulates in Nafion film was reported [334].

Third-generation biosensors, based on direct electron transfer between myoglobin-doped mesoporous oxides and the supporting electrode, was recently demonstrated by several authors. Zhang et al. [335] used layered nanostructured titania because the interlayer distance of these materials can be changed to allow the incorporation of the myoglobin without sacrificing the long-range order. Other researchers [329,336] have demonstrated direct charge transfer between the glassy carbon electrodes and myoglobin entrapped in silica and tungsten oxide, respectively. Sensing of hydrogen peroxide and nitrite was demonstrated in both cases.

Similar direct charge transfer hydrogen peroxide sensor based on immobilization of hemoglobin modified mesoporous silicate on glassy carbon electrode has also been reported [328]. Xian et al. [337] demonstrated the same for mesoporous material doped with gold nanoparticles. Yu et al. [338] demonstrated successful immobilization of hemoglobin in (a room temperature ionic liquid) 1-butyl-3-methylimidazolium hexafluorophosphate–doped mesoporous silicate deposited on glassy carbon. The authors demonstrated that the ionic liquid did not inactivate the enzyme but, on the contrary, a more efficient direct charge transfer was obtained. This is in line with several recent reports suggesting that ionic liquids may enhance heme-containing enzymes stability and activity [339].

In all these electrochemical studies, the mesoporous material is not conductive, and direct charge transfer is therefore limited to those enzymes that are adjacent to the electrode itself with the mesoporous material probably exerting little or no influence. The enzymes that are located a few angstroms away from the electrode do not contribute to the observed current. However, accessibility to the active site of the enzymes is highly accelerated by the presence of the mesoporous materials relative to amorphous silicate film.

4.4.2.1.2 Direct Deposition of Mesoporous Materials on eLectrodes by EISA

A more versatile method to produce mesoporous film-modified electrodes is by synthesizing the mesoporous material directly onto the electrode surface by EISA (evaporation-induced self-assembly). On the positive side, this allows formation of much richer structures, more uniform pore size distribution, and eliminates heterogeneity imposed by the polymeric binders, be it of inorganic or organic nature. On the negative side, the ability to remove the organic structure induced by thermal treatment is substantially reduced, the film is prone to aging processes, and too often, hydrolytic degradation is reported to explain the electrochemical observations. Moreover, it was noted [340] that electrochemistry conducted without removing the template formers from the mesoporous material results in a poor or at best sluggish response due to blockage of the electrode surface by the surfactants.

Etienne et al. [341] studied the mass transport of negatively charged ferricyanide and iodide, positively charged $Ru(bpy)_3^{2+}$, and ferrocene methanol in different EISA-derived silicates of comparable pore size. The studies were conducted with phthalate buffer, pH 4.1, to prevent slow mesoporous material hydrolysis, which was reported to take place at higher pH. As for other sol-gel materials, transport through the film is influenced by charge permselectivity determined by the charge of the silicate and the analyte as well as by the sizes of the pores and the analyte. As a side remark, unconnected to the previous report, it should be noted that mobility through the film can be reversed by grafting of the silicate by positively charged amino groups (e.g, [342]). Etienne et al. [340] showed that the preferred mass transport through the mesoporous materials followed the ranking cubic > 3-D hexagonal > 2-D hexagonal. However, in this particular study, the low response of the 2-D hexagonal structure was attributed to degradation of the 2-D structure, which partly blocked the electrode rather than to the parallel-to-the surface orientation of the dip-coated cylindrical mesostructures. Surprisingly, the cubic structures made of Pluronic polymer had poor permeability, which was also attributed to hydrolytic instability.

The field of EISA-modified electrodes is one of the most vibrant in contemporary sol-gel electrochemistry, with applications in metal ion sensing [340,341,343], gas sensing [344], lithium intercalation anodes [345], electrochromic windows

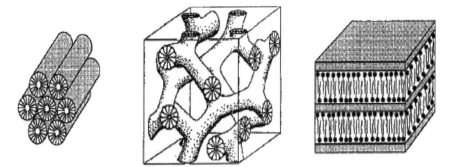

FIGURE 4.19 Typical forms of hexagonal (left), cubic (center), and bicontinuous mesoporous material. (From Raman, N. K., M. T. Anderson, and C. J. Brinker, 1996. *Chem Mater* 8:1682–1701. Used with permission.)

[346], dye-sensitized solar cells [347,348] as well as other interesting fundamental phenomena such as charge hopping in redox-grafted mesoporous material [349]. There is no doubt that EISA will play a dominant role in electroanalysis. The only reason that we do not dwell upon it further in this account is because findings are to be anticipated based on what is already known from silicate electrochemistry, with the obvious difference that mesoporosity increases accessibility and specific surface area. Improved ways to gain stability are still needed.

4.4.2.1.3 Electrochemical Deposition of Mesostructured Electrodes (EASA)

The Walcarius team [45] advanced mesoporous silicate-modified electrode formation by showing that cathodic deposition of mesoporous silicate film from a solution containing CTAB and TEOS at pH 3 yields predominantly hexagonal 2-D structures with pores oriented perpendicular to the supporting electrode. Figure 4.20 (modified from Reference 45) depicts a cross-sectional view and top view of the highly oriented porous structure obtained by this method. The method was termed EASA for electroassisted self-assembly.

The method is simple to perform. In a typical application, electrodeposition on the Au electrode was carried out in potentiostatic mode, applying −1.2 V. A typical starting solution contained 13.6 mmol TEOS, 20 mL ethanol, 20 mL aqueous solution of 0.1M $NaNO_3$, and 1 mmol HCl, to which approximately 4.35 mmol CTAB was added under stirring. The authors suggest that H^+ and water reduction (with possible contributions from nitrate reduction) increases the pH in the vicinity of the electrode and stimulates silicate deposition.

4.4.3 Hierarchical Networks

Most of the sol-gel-derived materials described thus far pertain to wide-size-distribution amorphous material or to porous materials having uniform size distribution. However, for chromatography as well as for several electrochemical

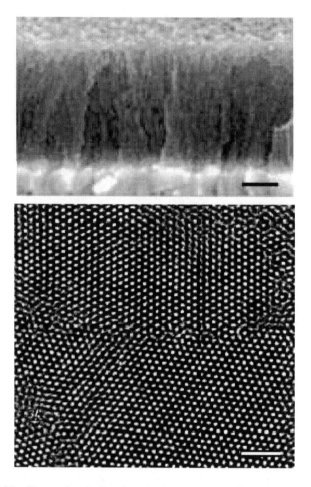

FIGURE 4.20 Electrochemically deposited mesoporous film. Top: Cross-sectional FE-SEM images for a film deposited on ITO after vertical cleaving. The scale bar corresponds to 100 nm. Bottom: TEM image of the top view of a surfactant-templated mesoporous silica film electrodeposited on ITO. The scale bar corresponds to 20 nm. (From Walcarius, A., E. Sibottier, M. Etienne, and J. Ghanbaja, 2007. *Nat Mater* 6:602–608. Used with permission.)

applications, it is desirable to have bimodal (or even multimodal) pore-size distributions. A material having macroporosity and microporosity within the same domain will allow high accessibility, for example, by convection, and high surface area for catalysis or adsorption of analytes. Although, as outlined later, we could find only one example of the use of such templated hierarchical material in electrochemistry, it is easy to conceive how flow-through electrodes, gas sensors, and other applications can benefit from such materials.

An example of hierarchical materials is the macroporous materials produced by Nakanishi et al. [286] by copolymerization of polyols (e.g., poly(vinyl alcohol) or poly(ethylene oxide)) and tetraethoxysilane.

Polymerization induced phase separation, and thermal removal of the segregated organic phase resulted in the formation of macroporous materials with large surface areas due to microporosity. The ability to produce monolithic rods of these materials and to seal them in PEEK opened a new field in liquid chromatography, and nowadays low-pressure drop monolithic chromatography columns are marketed by Merck under the trade name Chromlith™. Grafting allows the production of monoliths for reversed-phase, chiral, ion exchange, and hydrophilic interaction chromatographies [350]. Figure 4.21 shows the porous structure and

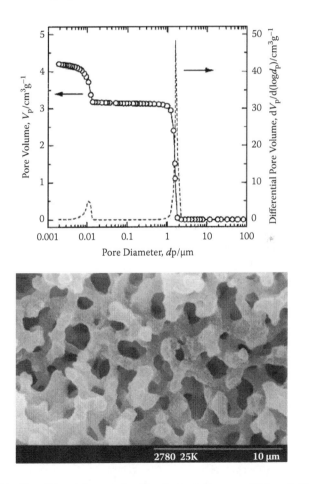

FIGURE 4.21   Top: Bimodal pore size distribution of a monolithic gel. Bottom: The corresponding electron micrograph of a fractured section of the corresponding gel. (From Nakanishi, K., H. Minakuchi, N. Soga, and N. Tanaka, 1997. *J Sol-Gel Sci Technol*, 8:547–552. Used with permission.)

the pore-size distribution of a monolithic chromatography column as obtained by superposition of mercury porosimetry and nitrogen adsorption (BET) techniques [351]. The combination of Nakanishi's phase separation approach and surfactant templating was also recently reported using a combination of poly(ethylene glycol) for phase separation and alkylammonium surfactant for templating.

Hierarchical films are still seldom used in electrochemistry. Recently, however, Sel et al. [352] examined the cyclic voltammetry response of ferrocenyl-grafted hierarchical macroporous–mesoporous silicate-modified electrodes, which were produced by a combination of three methods: (1) block polymer templating induced by the block polymer (poly(ω-hydroxypoly(ethylene-co-butylene)-co-poly(ethyleneoxide), which gave large mesoporosity (about 14 nm); (2) ionic liquid (1-hexadecyl-3-methylimidazolium chloride) templating inducing small mesoporosity (about 3 nm); and (3) induced macroporosity by addition of polymethylmthacrylate beads of about 120 nm size. They examined all unimodal, bimodal, and trimodal pore-size distribution combinations of materials produced with the aid of the different porogens. High coverage of the surface by ferrocenyl grafting allowed them to obtain charge transport by electron self-exchange through the modified silicate films. The authors demonstrated that films with bimodal pore-size distribution with the small pores or films with the trimodal pore-size distribution are optimal combinations. Small pores contributed by the ionic liquid templating were vital for obtaining good connectivity between the larger pores and high-rate self-exchange transport.

We cannot judge at this point whether the importance of small mesoporosity for charge transport is really general or specific to the investigated film, but the authors have clearly demonstrated that films with bimodal and trimodal pore-size distributions are useful for electrochemists, because such combinations may at times alter the obtained electrochemical response in unexpected ways. We anticipate that multimodal pore-size distribution film-modified electrodes and membranes (i.e., hierarchical networks) will become widespread in electrochemical sensing, catalysis, and energy storage applications.

4.5　CONCLUDING REMARKS

The seemingly dull chemistry of silicates, which is governed by the predominantly tetracoordinated, redox-poor silicon atom, continues to be an endless source of exciting scientific revelations, and, because silicates are so abundant and cheap, new discoveries often have far-reaching practical consequences.

Silicate chemistry is undergoing an exciting period. The unification of silane and silicone chemistry with inorganic silicate chemistry occurred just before the current boom in nanotechnology research. Sol-gel processing of hybrids and composites, which were largely developed over the past few decades, were ready with technological solutions to the atom-up (bottom-up) approach of modern nanotechnology. Thorough knowledge of the synthesis of organic–inorganic hybrids and composites, and specialized nanoparticle- and nanometer-thick films, was ready for integration in nanotechnology packages.

From an electrochemical point of view, the recent developments in silicate science are somewhat different from older ones. The electrochemical community is more eager to embrace and put to use recent advancements in silicate and sol-gel science. Simply put, today many more electrochemists are familiar with sol-gel processing, and the availability of comprehensive and rapidly updated reviews in the field makes sol-gel and silicate know-how readily available for potential users. In addition, sol-gel tools are most amenable to modern nanotechnology with its bottom-up architectural approaches. This also holds true for electrochemistry, which broadens and diversifies the scientific community engaged in and contributing to sol-gel electrochemistry.

We do not wish to predict future hot fields in sol-gel electrochemistry because new unexpected revelations are likely, based on the past trends in the recent and not-so-recent evolution of sol-gel science.

ACKNOWLEDGMENT

Ovadia Lev thanks the Israel Science Foundation and the Ministry of Science for financial support.

REFERENCES

1. Lev, O., Z. Wu, S. Bharathi, V. Glezer, A. Modestov, J. Gun, L. Rabinovich, and S. Sampath, 1997. Sol-gel materials in electrochemistry. *Chem Mater* 9:2354–75.
2. Collinson, M. M., 2002. Recent trends in analytical applications of organically modified silicate materials. *Trac-Trends Anal Chem* 21:30–38.
3. Walcarius, A., M. Etienne, S. Sayen, and B. Lebeau, 2003. Grafted silicas in electroanalysis: Amorphous versus ordered mesoporous materials. *Electroanalysis* 15:414–21.
4. Long, J. W., B. Dunn, D. R. Rolison, and H. S. White, 2004. Three-dimensional battery architectures. *Chem Rev* 104:4463–92.
5. Walcarius, A., D. Mandler, J. A. Cox, M. Collinson, and O. Lev, 2005. Exciting new directions in the intersection of functionalized sol-gel materials with electrochemistry. *J Mater Chem* 15:3663–89.
6. Walcarius, A., 2005. Impact of mesoporous silica-based materials on electrochemistry and feedback from electrochemical science to the characterization of these ordered materials. *C R Chim* 8:693–712.
7. Collinson, M. M., 2007. Electrochemistry: An important tool to study and create new sol-gel-derived materials. *Acc Chem Res* 40:777–83.
8. Zhitomirsky, I., 2007. Electrosynthesis of organic-inorganic nanocomposites. *J Alloys Comp* 434:823–25.
9. Walcarius, A., and A. Kuhn, 2008. Ordered porous thin films in electrochemical analysis. *Trac-Trends Anal Chem* 27:593–603.
10. Walcarius, A., 2008. Electroanalytical applications of microporous zeolites and mesoporous (Organo) silicas: Recent trends. *Electroanalysis* 20:711–38.
11. Avnir, D., 1995. Organic-chemistry within ceramic matrices—Doped sol-gel materials. *Acc Chem Res* 28:328–34.
12. Marx, S., and D. Avnir, 2007. The induction of chirality in sol-gel materials. *Acc Chem Res* 40:768–76.

13 Avnir, D., T. Coradin, O. Lev, and Livage, J., 2006. Recent bio-applications of sol-gel materials. *J Mater Chem* 16:1013–30.

14. Betancor, L., and H. R. Luckarift, 2008. Bioinspired enzyme encapsulation for bio-catalysis. *Trends Biotechnol* 26:566–72.

15. Sanchez, C., C. Boissiere, D. Grosso, C. Laberty, and L. Nicole, 2008. Design, synthesis, and properties of inorganic and hybrid thin films having periodically organized nanoporosity. *Chem Mater* 20:682–737.

16. Coradin, T., and J. Livage, 2007. Aqueous silicates in biological sol–gel applications: New perspectives for old precursors. *Acc Chem Res* 40:819–26.

17. Shriver, D. F., P. W. Atkins, and C. H. Langford, 1994. *Inorganic chemistry*, 2nd edition. Oxford University Press: Oxford.

18. Brinker, C. J., and G. W. Scherer, 1990. *Sol-gel science: The physics and chemistry of sol-gel processing*. Academic Press: San Diego, CA.

19. Iler, I. K., 1979. *The chemistry of silica*. John Wiley & Sons: New York.

20. Bergna, H. E., and W. O. Roberts, eds., 2006. *Colloidal silica fundamentals and applications*. CRC Press, Taylor and Francis Group: Boca Raton, FL.

21. Sakka, S., 2005. *Handbook of sol-gel science and technology, Volume I: Sol-gel processing*. Kluwer Academic Pub.: Norwell, MA.

22. Artaki, I., T. Zerda, and J. Jonas, 1985. Solvent effects on hydrolysis stage of the sol-gel process. *Mater Lett* 3:493–96.

23. Schmidt, H., and H. Scholze, 1985. In *Glass. 8 Current Issues*, ed. A. F. Wright and A. F. Dupuy. Martinus Nijhojj: Dordrecht, Netherlands, 253–56.

24. Schmidt, H., A. Kaiser, M. Rudolph, and A. Lentz, 1986. In *Sci Ceramic Chem Proc*, ed. L. L. Hench and D. R. Ulrich. Wiley: New York, 87–90.

25. Sommer, L. H., G. A. Parker, N. C. Lloyd, C. L. Frye, and K. W. Michael, 1967. Stereochemistry of asymmetric silicon .4. SN_2-Si Stereochemistry rule for good leaving groups. *J Am Chem Soc* 89:857–62.

26. Brinker, C. J., K. D. Keefer, D. W. Schaefer, and C. S. Ashley, 1982. Sol-gel transition in simple silicates. *J Non-Cryst Solids* 48:47–64.

27. Szekeres, M., O. Kamalin, P. G. Grobet, R. A. Schoonheydt, K. Wostyn, K. Clays, A. Persoons, and I. Dékány, 2003. Two-dimensional ordering of Stöber silica particles at the air/water interface. *Coll Surf A: Physicochem Eng Aspects* 227:77–83.

28. Mackenzie, J. D., 1986. In *Sci ceramic chem processing*, ed. L. L.Hench and D. R. Ulrich. Wiley: New York, 113–15.

29. Scherer, W., 1988. In *Better ceramics through chemistry III*. ed. C. J. Brinker, D. E. Clark, and D. R. Ulrich. Materials Research Society: Pittsburg, PA, 121:179–81.

30. Kistler, S. S., 1931. Coherent expanded aerogels and jellies. *Nature* 3211:741.

31. Brinker, C. J., G. C. Frye, A. J. Hurd, and C. S. Ashley, 1991. Fundamentals of sol–gel coating, *Thin Solid Films* 201:97–108.

32. Strawbridge, I., and P. F. James, 1986. The factors affecting the thickness of sol-gel derived silica coatings prepared by dipping. *J Non-Cryst Solids* 86:381–393.

33. Strawbridge, I., and P. F. James, 1986. Thin silica films prepared by dip coating. *J Non-Cryst Solids* 82:366–72.

34. Dislich, J., 1986. Sol–gel: science, processes and products. *J Non-Cryst Solids* 80:115–21.

35. Innocenzi, P., M. O. Abdirashid, and M. Guglielmi, 1994. Structure and properties of sol-gel coatings from methyltriethoxysilane and tetraethoxysilane. *J Sol-Gel Sci Technol* 3:47–55.

36. Sakka, S., K. Kamiya, K. Makita, and Y. Yamamoto, 1984. Formation of sheets and coating films from alkoxide solutions. *J Non-Cryst Solids* 63:223–25.

37. Yang, C. C., J. Y. Josefowicz, and L. Alexandru, 1980. Deposition of ultrathin films by a withdrawal method. *Thin Solid Films* 74:117–27.
38. Dislich, H., and E. Hussmann, 1981. Amorphous and crystalline dip coatings obtained from organometallic solutions: Procedures, chemical processes and products. *Thin Solid Films* 77:129–39.
39. Guglielmi, M., and S. Zenezini, 1990. The thickness of sol–gel silica coatings obtained by dipping. *J Non-Cryst Solids* 121:303–9.
40. Meyerhofer, D., 1978. Characteristics of resist films produced by spinning. *J Appl Phys* 49:3993–97.
41. Hirano, S., and K. Kato, 1983. Preparation of crystalline LiNb0$_3$ films with preferred orientation by hydrolysis of metal. *Adv Ceram Mater* 3:503–6.
42. Kozuka, H., S. Takenaka, and S. Kimura, 2001. Nanoscale radiative striations of sol–gel derived spin coating films. *Scripta Mater* 44:1807–11.
43. Zhitomirsky, I., 2000. New developments in electrolytic deposition of ceramic films. *Am Ceramic Soc Bull* 79:57–97.
44. Zhitomirsky, I. 2002. Cathiodic electrodeposition of ceramic and organiceramic materials: Fundamental aspects. *Adv Coll Interface Sci* 97:279–317.
45 Walcarius, A., E. Sibottier, M. Etienne, and J. Ghanbaja, 2007. Electrochemically assisted self-assembly of mesoporous silica thin films. *Nat Mater* 6:602–608.
46. Rabinovich, L., V. Glezer, Z. B. Wu, and O. Lev, 2001. Naphthoquinone-silicate based gas electrodes: chemical-electrochemical mode of operation. *J Electroanal Chem* 504: 146–59.
47. Rabinovich, L., and O. Lev, 2001. Sol-gel derived carbon electrodes. Electroanalysis 13:265–275.
48. Bharathi, S., and O. Lev, 1998. Sol-gel-derived nanocrystalline gold-silicate composite biosensor. *Anal Commun* 35:29–31.
49. Leventis, N., and M. Chen, 1997. Electrochemically assisted sol-gel process for the synthesis of polysiloxane films incorporating phenothiazine dyes analogous to methylene blue. Structure and ion-transport properties of the films via spectroscopic and electrochemical characterization. *Chem Mater* 9:2621–31.
50. Shacham, R., D. Avnir, and D. Mandler, 1999. Electrodeposition of methylated sol-gel films on conducting surfaces. *Adv Mater* 11:384–88.
51. Collinson, M. M., 2007. Electrochemistry: An important tool to study and create new sol-gel-derived materials. *Acc Chem Res* 40:777–83.
52. Walcarius, A., 2005. Impact of mesoporous silica-based materials on electrochemistry and feedback from electrochemical science to the characterization of these ordered materials. *C R Chim* 8:693–712.
53. Deepa, P. N., M. Kanungo, G. Claycomb, P. M. A. Sherwood, and M. M. Collinson, 2003. Electrochemically deposited sol-gel-derived silicate films as a viable alternative in thin-film design. *Anal Chem* 75:5399–405.
54. Collinson, M. M., N. Moore, P. N. Deepa, and M. Kanungo, 2003. Electrodeposition of porous silicate films from ludox colloidal silicae. *Langmuir* 19:7669–72.
55. Collinson, M. M., D. A. Higgins, R. Kommidi, and D. Campbell-Rance, 2008. Electrodeposited silicate films: Importance of supporting electrolyte. *Anal Chem* 80:651–56.
56. Jia, W. Z., K. Wang, Z. J. Zhu, H. T. Song, and X. H. Xia, 2007. One-step immobilization of glucose oxidase in a silica matrix on a Pt electrode by an electrochemically induced sol-gel process. *Langmuir* 23:11896–900.
57. Kanungo, M., H. S. Isaacs, and S. S. Wong, 2007. Quantitative control over lectrodeposition of silica films onto single-walled carbon nanotube. *J Phys Chem C* 111:17730–742.

58. Walcarius, A., M. Etienne, S. Sayen, and B. Lebeau, 2003. Grafted silicas in electroanalysis: Amorphous versus ordered mesoporous materials. *Electroanalysis* 15:414–21.

59. Carrington, N. A., L. Yong, and Z. L. Xue, 2006. Electrochemical deposition of sol-gel films for enhanced chromium(VI) determination in aqueous solutions. *Anal Chim Acta* 572:17–24.

60. Zhang, Z. H., L. H. Nie, and S. Z. Yao, 2006. Electrodeposited sol-gel-imprinted sensing film for cytidine recognition on Au-electrode surface. *Talanta* 69:435–42.

61. Herlem, G., O. Segut, and A. Antoniou, 2008. Electrodeposition and characterization of silane thin films from 3-(aminopropyl)triethoxysilane. *Surf. Coatings Technol* 202:1437–42.

62. Tian, F., E. Llaudet, and Dale, N. 2007. Ruthenium purple-mediated microelectrode biosensors based on sol-gel. *Anal Chem* 79:6760–66.

63. Nadzhafova, O., M. Etienne, and A. Walcarius, 2007. Direct electrochemistry of hemoglobin and glucose oxidase in electrodeposited sol-gel silica thin films on glassy carbon. *Electrochem Commun* 9:1189–95.

64. Some 1200 out of 3200 ISE search engine hits generated by the keywords "sol-gel and electro* and silica*" contain the phrase hybrid or composite http://apps.isiknowledge.com/summary.do.

65. Schmidt, H., 1998. In *Inorganic and organometallic polymers*, ed. K. J. Wyne and H. R. Allcock. *ACS Symp Ser* 360, 333–35, *Am Chem Soc* Washington, DC.

66. Gautier-Luneau, I., A. Denoyelle, J. Y. Sanchez, and C. Poinsignon, 1992. Organic-inorganic protonic polymer electrolytes for low-temperature fuel cell. *Electrochim Acta* 37:1615–18.

67. Gupta, R. K., H. Y. Jung, C. J. Wi, and C. M.. Whang, 2002. Electrical and structural properties of new Li^+ ion conducting sol-gel derived ormolytes: $SiO(2)(3)(-PEG-LiCF)(SO)3$. *Solid State Ionics: Trends in the new millenium, proceedings*: 369–76.

68. Bermudez, V. D., L. Alcacer, and J. L. Acosta, 1999. Synthesis and characterization of novel urethane cross-linked ormolytes for solid-state lithium batteries. *Solid State Ionics* 116:197–209.

69. Chang, T. C., C. W. Yang, and K. H. Wu, 2000. Organic-inorganic hybrid materials 3: Characterization and degradation of poly(imide-silica) hybrids doped with $LiCF_3SO_3$. *Polym Degrad Stab* 68:103–9.

70. Fujinami, T., M. A. Mehta, K. Sugie, and K. Mori, 2000. Molecular design of inorganic–organic hybrid polyelectrolytes to enhance lithium ion conductivity. *Electrochem Acta* 45:1181–86.

71. Nishio, K., K. Okubo, Y. Watanabe, and T. Tsuchiya, 2000. Structural analysis and properties of organic-inorganic hybrid ionic conductor prepared by sol-gel process. *J Sol-Gel Sci Technol* 19:187–91.

72. Kweon, J. O., and Noh, S. T. 2001. Thermal, thermomechanical, and electrochemical characterization of the organic–inorganic hybrids poly(ethylene oxide) (PEO)–silica and PEO–Silica–$LiClO_4$. *J Appl Polym Sci* 81:2471–80.

73. Philipp, G., and H. Schmidt, 1984. New materials for contact lenses prepared from Si- and Ti-alkoxides by the sol–gel process. *J Non-Cryst Solids* 63:283–92.

74. Avnir, D., D. Levy, and R. Reisfeld, 1984. The nature of the silica cage as reflected by spectral changes and enhanced photostability of trapped Rhodamine 6G. *J Phys Chem* 88:5956–59.

75. Zusman, R., C. Rottman, M. Ottolenghi, and D. Avnir, 1990. Doped sol-gel glasses as chemical sensors. *J Non-Crystal Solids* 122:107–9.

76. Rottman, C., M. Ottolenghi, R. Zusman, O. Lev, M. Smith, G. Gong, M. I. Kagan, and D. Avnir, 1992. Doped sol–gel; glasses as pH sensors. *Mater Lett* 13:293–298.
77. Iosefzon-Kuyavskaya, B., I. Gigozin, M. Ottolenghi, D. Avnir, and O. Lev, 1992. Spectrophotometric detection of heavy-metals by doped sol–gel glass detyectors. *J Non-Crystal Solids* 147:808–12.
78. Avnir, D., M. Braun, O. Lev, and M. Ottolenghi, 1994. Enzymes and other proteins entrapped in sol-gel materials. *Chem Mater* 6:1605–14.
79. Dickey, F. H., 1949. The preparation of specific adsorbents. *Proc Natl Acad Sci* 35:227–29.
80. Dickey, F. H., 1955. Specific adsorption. *J Phys Chem* 50:695–707.
81. Braun, S., S. Rappoport, R. Zusman, D. Avnir, and M. Ottolenghi, 1990. Biochemically active sol–gel glasses: The trapping of enzymes. *Mater Lett* 10:1–5.
82. Braun, S., S. Rappoport, R. Zusman, S. Shteltzer, S. Drukman, D. Avnir, and M. Ottolenghi, 1991. *Biotechnol: Bridging research and* applications, ed. D. Kamely, A. Chakrabarty, and S. E. Kornguth, 205–18. Kluwer Academic Publishers: Amsterdam.
83. Frenkel-Mullerad, H., and Avnir, D. 2005. Sol-gel materials as efficient protectors: Preserving the activity of phosphatases under extreme pH conditions. *J Am Chem Soc* 127:8077–81.
84. Nguyen, D. T., M. Smit, B. Dunn, and J. I. Zink, 2002. Stabilization of creatine kinase encapsulated in silica sol–gel materials and unusual temperature effects on its activity. *Chem Mater* 14:4300–306.
85. Reetz, M. T., A. Zonta, and J. Simpelkamp, 1995. Efficient heterogeneous biocatalysts by entrapment of lipases in hydrophobic sol-gel materials. *Angew Chem Int Ed Engl* 34:301–3.
86. Lim, F., and A. M. Sun, 1980. Microencapsulated islets as bioartificial endocrine pancreas. *Science* 210:908–10.
87. Premkumar, J. R., E. Sagi, R. Rozen, S. Belkin, A. D. Modestov, and O. Lev, 2002. Fluorescent bacteria encapsulated in sol–gel derived silicate films. *Chem Mater* 14:2676–86.
88. Premkumar, J. R., O. Lev, R. Rosen, and S. Belkin, 2001. Encapsulation of luminous recombinant *E-coli* in sol–gel silicate films. *Adv Mater* 13:1773–75.
89. Premkumar, J. R., O. Lev, R. Rosen, and S. Belkin, 2002. Sol-gel luminescence biosensors: Encapsulation of recombinant *E-coli* reporters in thick silicate films. *Anal Chim Acta* 462:11–23.
90. Liu, M. C., G. Y. Shi, L. Zhang, G. H. Zhao, and L. T. Jin, 2008. Electrode modified with toluidine blue-doped silica nanoparticles, and its use for enhanced amperometric sensing of hemoglobin. *Anal Bioanal Chem* 391:1951–59.
91. Li, W. J., R. Yuan, Y. Q. Chai, L. Zhou, S. H. Chen, and N. Li, 2008. Immobilization of horseradish peroxidase on chitosan/silica sol-gel hybrid membranes for the preparation of hydrogen peroxide biosensor. *J Biochem Biophysic Methods* 70:830–37.
92. Liang, R. P., H. Z. Peng, and J. D. Qiu, 2008. Fabrication, characterization, and application of potentiometric immunosensor based on biocompatible and controllable three-dimensional porous chitosan membranes. *J Coll Interface Sci* 320:125–31.
93. Kang, X. H., Z. B. Mai, X. Y. Zou, P. X. Cai, and J. Y. Mo, 2008. Glucose biosensors based on platinum nanoparticles-deposited carbon nanotubes in sol–gel chitosan/silica hybrid. *Talanta* 74:879–86.

94. Sahu, A. K., G. Selvarani, S. Pitchumani, P. Sridhar, and A. K. Shukla, 2007. Ameliorating effect of silica addition in the anode-catalyst layer of the membrane electrode assemblies for polymer electrolyte fuel cells. *J Appl Electrochem* 37:913–19.

95. Qian, L., and X. R. Yang, 2007. One-step synthesis of Ru(2,2′-bipyridine)(3)Cl-2-immobilized silica nanoparticles for use in electrogenerated chemiluminescence detection. *Adv Func Mater* 17:1353–58.

96. Chen, C. Y., J. I. Garnica-Rodriguez, M. C. Duke, R. F. Dalla Costa, A. L. Dicks, and J. C. D. da Costa, 2007. Nafion/polyaniline/silica composite membranes for direct methanol fuel cell application. *J Power Sour* 166:324–30.

97. Lei, C. H., M. M. Valenta, K. P. Saripalli, E. J. Ackerman, 2007. Biosensing paraoxon in simulated environmental samples by immobilized organophosphorus hydrolase in functionalized mesoporous silica. *J Environ Quality* 36:233–38.

98. Wu, H. L., P. Y. Yang, G. R. Fan, Y. P. Tian, H. J. Lu, and H. Jin, 2006. Sol-gel-derived poly(dimethylsiloxane) enzymatic reactor for microfluidic peptide mapping. *Chin J Chem* 24:903–9.

99. Ho, W. J., C. J. Yuan, and O. Reiko, 2006. Application of SiO2-poly(dimethylsiloxane) hybrid material in the fabrication of amperometric biosensor. *Anal Chim Acta* 572:248–52.

100. Pauliukaite, R., M. E. Ghica, M. Barsan, and C. M. A. Brett, 2007. Characterisation of poly(neutral red) modified carbon film electrodes; application as a redox mediator for biosensors. *J Solid State Electrochem* 11:899–908.

101. Park, T. M., E. I. Iwuoha, M. R. Smyth, R. Freaney, and A. J. McShane, 1997. Sol-gel based amperometric biosensor incorporating an osmium redox polymer as mediator for detection of l-lactate. *Talanta* 44:973–78.

102. Novak, B. N., 1993. Hybrid nanocomposite materials—between inorganic glasses and organic polymers. *Adv Mater* 5:422–33.

103. Donescu, D., 2001. Polymer-inorganic nanocomposites. *Mater Plastice* 38:3–16.

104. Chulhee, K., J. S. Kim, and M. H. Lee, 1998. Ionic conduction of sol-gel derived polyphosphazene/silicate hybrid network. *Synth Met* 98:153–56.

105. Dag, O., A. Verma, G. A. Ozin, and C. T. Kresge, 1999. Salted mesostructures: salt-liquid crystal templating of lithium triflate-oligo(ethylene oxide) surfactant-mesoporous silica nanocomposite films and monoliths. *J Mater Chem* 9:1475–82.

106. Mello, N. C., T. J. Bonagamba, H. Panepucci, K. Dahmouche, P. Judeinstein, M. A. Aegerter, 2000. NMR Study of ion-conducting organic-inorganic and composites poly(ethylene glycol)-silica-LiClO$_4$. *Macromolecules* 33:1280–88.

107. Judeinstein, P., and C. Sanchez, 1996. Hybrid organic–inorganic materials: A land of Multidisciplinarity. *J Mater Chem* 6:511–25.

108. Judeinstein, P., J. Titman, M. Stamm, and H. Schmidt, 1994. Investigation of ion-conducting ormolytes: Structure–property relationships. *Chem Mater* 6:127–34.

109. Dahmouche, K., M. Atik, N. C. Mello, T. J. Bonagamba, H. Panepucci, M. Aegerter, and P. Judeinstein, 1996. Preparation, characterization and properties of new ion-conducting ormolytes. *Mater Res Symp Proc* 435:363–64.

110. Dahmouche, K., P. H. De Souza, T. J. Bonagamba, H. Paneppucci, P. Judeinstein, S. H. Pulcinelli, and C. V. Santilli, 1998. Investigation of new ion conducting ormo-lytes silica-polypropyleneglycol. *J Sol-Gel Sci Technol* 13:909–13.

111. Dahmouche, K., M. Atik, N. C. Mello, T. J. Bonagamba, H. Panepucci, P. Judeinstein, and M. A. Aegerter, 1998. New Li$^+$ ion-conducting ormolytes. *Sol Energy Mater Sol Cells* 54:1–8.

112. Dahmouche, K., C. V. Santilli, M. Da Silva, C. A. Ribeiro, S. H. Pulcinelli, and A. F. Craievich, 1999. Silica-PEG hybrid ormolytes: structure and properties. *J Non-Cryst Solids* 247:108–13.

113. Sagar, M., and S. Sampath, 2002. Sol-gel derived, magnesium based ionically conducting composites *J Mater Chem* 12:2351–37.

114. Sagar, M., and S. Sampath, 2005. Alternating current conductivity and spectroscopic studies on sol-gel derived, trivalent ion containing silicate-tetra(ethylene glycol)-based composites. *Macromolecules* 38:134–44.

115. Ravaine, D., A. Seminet, and Y. Charbonillit, 1986. A new family of organically modified silicates prepared from gels. *J Non-Cryst Solids* 82:210–19.

116. Hench, L. L., and J. K. West, 1990. The sol–gel process. *Chem Rev* 90:33–72.

117. Judeinstein, P., J. Livage, A. Zarudiansky et al., 1988. Fractal interpretation of scatteringon $AgI:Ag_2O:B_2O_3$. *Solid State Ionics* 28–30:722–25.

118. Chen, X., B. Q. Wang, and S. J. Dong, 2001. Amperometric biosensor for hydrogen peroxide based on sol–gel/hydrogel composite thin film. *Electroanal Chem* 13:1149–52.

119. Kros, A., M. Gerritsen, V. S. I. Sprakel, N. A. Sommerdijk, J. A. Jansen, and R. J. Nolte, 2001. Silica-based hybrid materials as biocompatible coatings for glucose sensors. *Sensor Actuat B-Chem* 81:68–75.

120. Gao, Z. M., J. S. Nahrup, J. E. Mark, and A. Sakr, 2003. Poly(dimethylsiloxane) coatings for controlled drug release. I. Preparation and characterization of pharmaceutically acceptable materials. *J Appl Polym Sci* 90:658–66.

121. Novak, B. M., D. Auerbach, and C. Verrier, 1994. Low-density, mutually interpenetrating organic-inorganic composite-materials via supercritical drying techniques. *Chem Mater* 6:282–86.

122. Deng, Q., K. M. Cable, R. B. Moore, and K. A. Mauritz, 1996. Small-angle x-ray scattering studies of Nafion(R)/[silicon oxide] and Nafion(R)/ORMOSIL nanocomposites. *J Polym Sci Part B-Polym Phys* 34:1917–23.

123. Watanabe, M., H. Uchida, Y. Seki, M. Emori, and P. Stonehart, 1996. Self-humidifying polymer electrolyte membranes for fuel cells. *J Electrochem Soc* 143:1700–704.

124. Adjemian, K. T., S. J. Lee, S. Srinivasan, J. Benziger, and A. B. Bocarsly, 2002a. Silicon oxide Nafion composite membranes for proton-exchange membrane fuel cell operation at 80–140 degrees C. *J Electrochem Soc* 149:A256–A61.

125. Adjemian, K. T., S. Srinivasan, J. Benziger, and A. B. Bocarsly, 2002b. Investigation of PEMFC operation above 100 degrees C employing perfluorosulfonic acid silicon oxide composite membranes. *J Power Sources* 109:2, 356–64.

126. Hartmann-Thompson, C., A. Merrington, P. I. Carver, D. L. Keeley, J. L. Rousseau, D. Hucul, K. J. Bruza, L. S. Thomas, S. E. Keinath, R. M. Nowak, D. M. Katona, and P. R. Santurri, 2008. Hyperbranched polyesters with internal and exo-presented hydrogen-bond acidic sensor groups for surface acoustic wave sensors. *J Appl Polym Sci* 110:958–74.

127. Tay, S. W., X. Zhang, Z. Liu, L. Hong, and S. H. Chan, 2008. Composite Nafion (R) membrane embedded with hybrid nanofillers for promoting direct methanol fuel cell performance. *J Membrane Sci* 321:139–45.

128. Tang, H. L., and M. Pan, 2008. Synthesis and characterization of a self-assembled nafion/silica nanocomposite membrane for polymer electrolyte membrane fuel cells. *J Phys Chem* C 112:11556–68.

129. Jung, G. B., K. F. Lo, A. Su, F. B. Weng, C. H. Tu, T. F. Yang, and S. H. Chan, 2008. Experimental evaluation of ambient forced-feed air-supply PEM fuel cell. *Int J Hydrogen Energy* 33:2980–85.

130. Pereira, F., K. Valle, P. Belleville, A. Morin, S. Lambert, and C. Sanchez, 2008. Advanced mesostructured hybrid silica-nafion membranes for high-performance PEM fuel cell. *Chem Mater* 20:1710–18.

131. Vengatesan, S., H. J. Kim, S. Y. Lee, E. Cho, H. Y. Ha, I. H. Oh, S. A. Hong, and T. H. Lim, 2008. High temperature operation of PEMFC: A novel approach using MEA with silica in catalyst layer. *Int J Hydrogen Energy* 33:171–78.

132. Wang, Q. L., G. X., Lu, and B. J. Yang, 2004. Direct electrochemistry and electrocatalysis of hemoglobin immobilized on carbon paste electrode by silica sol–gel film. *Biosens Bioelectron* 19:1269–75.

133. Wang. L., and E. K. Wang, 2004. Direct electron transfer between cytochrome c and a gold nanoparticles modified electrode. *Electrochem Commun* 6:49–54.

134. Nicotera, I., A. Khalfan, G. Goenaga, T. Zhang, A. Bocarsly, and S. Greenbaum, 2008. NMR investigation of water and methanol mobility in nanocomposite fuel cell membranes. *Ionics* 14:243–53.

135. Matos, B. R., E. I. Santiago, F. C. Fonseca, M. Linardi, V. Lavayen, R. G. Lacerda, L. O. Ladeira, and A. S. Ferlauto, 2008. Nafion-Titanate nanotube composite membranes for PEMFC operating at high temperature. *J Electrochem Soc* 154:B1358–61.

136. Sunarso, J., C. Y. Chen, L. Z. Wang, R. F. D. Costa, G. Q. Lu, and J. C. D. da Costa, 2008. Characterization of hybrid organic and inorganic functionalised membranes for proton conduction. *Solid State Ionics* 179:477–82.

137. Lepiller, C., V. Gauthier, J. Gaudet, A. Pereira, M. Lefevre, D. Guay, and A. Hitchcock, 2008. Studies of Nafion-RuO$_2$ center dot xH(2)O composite membranes. *J Electrochem Soc* 155:B70–B78.

138. Shao, Z. G., H. F. Xu, I. M. Hsing, and H. M. Zhang, 2007. Tungsten trioxide hydrate incorporated Nafion composite membrane for proton exchange membrane fuel cells operated above 100 degrees C. *Chem Eng Commun* 194:667–74.

139. Sacca, A., I. Gatto, A. Carbone, R. Pedicini, and E. Passalacqua, 2006. ZrO2-Nafion composite membranes for polymer electrolyte fuel cells (PEFCs) at intermediate temperature. *J Power Sources* 163:47–51.

140. Di Noto, V., R. Gliubizzi, E. Negro, M. Vittadello, and G. Pace, 2007. Hybrid inorganic-organic proton conducting membranes based on Nafion and 5 wt.% of MxOy (M = Ti, Zr, Hf, Ta and W). Part I. Synthesis, properties and vibrational. *Electrochim Acta* 53:1618–27.

141. Valle, K., P. Belleville, F. Pereira, and C. Sanchez, 2006. Hierarchically structured transparent hybrid membranes by in situ growth of mesostructured organosilica in host polymer. *Nat Mater* 5:107–11.

142. Croce, F., G. B. Appetecchi, and B. Scrosati, 1998. Nanocomposite polymer electrolytes for lithium batteries, *Nature* 394:456–58.

143. Walls, H. J., M. W. Riley, R. R., Singhal, R. J. Spontak, P. S. Fedwik, and S. A. Khan, 2003. Nanocomposite electrolytes with fumed silica and hectorite clay networks: Passive versus active filters. *Adv Func Mater* 13:710–17.

144. Mello, N. C., T. J. Bonagamba, H. Panepucci, K. Dahmouche, P. Judeinstein, and M. A. Aegerter, 2000. NMR study of ion-conducting organic-inorganic nanocomposites poly(ethylene glycol)—Silica—LiClO$_4$. *Macromolecules* 33:1280–88.

145. de Souza, P. H., R. F. Bianchi, K. Dahmouche, P. Judeinstein, R. M. Faria, and T. J. Bonagamba, 2001. Solid-state NMR, ionic conductivity, and thermal studies of lithium-doped siloxane-poly(propylene glycol) organic-inorganic nanocomposites. *Chem Mater* 13:3685–92.

146. Liu, Y., J. Y. Lee, and L. Hong, 2004. In situ preparation of poly(ethylene oxide)-SiO$_2$ composite polymer electrolytes. *J Power Sources* 129:303–11.

147. Popall, M., and X. M. Du, 1995. Inorganic-organic copolymers as solid-state ionic conductors with grafted anions. *Electrochim Acta* 40:2305–8.

148. Yeh, J. M., C. J. Weng, K. Y. Huang, and C. C. Lin, 2006. Effect of baking treatment and materials composition on the properties of bulky PMMA-silica hybrid sol-gel materials with low volume shrinkage. *J Appl Polym Sci* 101:1151–59.

149. Ita, M., Y. Uchida, and K. Matsui, 2003. Polyaniline/silica hybrid composite gels prepared by the sol-gel. *J. Sol-Gel Sci Technol* 26:479–82.

150. Widera, J., and J. A. Cox, 2002. Electrochemical oxidation of aniline in a silica sol-gel matrix. *Electrochem Commun* 4:118–22.

151. Verghese, M. M., K. Ramanathan, S. M. Ashraf, M. N. Kamalasanan, and B. D. Malhotra, 1996. Electrochemical growth of polyaniline in porous sol-gel. *Chem Mater* 8:822–24.

152. Neves, S., and C. P. Fonseca, 2002. Influence of template synthesis on the performance of polyaniline cathodes. *J Power Sources* 107:13–17.

153. Chowdhury, A. N., M. R. Rahman, D. S. Islam, and F. S. Saleh, 2008. Electrochemical preparation and characterization of conducting copolymer/silica. *J Appl Polym Sci* 110:808–16.

154. Tan, X. C., M. J. Ll, P. X. Cai, L. J. Luo, and X. Y. Zou, 2005. An amperometric cholesterol biosensor based on multiwalled carbon nanotubes and organically modified sol-gel/chitosan hybrid composite film. *Anal Biochem* 337:111–20.

155. Tan, X. C., Y. X. Tian, P. X. Cai, and X. Y. Zou, 2005. Glucose biosensor based on glucose oxidase immobilized in sol-gel chitosan/silica hybrid composite film on Prussian blue modified glass carbon electrode. *Anal Bioanal Chem* 381:500–507.

156. Tang, D. P., and J. J. Ren, 2005. Direct and rapid detection of diphtherotoxin via potentiometric immunosensor based on nanoparticles mixture and polyvinyl butyral as matrixes. *Electroanalysis* 17:2208–16.

157. Wang, Q. L., G. X. Lu, and B. J. Yang, 2004. Hydrogen peroxide biosensor based on direct electrochemistry of hemoglobin immobilized on carbon paste electrode by a silica sol-gel film. *Sens Actuat B-Chem* 99:50–57.

158. Tang, D. P., R. Yuan, Y. Q. Chai, X. Zhong, Y. Liu, and J. Y. Dai, 2006. Electrochemical detection of hepatitis B surface antigen using colloidal gold nanoparticles modified by a sol-gel network interface. *Clin Biochem* 39:309–14.

159. Jena, B. K., and C. R. Raj, 2006. Enzyme-free amperometric sensing of glucose by using gold nanoparticles. *Chem-A Eur J* 12:2702–8.

160. Zhang, S. X., N. Wang, Y. M. Niu, and C. Q. Sun, 2005. Immobiliization of glucose oxidase on gold nanoparticles modified Au electrode for the construction of biosensor. *Sens Actuat B-Chem* 109:367–74.

161. Geng, R., G. H. Zhao, M. C. Liu, and M. F. Li, 2008. A sandwich structured SiO_2/cytochrome c/SiO_2 on a boron-doped diamond film electrode as an electrochemical nitrite biosensor. *Biomaterials* 29:2794–801.

162. Bharathi, S., and O. Lev, 2000. Sol-gel-derived prussian blue-silicate amperometric glucose biosensor. *Appl Biochem Biotechnol* 89:209–16.

163. Li, J. P., and T. Z. Peng, 2003. Cholesterol biosensors based on cholesterol oxidase immobilized in a silica sol-gel matrix on a Prussian Blue modified electrode. *Chem J Chinese Universities-Chinese* 24:798–802.

164. Zamponi, S., A. M. Kijak, A. J. Sommer, R. Marassi, P. J. Kulesza, and J. A. Cox, 2002. Electrochemistry of Prussian Blue in silica sol-gel electrolytes doped with polyamidoamine dendrimers. *J Solid State Electrochem* 6:528–33.

165. Wang, P., and G. Y. Zhu, 2002. Cupric hexacyanoferrate nanoparticle modified carbon ceramic composite electrodes. *Chinese J Chem* 20:374–80.

166. Zuo, S. H., Y. J. Teng, H. H. Yuan, and M. B. Lan, 2008. Development of a novel silver nanoparticles-enhanced screen-printed amperometric glucose biosensor. *Anal Lett* 41:1158–72.

167. Li, T., Z. H. Yao, and L. Ding, 2004. Development of an amperometric biosensor based on glucose oxidase immobilized through silica sol-gel film onto Prussian Blue modified electrode. *Sens Actuat B-Chem* 101:155–160.

168. Miecznikowski, K., J. A. Cox, A. Lewera, and P. J. Kulesza, 2000. Solid state voltammetric characterization of iron hexacyanoferrate encapsulated in silica. *J Solid State Electrochem* 4:199–204.

169. Stoch, J., M. Klisch, and I. Babytch, 1995. The structure of x-center-dot-wo3-center-dot(1-x)center-dot-sio2 sol-gel thin-films. *Bull Polish Acad Sci-Chem* 43:173–80.

170. Klisch, M., 1998. 12-tungstosilicic acid (12-TSA) as a tungsten precursor in alcoholic solution for deposition of $xWO(3)(1-x)SiO_2$ thin films (x <= 0.7) exhibiting electrochromic coloration ability. *J Sol-Gel Sci Technol* 12:21–33.

171. Stangar, U. L., B. Orel, N. Groselj, P. Judeinstein, F. Decker, and P. Lianos, 2001. Organic-inorganic sol–gel hybrids with ionic properties. *Monatshefte Chem* 132:103–12.

172. Song, W. B., Y. Liu, N. Lu, H. D. Xu, and C. Q. Sun, 2000. Application of the sol–gel technique to polyoxometalates: Towards a new chemically modified. *Electrochim Acta* 45:1639–44.

173. Aliev, A. E., and H. W. Shin, 2002. Nanostructured materials for electrochromic devices. *Solid State Ionics* 154:425–31.

174. Tsionsky, M., G. Gun, V. Glezer, and O. Lev, 1994. Sol-gel-derived ceramic-carbon composite electrodes—introduction and scope of applications. *Anal Chem* 66:1747–53.

175. Ramanathan, K., B. R. Jonsson, and B. Danielsson, 2001. Sol-gel based thermal biosensor for glucose. *Anal Chim Acta* 427:1–10.

176. Thangamuthu, R., and C. W. Lin, 2005. Membrane electrode assemblies based on sol-gel hybrid membranes—a preliminary investigation on fabrication aspects. *J. Power Sources* 150:48–56.

177. Cordero-Rando, M. D., I. Naranjo-Rodriguez, J. M. Palacios-Santander, L. M. Cubillana-Aguilera, and J. L. Hidalgo-Hidalgo-de-Cisneros, 2005. Study of the responses of a sonogel-carbon electrode towards phenolic compounds. *Electroanalysis* 17:806–14.

178. Cubillana-Aguilera, L. M., J. M. Palacios-Santander, I. Naranjo-Rodriguez, and J. L. Hidalgo-Hidalgo-de-Cisneros, 2006. Study of the influence of the graphite powder particle size on the structure of the Sonogel-Carbon materials. *J Sol-Gel Sci Technol* 40:55–64.

179. Bharathi, S., J. Joseph, and O. Lev, 1999. Electrodeposition of this gold films from an anminosilicate stabilized gold sold. *Electrochem Solid State Lett* 2:284–87.

180. Wang, J., P. V. A. Pamidi, and M. Jiang, 1998a. Low-potential stable detection of beta-NADH at sol-gel derived carbon composite electrodes. *Anal Chim Acta* 360:171–78.

181. Kim, H., and Popov, B. N., 2002. Characterization of hydrous ruthenium oxide/carbon nanocomposite supercapacitors prepared by a colloidal method. *J Power Sources* 104:52–61.

182. Dandekar, M. S., G. Arabale, and K. Vijayamohanan, 2005. Preparation and characterization of composite electrodes of coconut-shell-based activated carbon and hydrous ruthenium oxide for supercapacitors. *J Power Sources* 141:198–203.

183. Iwata, T., T. Hirose, A. Ueda, and N. Sawtari, 2001. Ruthenium oxide impregnated carbon pseudocapacitors. *Electrochemistry* 69:177–81.

184. Gun, G., M. Tsionsky, and O. Lev, 1994. Characterization of sol-gel derived composite silica carbon electrodes. 1994. *Better ceramics through chemistry, VI*, C. Sanchez, M. L. Mecartney, C. J. Brinker, A. Cheetham, eds, Materials Research Society Symposium Proceedings, 346, 1011–16.

185. Gun, G., M. Tsionsky, and O. Lev, 1994. Voltammetric studies of composite ceramic carbon working electrodes. *Anal Chim Acta* 294:261–70.

186. Wang, J., D. S. Park, and P. V. A. Pamidi, 1997. Tailoring the macroporosity and performance of sol–gel derived carbon composite glucose sensors. *J Electroanal Chem.* 434:185–89.

187. Macdonald, S. M., K. Szot, J. Niedziolka, F. Marken, and M. Opallo, 2008. Introducing hydrophilic carbon nanoparticles into hydrophilic sol–gel film film electrodes. *J Solid State Electrochem* 12:287–93.

188. Britto, P. J., K. S. V. Santhanam, and P. M. Ajayan, 1996. Carbon nanotube electrode for oxidation of dopamine. *Bioelectrochem Bioenergetics* 41:121–25.

189. Guo, Y. Z., and A. R. Guadalupe, 1998. Screen-printable surfactant-induced sol–gel graphite composites for electrochemical sensors. *Sensor Actuat B-Chem* 46:213–19.

190. Li, J. P., T. Z. Peng, and C. Fang, 2002. Screen-printable sol–gel ceramic carbon composite pH sensor with a receptor zeolite. *Anal Chim Acta* 455:53–60.

191. Wang, J., P. V. A. Pamidi, C. Parrado, D. S. Park, and J. Pingaron, 1997. Sol-gel-derived cobalt phthalocyanine-dispersed carbon composite electrodes for electrocatalysis and amperometric flow detection. *Electroanalysis* 9:908–11.

192. Wang, J., P. V. A. Pamidi, and D. S. Park, 1997. Sol-gel-derived metal-dispersed carbon composite amperometric biosensors. *Electroanalysis* 9:52–55.

193. Hua, L., and S. N. Tan, 2000. Capillary electrophoresis with an integrated on-capillary tubular detector based on a carbon sol-gel-derived platform. *Anal Chem.* 72:4821–25.

194. Li, J., L. S. Chia, N. K. Goh, and S. N. Tan, 1999. Renewable silica sol-gel derived carbon composite based glucose biosensor. *J Electroanal Chem* 460:234–41.

195. Gun, J., and O. Lev, 1996. Wiring of glucose oxidase to carbon matrices via sol-gel derived redox modified silicate. *Anal Lett* 29:1933–38.

196. Gun, J., and O. Lev, 1996. Sol-gel derived, ferrocenyl-modified silicate-graphite composite electrode: Wiring of glucose oxidase. *Anal Chim Acta* 336:95–106.

197. Wang, J., P. V. A., Pamidi, and D. S. Park, 1996. Screen-printable sol-gel enzyme-containing carbon inks. *Anal Chem* 68:2705–8.

198. Wang, J., and P. V. A. Pamidi, 1997. Sol-gel-derived gold composite electrodes. *Anal Chem* 69:4490–94.

199. Bharathi, S., N. Fishelson, and O. Lev, 1999. Direct synthesis and characterization of gold and other noble metal nanodispersions in sol-gel-derived organically modified silicates. *Langmuir* 15:1929–37.

200. Bharathi, S., and O. Lev, 1997. Direct synthesis of gold nanodispersions in sol-gel derived silicate sols, gels and films. *Chem Commun* 23:2303–4.

201. Chen, X. H., and G. S. Wilson, 2004. Electrochemical and spectroscopic characterization of surface sol–gel processes. *Langmuir* 20:8762–67.

202. Tang, D. P., and J. J. Ren, 2005. Direct and rapid detection of diphtherotoxin via potentiometric immunosensor based on nanoparticles mixture and polyvinyl butyral as matrixes. *Electroanalysis* 17:2208–16.

203. Wang, F. C., R. Yuan, and Y. Q. Chai, 2006. Direct electrochemical immunoassay based on a silica nanoparticles/sol-gel composite architecture for encapsulation of immunoconjugate. *Appl Microbiol Biotechnol* 72:671–75.

204. Wang, J., P. V. A. Pamidi, and K. R. Rogers, 1998. Sol-gel-derived thick-film amperometric immunosensors. *Anal Chem* 70:1171–75.

205. Cai, W. Y., Q. Xu, X. N. Zhao, J. H. Zhu, and H. Y. Chen, 2006. Porous gold-nanoparticle-CaCO$_3$ hybrid material: Preparation, characterization, and application for horseradish peroxidase assembly and direct electrochemistry. *Chem Mater* 18:279–84.

206. Xu, Q., C. Mao, N. N. Liu, J. Zhu, and J. Sheng, 2006. Direct electrochemistry of horseradish peroxidase based on biocompatible carboxymethyl chitosan-gold nanoparticle nanocomposite. *Biosens Bioelectron* 22:768–73.

207. Du, D., S. Chen, J. Cai, and A. Zhang, 2007. Immobilization of acetylcholinesterase on gold nanoparticles embedded in sol-gel film for amperometric detection of organophosphorous insecticide. *Biosens Bioelectron* 23:130–34.

208. Di, J. W., S. H. Peng, C. P. Shen, Y. S. Gao, and Y. F. Tu, 2007. One-step method embedding superoxide dismutase and gold nanoparticles in silica sol-gel network in the presence of cysteine for construction of third-generation biosensor. *Biosens Bioelectron* 23:88–94.

209. Iijima, S., 1991. Helical microtubules of graphitic carbon. *Nature* 354:56–58.

210. Britto, P. J., K. S. V. Santhanam, A. Rubio, J. A. Alonso, and P. M. Ajayan, 1999. Improved charge transfer at carbon nanotube electrodes. *Adv Mater* 11:154–57.

211. Wang, J. X., M. X. Li, Z. J. Shi, N. Q. Li, and Z. N. Gu, 2002. Electrocatalytic oxidation of norepinephrine at a glassy carbon electrode modified with single wall carbon nanotubes. *Electroanalysis* 14:225–30.

212 Luo, H. X., Z. J. Shi, N. Q. Li, Z. N. Gu, Q. K. Zhuang, 2001. Investigation of the electrochemical and electrocatalytic behavior of single-wall carbon nanotube film on a glassy carbon. *Anal Chem*. 73:915–20.

213. Gong, K. P., M. N. Zhang, Y. M. Yan, L. Su, L. Q. Mao, Z. X. Xiong, and Y. Chen, 2004. Sol-gel-derived ceramic-carbon nanotube nanocomposite electrodes: Tunable electrode dimension and potential electrochemical applications. *Anal Chem* 76:6500–505.

214. Yang, M. H., Y. H. Yang, Y. L. Liu, G. L. Shen, and R. Q. Yu, 2006. Platinum nanoparticles-doped sol-gel/carbon nanotubes composite electrochemical sensors and biosensors. *Biosens Bioactuat* 21:1125–1131.

215. Gavalas, V. G., S. A. Law, J. C. Ball, R. Andrews, and L. G. Bachas, 2004. Carbon nanotube aqueous sol–gel composites: Enzyme friendly platform for the development of stable biosensors. *Anal Biochem* 329:247–52.

216. Salimia, A., R. G. Compton, and R. Hallaja, 2004. Glucose biosensor prepared by glucose oxidase encapsulated sol-gel and carbon-nanotube-modified basal plane pyrolytic graphite electrode. *Anal Biochem* 333:49–56.

217 Yang, M., Y. Yang, Y. Liu, G. Shen, and R. Yu, 2006. Platinum nanoparticles-doped sol-gel/carbon nanotubes composite electrochemical sensors and biosensors. *Biosens Bioelectronics* 21:1125–31.

218. Kandimalla, V. B., V. S. Tripathi, and H. Ju, 2006. A conductive ormosil encapsulated with ferrocene conjugate and multiwall carbon nanotubes for biosensing application. *Biomaterials* 27:1167–74.

219 Shi, G. Y., Z. Y. Sun, M. C. Liu, L. Zhang, Y. Liu, Y. H. Qu, and L. T. Jin, 2007. Electrochemistry and electrocatalytic properties of hemoglobin in layer-by-layer films of SiO_2 with vapor-surface sol–gel deposition. *Anal Chem* 79:3581–88.

220. Gooding, J. J., R. Wibowo, J. Q. Liu, W. R., Yang, D. Losic, S. Orbons, F. J. Mearns, J. G. Shapter, and D. B. Hibbert, 2003. Protein electrochemistry using aligned carbon nanotube. *J Am Chem Soc* 125:9006–7.

221. Chen, H., and S. Dong, 2007. Direct electrochemistry and electrocatalysis of horseradish peroxidase immobilized in sol-gel-derived ceramic–carbon nanotube nanocomposite film. *Biosens Bioelectronics* 22:1811–15.

222. Wang, J., M. Musameh, and Y. Lin, 2003. Solubilization of carbon nanotubes by Nafion toward the preparation of amperometric biosensors. *J Am Chem Soc* 125:2408–9.

223. Chu, D. B., L. Y. Zhang, J. H. Zhang, and X. J. Yin, 2006. Heterogeneous electro-catalytic reduction of furfural on nanocrystalline TiO_2–CNT complex film electrode in DMF solution. *Acta Phys-Chim A Sinica* 22:373–77.

224. Choi, H. N., Y. K. Lyu, J. H. Han, and W. Y. Lee, 2007. Amperometric ethanol bio-sensor based on carbon nanotubes dispersed in sol-gel-derived titania–nafion com-posite film. *Electroanalysis* 19:1524–30.

225. Sadeh, A., S. Sladkevich, F. Gelman, P. Prikhodchenko, I. Baumberg, O. Berezin, and O. Lev, 2007. Sol-gel-derived composite antimony-doped, tin oxide-coated clay-silicate semitransparent and conductive electrodes. *Anal Chem* 70:5188–95.

226. Lipatov, Y. S., 2002. Polymer blends and interpenetrating polymer networks at the interface with solids. *Progress Polym Sci* 27:1721–1801.

227. Sampath, S., and O. Lev, 1996a. Inert metal-modified, composite ceramic-carbon, amper-ometric biosensors: renewable, controlled reactive layer. *Anal Chem* 68:2015–21.

228. Sampath, S., and O. Lev, 1996b. Renewable, reagentless glucose sensor based on a redox modified enzyme and carbon-silica composite. *Electroanalysis* 8:1112–16.

229. Rabinovich, L., J. Gun, M. Tsionsky, and O. Lev, 1997. Fuel-cell type ceramic-car-bon oxygen sensors. *J Sol-Gel Sci Technol* 8:1077–81.

230. Tess, M. E., and J. A. Cox, 1999. Chemical and biochemical sensors based on advances in materials chemistry. *J Pharm Biomed Anal* 19:55–68.

231. Etienne, M., and A. Walcarius, 2006. Evaporation induced self-assembly of tem-plated silica and organosilica thin films on various electrode surfaces. *Electrochem Commun* 7:1449–56.

232. Kubiaka, P., J. Geserick, N. Husing, and A. Wohfahrt-Mehrens, 2008. Electrochemical performance of mesoporous TiO_2 anatase. *J Power Sources* 175:510–16.

233. Deepa, M., M. Kar, D. P. Singh, A. K. Srivastava, and S. Ahmad, 2008. Influence of polyethylene glycol template on microstructure and electrochromic properties of tungsten oxide. *Solar Energy Mater Solar Cells* 92:170–78.

234. Tsionsky, M., A. Vanger, and O. Lev, 1994. Macroporous thin films for planar chro-matography. *J Sol-Gel Sci Technol* 2:595–99.

235. Minakuchi, H., K. Nakanishi, N. Soga, N. Ishizuka, and N. Tanaka, 1996. Octadecylsilylated porous silica rods as separation media for reversed-phase liquid chromatography. *Anal Chem* 68:3498–3501.

236. Kresge, C. T., M. E. Leonowicz, W. J. Roth, J. C. Vartuli, and J. S. Beck, 1992. Ordered mesoporous molecular sieves synthesized by a liquid-crystal template mechanism. *Nature* 359:710–12.

237. Lu, Y. F., R. Ganguli, C. A. Drewien, M. T. Anderson, C. J. Brinker, W. L. Gong, Y. X. Guo, H. Soyez, B. Dunn, M. H. Huang, and J. I. Zink, 1997. Continuous for-mation of supported cubic and hexagonal mesoporous films by sol–gel dip-coating. *Nature* 389:364–68.

238. Leddy, J., 1999. Characterizing flux through micro- and nanostructured composites. *Langmuir* 15:710–16.

239. Collinson, M. M., C. G. Rausch, and A. Voigt, 1997. Electroactivity of redox probes encapsulated within sol-gel-derived silicate films. *Langmuir* 13:7245–51.

240. Kanungo, M., and M. M. Collinson, 2003. Diffusion of redox probes in hydrated sol-gel-derived glasses: Effect of gel structure. *Anal Chem* 75:6555–59.

241. Hsueh, C. C., and M. M. Collinson, 1997. Permselectivities and ion-exchange prop-erties of organically modified sol–gel electrodes. *J Electroanal Chem* 420:243–49.

242. Barroso-Fernandez, B., M. T. Lee-Alvarez, C. J. Seliskar, and W. R. Heineman, 1998. Electrochemical behavior of methyl viologen at graphite electrodes modified with Nafion sol-gel composite. *Anal Chim Acta* 370:221–30.

243. Zoppi, R. A., and S. P. Nunes, 1998. Electrochemical impedance studies of hybrids of perfluorosulfonic acid ionomer and silicon oxide by sol-gel reaction from solution. *J Electroanal Chem* 445:39–45.

244. Zhao, C.-Z., N. Egashira, Y. Kurauchi, and K. Ohga, 1998. Electrochemiluminescence sensor having a Pt electrode coated with a Ru(bpy)$_3^{2+}$-modified chitosan/silica gel membrane. *Anal Sci* 14:439–41.

245. Brinker, C. J., T. L. Ward, R. Sehgal, N. K. Raman, S. L. Hietala, D. M. Smith, D.-W. Hua, and T. J. Headley, 1993. "Ultramicroporous" silica-based supported inorganic membranes. *J Membr Sci* 77:165–79.

246. Chia, S. Y., J. Urano, F. Tamanoi, B. Dunn, and J. I. Zink, 2000. Patterned hexagonal arrays of living cells in sol-gel silica films. *J Am Chem Soc* 122:6488–89.

247. Dickert, F. L., and O. Hayden, 2002. Bioimprinting of polymers and sol-gel phases. Selective detection of yeasts with imprinted polymers. *Anal Chem* 74:1302–6.

248. Kanungo, M., P. N. Deepa, and M. M. Collinson, 2004. Template-directed formation of hemispherical cavities of varying depth and diameter in a silicate matrix prepared by the sol-gel process. *Chem Mater* 16:5535–41.

249. Rolison, D. R., and B. Dunn, 2001. Electrically conductive oxide aerogels: New materials in electrochemistry. *J Mater Chem* 11:963–80.

250. Dong, W., D. R. Rolison, and B. Dunn, 2000. Electrochemical properties of high surface area vanadium oxide aerogels. *Electrochem Solid State Lett* 3:457–59.

251. Collinson, M. M., 1999. Sol-gel strategies for the preparation of selective materials for chemical analysis. *Crit Rev Anal Chem* 29:289–311.

252. Sayen, S., M. Etienne, J. Bessiere, and A. Walcarius, 2002. Tuning the sensitivity of electrodes modified with an organic-inorganic hybrid by tailoring the structures of the nanocomposite material. *Electroanalysis* 14:1521–25.

253. Bardo, A. M., M. M. Collinson, and D. A. Higgins, 2001. Nanoscale properties and matrix-dopant interactions in dye-doped organically modified silicate thin films. *Chem Mater* 13:2713–21.

254. Avnir, D., S. Braun, O. Lev, and M. Ottolenghi, 1994. Enzymes and other proteins entrapped in sol-gel materials. *Chem Mater* 6:1605–14.

255. Shtelzer, S., S. Rappoport, D. Avnir, M. Ottolenghi, and S. Braun, 1992. Properties of trypsin and of acid-phosphatase immobilized in sol-gel glass matrices. *Biotechnol Appl Biochem* 15:227–35.

256. Chen, Q., G. L. Kenausis, and A. Heller, 1998. Stability of oxidases immobilized in silica gels. *J Am Chem Soc* 120:4582–85.

257. Guo, W., H. Y. Lu, and N. F. Hu, 2006. Comparative bioelectrochemical study of two types of myoglobin layer-by-layer films with alumina: Vapor-surface sol-gel deposited Al$_2$O$_3$ films versus AlO$_3$ nanoparticle films. *Electrochim Acta* 52:123–32.

258. Xu, J. S., and G. C. Zhao, 2008. Direct electrochemistry of cytochrome c on a silica sol-gel film modified electrode. *Electroanalysis* 20:1200–203.

259. Kafi, A. K. M., D. Y. Lee, S. H. Park, and Y. S. Kwon, 2008. Potential application of hemoglobin as an alternative to peroxidase in a phenol biosensor. *Thin Solid Films* 516:2816–21.

260. Ray, A., M. L. Feng, and H. Tachikawa, 2005. Direct electrochemistry and Raman spectroscopy of sol-gel-encapsulated myoglobin. *Langmuir* 21:7456–60.

261. Gill, I., and A. Ballesteros, 1998. Encapsulation of biologicals within silicate, siloxane, and hybrid sol-gel polymers: an efficient and generic approach. *J Am Chem Soc* 120:8587–98.

262. Miao, Y., and S. N. Tan, 2001. Amperometric hydrogen peroxide biosensor with silica sol-gel/chitosan film as immobilization matrix. *Anal Chim Acta* 437:87–93.

263. Wang, G., J. J. Xu, H. Y. Chen, and Z. H. Lu, 2003. Amperometric hydrogen peroxide biosensor with sol-gel/chitosan network-like film as immobilization matrix. *Biosens Bioelectronics* 18:335–43.

264. Chen, M. H., Z. C. Huang, G. T. Wu, G. M. Zhu, J. K. You, and Z. G. Lin, 2003. Synthesis and characterization of SnO-carbon nanotube composite as anode material for lithium-ion batteries. *Mater Res Bull* 38:831–36.

265. Shyuan, L. K., L. Y. Heng, M. Ahmad, S. A. Aziz, and Z. Ishak, 2006. Screen-printed biosensor with alkaline phosphatase immobilized in sol-gel/chitosan film for the detection of 2,4-dichlorophenoxyacetic acid. *Sens Lett* 4:17–21.

266. Kim, M. A., and W.-Y. Lee, 2003. Amperometric phenol biosensor based on sol-gel silicate/Nafion composite film. *Anal Chim Acta* 479:143–50.

267. Prieto-Simon, B., G. S. Armatas, P. J. Pomonis, C. G. Nanos, and M. I. Prodromidis, 2004. Metal-dispersed xerogel-based composite films for the development of interference free oxidase-based. *Chem Mater* 16:1026–34.

268. Pankratov, I., and O. Lev, 1995. Sol-gel derived renewable-surface biosensors. *J Electroanal Chem* 393:35–41.

269. Sampath, S., I. Pankratov, J. Gun, and O. Lev, 1996. Sol-gel derived ceramic-carbon enzyme electrodes: Glucose oxidase as a test case. *J Sol-Gel Sci Technol* 7:123–128.

270. Zhu, L. D., Y. X. Li, F. M. Tian, B. Xu, and G. Y. Zhu, 2002. Electrochemiluminescent determination of glucose with a sol-gel derived ceramic-carbon composite electrode as a renewable optical fiber biosensor. *Sens Actuat B-Chem* 84:265–70.

271. CocheGuerente, L., S. Cosnier, and L. Labbe, 1997. Sol-gel derived composite materials for the construction of oxidase/peroxidase mediatorless biosensors. *Chem Mater* 9:1348–52.

272. Nogala, W., E. Rozniecka, I. Zawisza, J. Rogalski, and M. Opallo, 2006. Immobilization of ABTS—laccase system in silicate based electrode for biolectrocatalytic reduction of dioxygen. *Electrochem Commun* 8:1850–54.

273. Haghighi, B., A. Rahmati-Panah, S. Shleev, and L. Gorton, 2007. Carbon ceramic electrodes modified with laccase from Trametes hirsuta: Fabrication, characterization and their use for phenolic compounds. *Electroanalysis* 19:907–17.

274. Tian, F. M., and G. Y. Zhu, 2002. Bienzymatic amperometric biosensor for glucose based on polypyrrole/ceramic carbon as electrode material. *Anal Chim Acta* 451:251–58.

275. Wang, P., X. P. Wang, L. H. Bi, and G. Y. Zhu, 2000. Sol-gel-derived alpha(2)-$K_7P_2W_{17}VO_{62}$/graphite/organoceramic composite as the electrode material for a renewable amperometric hydrogen peroxide. *J Electroanal Chem* 495:51–56.

276. Yang, X. H., L. Hua, H. Q. Gong, and S. N. Tan, 2003. Covalent immobilization of an enzyme (glucose oxidase) onto a carbon sol-gel silicate composite surface as a biosensing platform. *Anal Chim Acta* 478:67–75.

277. Xu, Q., C. Mao, N. N. Liu, J. Zhu, and J. Sheng, 2006. Direct electrochemistry of horseradish peroxidase based on biocompatible carboxymethyl chitosan-gold nanoparticle nanocomposite. *Biosens Bioelectronics* 22:768–73.

278. Zou, Y. J., C. L. Xiang, L. X. Sun, and F. Xu, 2008. Glucose biosensor based on electrodeposition of platinum nanoparticles onto carbon nanotubes and immobilizing enzyme with chitosan-SiO_2 sol-gel. *Biosens Bioelectronics* 23:1010–16.

279. Opallo, M., 2003. Silicate solvated by an organic solvent as electrolyte or electrode material. *Mater Sci-Poland* 21:453–60.

280. Shul, G., M. Saczek-Maj, and M. Opallo, 2004. Electroactive ceramic carbon electrode modified with hydrophobic polar solvent. *Electroanalysis* 16:1254–61.

281. Rozniecka, E., J. Niedziolka, J. Sirieix-Plenet, L. Gaillon, M. A. Murphy, and F. Marken, 2006. Ion transfer processes at the room temperature ionic liquid vertical bar aqueous solution interface supported by a hydrophobic carbon nanofibers–silica composite film. *J Electroanal Chem* 587:133–39.

282. Lesniewski, A., J. Niedziolka, B. Palys, C. Rizzi, L. Gaillon, and M. Opallo, 2007. Electrode modified with ionic liquid covalently bonded to silicate matrix for accumulation of electroactive anions. *Electrochem Commun* 9:2580–84.

283. Zhludkevich, M. L., D. G. Shchukin, K. A. Yasakau, H. Mohwald, and M. G. S. Ferreira, 2007. Anticorrosion coatings with self-healing effect based on nanocontainers impregnated with corrosion. *Chem Mater* 19:402–11.

284. Wakamatsu, H., S. P. Szu, L. C. Klein, and M. Greenblatt, 1992. Effect of lithium-salts on the ionic-conductivity of lithium silicate gels. *J Non-Crystalline Solids* 147:668–71.

285. Wei, Y., J. M. Yeh, D. L. Jin, X. R. Jia, J. G. Wang, G. W. Jang, C. C. Chen, and R. W. Gumbs, 1995. Composites of electronically conductive polyaniline with polycrylate-silica hybrid sol-gel materials. *Chem Mater* 7:969–74.

286. Nakanishi, K., H. Komura, R. Takahashi, and N. Soga, 1994. Phase separation in silica sol-gel system containing poly(ethylene oxide). I. Phase relation and gel morphology. *Bull Chem Soc Jpn* 67:1327–35.

287. Srebnik, S., and O. Lev, 2003. Theoretical investigation of imprinted crosslinked silicates. *J Sol-Gel Sci Technol* 26:107–13.

288. Srebnik, S., and O. Lev, 2002. Toward establishing criteria for polymer imprinting using mean-field theory. *J Chem Phys* 116:10967–72.

289. Srebnik, S., O. Lev, and D. Avnir, 2001. Pore size distribution induced by microphase separation: Effect of the leaving group during polycondensation. *Chem Mater* 13:811–16.

290. Haupt, K., and K. Mosbach, 2000. Molecularly imprinted polymers and their use in biomimetic sensors. *Chem Rev* 100:2495–2504.

291. Wulff, G., 2002. Enzyme-like catalysis by molecularly imprinted polymers. *Chem Rev* 102:1–27.

292. Whitcombe, M. J., and E. N. Vulfson, 2001. Imprinted polymers. *Adv Mater* 13:467–73.

293. Zimmerman, S. C., and N. G. Lemcoff, 2004. Synthetic hosts via molecular imprinting—are universal synthetic antibodies realistically possible? *Chem Commun* 1:5–14.

294. Alexander, C., H. S. Andersson, L. I. Andersson, R. J. Ansell, N. Kirsch, I. A. Nicholls, J. O'Mahony, and M. J. Whitcombe, 2006. Molecular imprinting science and technology: A survey of the literature for the years up to and including 2003. *J Mol Recogn* 19:106–80.

295. Makote, R., and M. M. Collinson, 1998. Template recognition in inorganic-organic hybrid films prepared by the sol-gel process. *Chem Mater* 10:2440–45.

296. Holland, B. T., C. F. Blanford, T. Do, and A. Stein, 1999. Synthesis of highly ordered, three-dimensional, macroporous structures of amorphous or crystalline inorganic oxides, phosphates, and hybrid composites. *Chem Mater* 11:795–805.

297. Liu, S. T., L. F. Wood, D. E. Ohman, and M. M. Collinson, 2007. Creating aligned arrays of *Bacillus megaterium* in sol-gel matrixes. *Chem Mater* 19:2752–56.

298. Shustak, G., S. Marx, I. Turyan, and D. Mandler, 2003. Application of sol-gel technology for electroanalytical sensing, *Electroanalysis* 15:398–408.

299. Ling, T. R., Y. Z. Syu, Y. C. Tasi, T. C. Chou, and C. C. Liu, 2005. Size-selective recognition of catecholamines by molecular imprinting on silica-alumina gel. *Biosens Bioelectronics* 21:901–7.

300. Blanco-Lopez, M. C., S. Gutierrez-Fernandez, M. J. Lobo-Castanon, A. J. Miranda-Ordieres, and P. Tunon-Blanco, 2004. Electrochemical sensing with electrodes modified with molecularly imprinted polymer films. *Anal Bioanal Chem* 378:1922–28.

301. Glad, M., O. Norrlow, B. Sellergren, N. Siegbahn, and K. Mosbach, 1985. Use of silane monomers for molecular imprinting and enzyme entrapment in polysiloxane-coated porous silica. *J Chromatogr* 347:11–23.

302. Hsu, C. W., and M. C. Yang, 2008. Electrochemical epinephrine sensor using artificial receptor synthesized by sol-gel process. *Sens Actuat B-Chem* 134:680–86.

303. Marx, S., and Z. Liron, 2001. Molecular imprinting in thin films of organic-inorganic hybrid sol-gel and acrylic polymers. *Chem Mater* 10: 3624–29.

304. Marx, S., A. Zaltsman, I. Turyan, and D. Mandler, 2004. Parathion sensor based on molecularly imprinted sol-gel films. *Anal Chem* 76:120–26.

305. Hunnius, M., A. Rufinska, and W. F. Maier, 1999. Selective surface adsorption versus imprinting in amorphous microporous silicas. *Microporous Mesoporous Mater* 29:389–403.

306. Boury, B., R. J. P. Corriu, V. Le Strat, and P. Delord, 1999. Generation of porosity in a hybrid organic-inorganic xerogel by chemical treatment. *New J Chem* 23:531–38.

307. Mikes, F., G. Boshart, and E. Gil-av, 1976. Resolution of optical isomers by high-performance liquid chromatography, using coated and bonded chiral charge-transfer complexing agents as stationary phases. *J Chromatogr* 122:205–21.

308. Bied, C., J. J. E. Moreau, and M. Chi Man Wong, 2001. Chiral amino-urea derivatives of (1R,2R)-1,2-diaminocyclohexaneas ligands in the ruthenium catalyzed asymmetric reduction of aromatic ketones by hydride transfer. *Tetrahedron Asymmetry* 12:329–36.

309. Marx, S., and D. Avnir, 2007. The induction of chirality in sol-gel materials. *Acc Chem Res* 40:768–76.

310. Huq, R., L. Mercier, and P. J. Kooyman, 2001. Incorporation of cyclodextrin into mesostructured silica. *Chem Mater* 13:4512–19.

311. Lahav, M., A. B. Kharitonov, and I. Willner, 2001. Imprinting of chiral molecular recognition sites in thin TiO2 films associated with field-effect transistors: Novel functionalized devices for chiroselective and chirospecific analyses. *Chem-A Eur J* 7:3992–97.

312. Lahav, M., A. B. Kharitonov, E. Katz, T. Kunitake, and I. Willner, 2001. Tailored chemosensors for chloroaromatic acids using molecular imprinted TiO_2 thin films on ion-sensitive field-effect transistors. *Anal Chem* 73:720–23.

313. Fireman-Shoresh, S., I. Turyan, D. Mandler, D. Avnir, and S. Marx, 2005. Chiral electrochemical recognition by very thin molecularly imprinted sol-gel films. *Langmuir* 21:7842–47.

314. Huo, Q. S., D. I. Margolese, U. Ciesla, P. Y. Feng, T. E. Gier, P. Sieger, R. Leon, P. M. Petroff, F. Schuth, and G. D. Stucky, 1994. Generalized synthesis of periodic surfactant inorganic composite-materials. *Nature* 368:317–21.

315. Ogawa, M., 1994. Formation of novel oriented transparent films of layered silica-surfactant nanocomposites. *J Am Chem Soc* 116:7941–42.

316. Aksay, I. A., M. Trau, S. Manne, I. Honma, N. Yao, L. Zhou, P. Fenter, P. M. Eisenberger, and S. M. Gruner, 1996. Biomimetic pathways for assembling inorganic thin films. *Science* 273:892–98.

317. Yang, H., A. Kuperman, N. Coombs, S. Mamiche-Afara, and G. A. Ozin, 1996. Synthesis of oriented films of mesoporous silica on mica. *Nature* 379:703–5.

318. Brinker, C. J., Y. F. Lu, A. Sellinger, and H. Y. Fan, 1999. Evaporation-induced self-assembly: Nanostructures made easy. *Adv Mater* 11:579–85.

319. Huang, M. H., H. M. Soyez, B. S. Dunn, J. I. Zink, A. Sellinger, and C. J. Brinker, 2008. In situ fluorescence probing of the chemical and structural changes during formation of hexagonal phase cetyltrimethylammonium bromide and lamellar phase CTAB/Poly(dodecylmethacrylate) sol-gel silica thin films. *J Sol-Gel Sci Technol* 47:300–310.

320. Raman, N. K., M. T. Anderson, and C. J. Brinker, 1996. Template-based approaches to the preparation of amorphous, nanoporous silicas. *Chem Mater* 8:1682–1701.

321. Zhao, D. Y., J. L. Feng, Q. S. Huo, N. Melosh, G. H. Fredrickson, B. F. Chmelka, and G. D. Stucky, 1998. Triblock copolymer syntheses of mesoporous silica with periodic 50 to 300 angstrom pores. *Science* 279:548–52.

322. Yang, P. D., D. Y. Zhao, D. I. Margolese, B. F. Chmelka, and G. D. Stucky, 1999. Block copolymer templating syntheses of mesoporous metal oxides with large ordering lengths and semicrystalline framework. *Chem Mater* 11:2813–26.

323. Walcarius, A., 2008. Electroanalytical applications of microporous zeotites and mesoporous (Organo)silicas: Recent trends. *Electroanalysis* 20:711–38.

324. Walcarius, A., and A. Kuhn, 2008. Ordered porous thin films in electrochemical analysis. *Trac-Trends Anal Chem* 27:593–603.

325. Walcarius, A., N. Luthi, J. L. Blin, B. L. Su, and L. Lamberts, 1999. Electrochemical evaluation of polysiloxane-immobilized amine ligands for the accumulation of copper(II) species. *Electrochim Acta* 44:4601–10.

326. Delacote, C., J. P. Bouillon, and A. Walcarius, 2006. Voltammetric response of ferrocene-grafted mesoporous silica. *Electrochim Acta* 51:6373–83.

327. Lunsford, S. K., H. Choi, J. Stinson, A. Yeary, and D. D. Dionysiou, 2007. Voltammetric determination of catechol using a sonogel carbon electrode modified with nanostructured titanium dioxide. *Talanta* 73:172–77.

328. Liu, Y. G., Q. Xu, X. M. Feng, J. J. Zhu, and W. H. Hou, 2007. Immobilization of hemoglobin on SBA-15 applied to the electrocatalytic reduction of H_2O_2. *Anal Bioanal Chem* 387:1553–59.

329. Dai, Z. H., X. X. Xu, and H. X. Ju, 2004. Direct electrochemistry and electrocatalysis of myoglobin immobilized on a hexagonal mesoporous silica matrix. *Anal Biochem* 332:23–31.

330. Walcarius, A., S. Sayen, C. Gerardin, F. Hamdoune, and L. Rodehuuser, 2004. Dipeptide-functionalized mesoporous silica spheres. *Coll Surf A-Physicochem Eng Aspects* 234:145–51.

331. Lee, C. H., J. Lang, C. W. Yen, P. C. Shih, T. S. Lin, and C. Y. Mou, 2005. Enhancing stability and oxidation activity of cytochrome c by immobilization in the nanochannels of mesoporous aluminosilicates. *J Phys Chem B* 109:12277–286.

332. Itoh, T., K. Yano, T. Kajino, S. Itoh, Y. Shibata, H. Mino, R. Miyamoto, Y. Inada, S. Iwai, and Y. Fukushima, 2004. Organization of chlorophyll a in mesoporous silica: Efficient energy transfer and stabilized charge separation as in natural photosynthesis. *J Phys Chem B* 108:13683–87.

333. Ganesan, R., and B. Viswanathan, 2004. Redox properties of bis(8-hydroxyquinoline) manganese(II) encapsulated in various zeolites. *J Mol Catal A-Chem* 223:21–29.

334. Lei, C. H., Y. S. Shin, J. Liu, and E. J. Ackerman 2002. Entrapping enzyme in a functionalized nanoporous support. *J Am Chem Soc* 124:11242–43.

335. Zhang, L., Q. Zhang, and J. H. Li, 2007. Layered titanate nanosheets intercalated with myoglobin for direct electrochemistry. *Adv Func Mater* 17:1958–65.

336. Feng, J. J., J. J. Xu, and H. Y. Chen, 2006. Direct electron transfer and electrocatalysis of hemoglobin adsorbed onto electrodeposited mesoporous tungsten oxide. *Electrochem Commun* 8:77–82.

337. Xian, Y. Z., Y. Xian, L. H. Zhou, F. H. Wu, Y. Ling, and L. T. Jin, 2007. Encapsulation hemoglobin in ordered mesoporous silicas: Influence factors for immobilization and bio electrochemistry. *Electrochem Commun* 9:142–48.

338. Yu, J. J., T. Zhao, F. Q. Zhao, and B. Z. Zeng, 2008. Direct electron transfer of hemoglobin immobilized in a mesocellular siliceous foams supported room temperature ionic liquid matrix and the electrocatalytic reduction of H_2O_2. *Electrochim Acta* 53:5760–65.

339. Laszlo, J. A., and D. L. Compton, 2002. Comparison of peroxidase activities of hemin, cytochrome c and microperoxidase-11 in molecular solvents and imidazolium-based ionic liquids. *J Mol Catal B-Enzym* 18:109–20.

340. Etienne, M., A. Quach, D. Grosso, L. Nicole, C. Sanchez, and A. Walcarius, 2007. Molecular transport into mesostructured silica thin films: Electrochemical monitoring and comparison between p6m, P6(3)/mmc, and Pm3n structures. *Chem Mater* 19:844–56.

341. Etienne, M., J. Cortot, and A. Walcarius, 2007. Preconcentration electroanalysis at surfactant-templated thiol-functionalized silica thin films. *Electroanalysis* 19:129–138.

342. Rathousky, D., J. Rohlfing, Y. Bartels, and O. M. Wark, 2005. Functionalized mesoporous silica films as a matrix for anchoring electrochemically active guests Fattakhova-Rohlfing. *Langmuir* 21:11320–29.

343. Yantasee, W., Y. Lin, X. Li, G. E. Fryxell, T. S. Zemanian, and V. V. Viswanathan, 2003. Nanoengineered electrochemical sensor based on mesoporous silica thin-film functionalized with thiol-terminated monolayer. *Analyst* 128:899–904.

344. Palaniappan, A., X. Li, F. E. H. Tay, J. Li, and X. D. Su, 2006. Cyclodextrin functionalized mesoporous silica films on quartz crystal microbalance for enhanced gas sensing. *Sens Actuat B-Chem* 119:220–26.

345. Shi, Z. C., Q. Wang, W. L. Ye, Y. X. Li, and Y. Yang, 2006. Synthesis and characterization of mesoporous titanium pyrophosphate as lithium intercalation electrode materials. *Microporous Mesoporous Mater* 88:232–37.

346. Jheong, H. K., Y. J. Kim, J. H. Pan, T. Y. Won, and W. I. Lee, 2006. Electrochromic property of the viologen-anchored mesoporous TiO_2 films. *J Electroceramics* 17:929–32.

347. Kitiyanan, A., S. Ngamsinlapasathian, and S. Pavasupree, 2005. The preparation and characterization of nanostructured TiO2-ZrO2 mixed oxide electrode for efficient dye-sensitized solar cells. *J Solid State Chem* 178:1044–48.

348. Imahori, H., S. Hayashi, T. Umeyama, S. Eu, A. Oguro, S. Kang, Y. Matano, T. Shishido, S. Ngamsinlapasathian, and S. Yoshikawa, 2006. Comparison of electrode structures and photovoltaic properties of porphyrin-sensitized solar cells with TiO_2 and Nb, Ge, Zr-added TiO_2 composite electrodes. *Langmuir* 22:11405–11.

349. Martinez-Ferrero, E., D. Grosso, C. Boissiere, C. Sanchez, O. Oms, D. Leclercq, A. Vioux, F. Miomandre, and Audebert, P., 2006. Electrochemical investigations into ferrocenylphosphonic acid functionalized mesostructured porous nanocrystalline titanium oxide films. *J Mater Chem* 16:3762–67.

350. Nunez, O., K. Nakanishi, and N. Tanaka, 2008. Preparation of monolithic silica columns for high-performance liquid chromatography. *J Chromtography A* 1191:231–52.

351. Nakanishi, K., H. Minakuchi, N. Soga, and N. Tanaka, 1997. Double pore silica gel monolith applied to liquid chromatography. *J Sol-Gel Sci Technol*, 8:547–52.

352. Sel, O., S. Sallard, T. Brezesinski, J. Rathousky, D. R. Dunphy, A. Collord, and B. M. Smarsly, 2007. Periodically ordered meso- and macroporous SiO_2 thin films and their induced electrochemical activity as a function of pore hierarchy. *Adv Func Mater* 17:3241–50.

353 Zhitominsky, I., 2007. Electrosynthesis of organ-inorganic nanocomposites: *J Alloys Comp* 434:823–825.

354 Bharathi, S., M. Nogami, and O. Lev, 2001. Electrochemical organization of gold sol and their characterization. *Langmuir* 17:2602–09.

Index